CLASSICAL TOPOLOGY
AND QUANTUM STATES

CLASSICAL TOPOLOGY AND QUANTUM STATES

A. P. Balachandran
Physics Department
Syracuse University
Syracuse, N.Y. 13244-1130
USA

G. Marmo
Department of Physical Sciences
Naples University and INFN, Naples Section
I-80125 Naples
Italy

B. S. Skagerstam
Chalmers University of Technology
University of Göteborg
S-41296, Göteborg
Sweden

A. Stern
Department of Physics and Astronomy
University of Alabama
Tuscaloosa, AL 35487-0324
USA

World Scientific
Singapore • New Jersey • London • Hong Kong

Published by

World Scientific Publishing Co. Pte. Ltd.

P O Box 128, Farrer Road, Singapore 9128

USA office: 687 Hartwell Street, Teaneck, NJ 07666

UK office: 73 Lynton Mead, Totteridge, London N20 8DH

Library of Congress Cataloging-in-Publication Data

Balachandran, A. P., 1938-
    Classical topology and quantum states/A.P. Balachandran . . . [et al.].
        p.   cm.
    Includes bibliographical references and index.
    ISBN 9810203292. -- ISBN 9810203306 (pbk.)
    1. Topology. 2. Quantum field theory. 3. Nuclear forces
(Physics) I. Title.
    QC20.7.T65B35 1991
    514--dc20   91-10473
                                                 CIP

Printed in Singapore by Loi Printing Pte. Ltd.

We dedicate this book to

Appa, Indra and Vinod;
Patrizia;
Helene, Emilie and Cecilia;
and
Lore, Max and Phoebe.

PREFACE

The topology of the configuration space for many classical systems of physical importance, such as the free particle and the harmonic oscillator in $\mathbf{R}^N$, happens to be very simple. The conventional exposition of classical and quantum mechanics generally concentrates on such systems, explains their quantization and goes on to show that there is a representation of wave functions as functions on the configuration space Q. However, already in 1931, Dirac understood in his work on magnetic monopoles that there are problems of interest where it is not natural to regard wave functions as functions on Q. Subsequent developments, especially in recent years, have made it clear that there are numerous important physical theories, all distinguished by a nontrivial topology of Q, where too it is not appropriate to regard wave functions as functions on Q. Rather they are to be thought of as special kinds of functions on a principal fibre bundle over Q or as sections of an associated vector bundle.

The nature of wave functions is only one aspect of the properties of a physical system subject to the influence of the configuration space topology. It is now appreciated that there are several other attributes which may show the effects of this topology. In particular, for appropriate topologies of Q, already the classical theory can predict the existence of novel sorts of stable configurations, such as line and point defects in condensed matter systems, solitons and monopoles. It happens that the nature of wave functions of the quantal version of many of these configurations is in turn influenced by the topology of Q. The latter can thus influence quantum dynamics both by leading to the existence of new states such as those describing solitons and by affecting their qualitative properties.

This book is an introduction to the role of topology in the quantization of classical systems. It is also an introduction to topological solitons with special emphasis on Skyrmions. As regards the first aspect, several issues of current interest are dealt with at a reasonably elementary level. Examples of such topics we cover are principal fibre bundles and their role in quantum physics, the possibility of spinorial quantum states in a Lagrangian theory based on tensorial variables, and multiply connected configuration spaces and associated quantum phenomena like the QCD θ angle and exotic statistics. The ideas are also illustrated by simple examples such as the spinning particle, the charge-monopole system and strings in $3 + 1$ dimensions. The application of these ideas to quantum gravity is another subject treated at an introductory level. In the field of topological solitons, our main interest has been in the exposition of Skyrmion physics. For this reason, we have limited ourselves to a comparatively brief treatment of the general theory of solitons, adequate to follow the subsequent chapters on Skyrmion physics. Some Skyrmion phenomenology is also discussed although it is far from being exhaustive. A chapter on electroweak Skyrmions has also been included. Although these are not topological solitons, they do resemble Skyrmions in many ways so that such a chapter seemed appropriate. There is also another class of solitons called topological geons to which this book may serve as an introduction. They were discovered by Friedman and Sorkin and possess many remarkable properties because of the rich topological complexities to be found in gravitational models.

An attempt has been made in this book to introduce the reader to the significance of topology for many distinct physical systems such as spinning particles, the charge-monopole system, strings, Skyrmions, QCD and gravity. It is our hope that it will contribute to a wider appreciation of the profound role of topology in classical and quantum dynamics.

There are several important aspects of the role of topology in quantization and soliton physics that we have not dealt with in this book. A major omission has been the subject of anomalies. As indicated in the text, this is a topic which can be naturally approached using the concepts that we develop. Limitations on our time have forced us into this omission and we are obliged to refer the interested reader to the several excellent reviews which exist today. Another interesting aspect of quantization which we shall not discuss is the relation between our approach to quantization with its emphasis on topology and the one which concentrates instead on domain and extension problems of operators. As alluded to previously, our treatment of general soliton theory and Skyrmion phenomenology has also been rather brief. In Skyrmion physics, there exists in particular a substantial body of research which considers the applications of Skyrme's model and its variants to low energy physics and which we have not attempted to cover adequately because of our limitations. Fortunately detailed reviews of these developments are also available to which the interested reader can refer.

With regard to references, no attempt has been made to give an exhaustive bibliography. We shall list essentially only those publications which we have frequently used during the preparation of this book. We shall also list a few representative review articles which cover material not treated by us here. We apologize to those authors whose work we have overlooked and to those who feel that their work should have been referred to.

The book is an outgrowth of lectures given by the authors at various institutions and conferences. We thank the audience at these lectures as well as our collegues at Syracuse and elsewhere for their suggestions and criticism. We are especially grateful to Rafael Sorkin for numerous discussions about the material treated in this book and for collaboration on its title. We thank Ted Allen for carefully proofreading the manuscript and several useful suggestions, and David Dallman and Kumar Gupta for their generous help in collecting references. The typing of the several versions of the manuscript of this book was done by Jane Boyd at Tuscaloosa, by Guido Celentano at Naples and by Annika Hofling at Göteborg. We also gratefully record our appreciation of their patience and accurate work here. Finally, the TH-division at CERN is acknowledged for hospitality while this book was completed.

A.P.B. was supported by the U.S. Department of Energy under contract

CLASSICAL TOPOLOGY

AND

QUANTUM STATES

CONTENTS

PART II

TOPOLOGICAL SOLITONS AND NONLINEAR MODELS

PART III

SKYRMIONS

PART IV

GAUGE, GRAVITY AND STRING THEORIES

Chapter 1

INTRODUCTION

The dynamics of a system in classical mechanics can be described by equations of motion on a configuration space Q. These equations are generally of second order in time. Thus if the position $q(t_0)$ of the system in Q and its velocity $\dot{q}(t_0)$ are known at some time t_0, then the equations of motion uniquely determine the trajectory $q(t)$ for all time t.

When the classical system is quantized, the state of a system at time t_0 is not specified by a position in Q and a velocity. Rather, it is described by a wave function ψ which in elementary quantum mechanics is a (normalized) function on Q. The correspondence between the quantum states and wave functions however is not one to one since two wave functions which differ by a phase describe the same state. The quantum state of a system is thus an equivalence class $\{e^{i\alpha}\psi | \alpha \text{ real}\}$ of normalized wave functions. The physical reason for this circumstance is that experimental observables correspond to functions like $\psi^{\star}\psi$ which are insensitive to this phase.

In discussing the transformation properties of wave functions, it is often convenient to enlarge the domain of definition of wave functions in elementary quantum mechanics in such a way as to naturally describe all the wave functions of an equivalence class. Thus instead of considering wave functions as functions on Q, we can regard them as functions on a larger space $\hat{Q} = Q \times S^1 \equiv \{(q, e^{i\alpha})\}$. The space $\hat{Q}$ is obtained by associating circles S^1

to each point of Q and is said to be a $U(1)$ bundle on Q. Wave functions on $\hat{Q}$ are not completely general functions on $\hat{Q}$, rather they are functions with the property $\psi(q, e^{i(\alpha+\theta)}) = \psi(q, e^{i\alpha})e^{i\theta}$. [Here we can also replace $e^{i\theta}$ by $e^{ni\theta}$ where n is a fixed integer]. Because of this property, experimental observables like $\psi^\star\psi$ are independent of the extra phase and are functions on Q as they should be. The standard elementary treatment which deals with functions on Q is recovered by restricting the wave functions to a surface $\{(q, e^{i\alpha_0})|q \in Q\}$ in Q where α_0 has a fixed value. Such a choice α_0 of α corresponds to a phase convention in the elementary approach.

When the topology of Q is nontrivial, it is often possible to associate circles S^1 to each point of Q so that the resultant space $\hat{Q} = \{\hat{q}\}$ is not $Q \times S^1$ although there is still an action of $U(1)$ on $\hat{Q}$. We shall indicate this action by $\hat{q} \to \hat{q}e^{i\theta}$. It is the analogue of the transformation $(q, e^{i\alpha}) \to (q, e^{i\alpha}e^{i\theta})$ we considered earlier. We shall require this action to be free, which means that $\hat{q}e^{i\theta} = \hat{q}$ if and only if $e^{i\theta}$ is the identity of $U(1)$. When $\hat{Q} \neq Q \times S^1$, the $U(1)$ bundle $\hat{Q}$ over Q is said to be twisted. It is possible to contemplate wave functions which are functions on $\hat{Q}$ even when this bundle is twisted provided they satisfy the constraint $\psi(\hat{q}e^{i\theta}) = \psi(\hat{q})e^{ni\theta}$ for some fixed integer n. If this constraint is satisfied, experimental observables being invariant under the $U(1)$ action are functions on Q as we require. However, when the bundle is twisted, it does not admit globally valid coordinates of the form $(q, e^{i\alpha})$ so that it is not possible (modulo certain technical qualifications) to make a global phase choice, as we did ealier. In other words, it is not possible to regard wave functions as functions on Q when $\hat{Q}$ is twisted.

The classical Lagrangian L often contains complete information on the nature of the bundle $\hat{Q}$. We can regard the classical Lagrangian as a function on the tangent bundle $T\hat{Q}$ of $\hat{Q}$. The space $T\hat{Q}$ is the space of positions in $\hat{Q}$ and the associated velocities. When $\hat{Q}$ is trivial, it is possible to reduce any such Lagrangian to a Lagrangian on the space TQ of positions and velocities associated with Q thereby obtaining the familiar description. On the other hand, when $\hat{Q}$ is twisted, such a reduction is in general impossible. Since the equations of motion deal with trajectories on Q and not on $\hat{Q}$, it is necessary that there is some principle which renders the additional $U(1)$ degrees of freedom in such a Lagrangian nondynamical. This principle is the principle of gauge invariance for the gauged group $U(1)$. Thus under the gauge transformation $\hat{q}(t) \rightarrow \hat{q}(t)e^{i\theta(t)}$, these Lagrangians change by constant times $d\theta/dt$, where t is the time. Since the equations of motion therefore involve only gauge invariant quantities which can be regarded as functions of positions and velocities associated with Q, these equations describe dynamics on Q. The Lagrangians we deal with in this book split into two terms L_0 and L_{WZ}, where L_0 is gauge invariant while L_{WZ} changes as indicated above. This term L_{WZ} has a geometrical interpretation. It is the one which is associated with the nature of the bundle $\hat{Q}$.

In particle physics, such a topological term was first discovered by Wess and Zumino [1] in their investigation of nonabelian anomalies in gauge theories. The importance and remarkable properties of such "Wess-Zumino terms" have been forcefully brought to the attention of particle physicists in recent years because of the realization that they play a critical role in creating fermionic states in a theory with bosonic fields and in determining the anomaly structure of effective field theories.

In point particle mechanics, the existence and significance of Wess-Zumino terms have long been understood. For example, such terms play an essential role in the program of geometric quantization [2] and related investigations which study the Hamiltonian or Lagrangian description of particles of fixed spin [3-18]. A similar term occurs in the description of the charge-monopole system [19-24] and has also been discussed in the literature. Recently such terms have been found in dual string models as well [25,26].

The Wess-Zumino term affects the equations of motion and has significant dynamical consequences already at the classical level. Its impact however is most dramatic in quantum theory where as was indicated above it affects the structure of the state space. For example, in the $SU(3)$ chiral model it is this term which is responsible for the fermionic nature of the Skyrmion [27].

There are many interesting examples in classical mechanics and in field theory where a Wess-Zumino term occurs. As indicated above, two such examples in classical mechanics are particles of fixed spin and the charge-monopole system. The $SU(3)$ Skyrme model provides a physically important field theoretic example. In the succeeding Chapters, we will try to explain principal fibre bundles and Wess-Zumino terms using these two classical mechanical examples. We shall then describe a general method for the construction of the space $\hat{Q}$ and of wave functions in the presence of a Wess-Zumino term. We shall also describe some of the implications of this construction for the symmetry properties of the quantum system. It has been understood for some time that these implications can be far reaching. In many of the systems of interest, there is a group G which operates on the configuration space Q, and the action, including the contribution of the Wess-Zumino term, is invariant under G. The wave functions in the presence of topologically nontrivial Wess-Zumino terms are functions on $\hat{Q}$ and not on Q. In general it is not the group G but a different group $\hat{G}$ which acts on $\hat{Q}$ and hence on these wave functions. For the charge-monopole system, for example, G is the group $SO(3)$ of spatial rotations while $\hat{G}$ can be taken to be $U(2)$ when the product eg of electric and magnetic charges is one half of an odd integer. The charge-monopole system in this case is therefore spinorial, which is clearly an important physical effect.

Wess-Zumino terms also play a crucial role in the theory of anomalies. Anomalies arise for example when the Wess-Zumino term L_{WZ} (suitably interpreted) is not invariant under the action of the group G although L_0 is. In that case, there is no group $\hat{G}$ which acts on $\hat{Q}$ and on wave functions. The group G therefore becomes anomalous. In this book we will not be dealing with the theory of anomalies. Rather we refer the reader to the many excellent reviews [28-32].

The exposition of the subject matter starts with a brief description of the Dirac-Bergmann theory of constraints, which is an important tool to deal with gauge theories. This is so since constraints always appear in the Hamiltonian formulation of gauge theories. Particles of fixed spin and the charge-monopole system are then treated and used to explain principle fibre bundles and Wess-Zumino terms. The general construction of $\hat{Q}$ and $\hat{G}$ are then described. This construction is then used for the charge-monopole system to show how the classical symmetry group $SO(3)$ becomes $U(2)$ in quantum mechanics when eg is half an odd integer.

The preceding remarks on the nature of wave functions in quantum theory can be generalized by replacing the group $U(1)$ by more general abelian or nonabelian groups. We shall explain this generalization in Chapter 8. A particularly important class of physical systems where such groups are discrete are those with multiply connected configuration spaces. Chapter 8 also discusses the quantization of such systems in detail. As we shall see, such configuration spaces occur in many physical situations, including molecular systems, collective models of nuclei, nonabelian gauge theories, gravity and $3+1$ dimensional string models.

The succeeding Chapters describe several models which exhibit interesting topological and bundle theoretic features, many of these models being of physical importance as well. Chapter 9 initiates the study of topological solitons and describes these solitons in one and two dimensions, while in Chapter 10 we describe a formulation of "nonlinear" models as gauge theories. In Chapter 11 we describe the Chern-Simons term and its role in fixing the properties of certain $2 + 1$ dimensional solitons. The next seven Chapters are devoted to the QCD effective Lagrangian and Skyrmions. Chapter 19 deals with electroweak Skyrmions. Although these are not topological solitons, they resemble Skyrmions in many ways, and because of this and because of their physical interest it has seemed to us appropriate to include this Chapter. The next three Chapters discuss certain topological aspects of gauge, gravity and string theories. In Chapter 20, we explain the θ states of gauge theories and their gravitational analogues, and how they are related to the connectivity of the appropriate configuration spaces. Chapter 21 deals with a family of very interesting topological excitations called geons in quantum gravity, which seem to admit an interpretation as particles. Many of the rich structures allowed in principle for quantum wave functions, including certain exotic ones, are seen to be realized in the geonic context. In particular, multiply connected configuration spaces (with both abelian and nonabelian fundamental groups) play a role in the description of these geons, whereas configuration spaces with non-abelian fundamental groups do not as a rule occur in Poincaré invariant gauge theories. Chapter 22 discusses certain remarkable statistical possibilities which exist for identical strings and geons in $3 + 1$ dimensions. The book concludes with some final remarks in Chapter 23.

PART I

CLASSICAL MECHANICS

AND

QUANTUM STATES

Chapter 2

THE DIRAC-BERGMANN THEORY OF CONSTRAINTS

2.1 INTRODUCTION

Constraints appear in the Hamiltonian formulation of almost all of the systems which we will discuss in this text. For this purpose we shall be applying the Dirac-Bergmann constraint theory. For readers unfamilar with the subject, we therefore give a very brief summary of this theory of constraints in the discussion which follows. For further expositions, along with many examples many of the excellent and detailed reviews can be resulted [33-37].

Let M be the configuration space appropriate to a Lagrangian L. It is the space Q on which equations of motion give trajectories if the Lagrangian is of the sort treated in elementary classical mechanics. More generally, it can be different from Q as indicated by our previous discussion in the Introduction. We denote the points of M by $m = (m_1, m_2, \ldots)$.

Now given any manifold M, it is possible to associate two spaces TM and T^*M to M. The space TM is called the tangent bundle over M. The coordinate of a point $(m, \dot{m})$ $[\dot{m} = (\dot{m}_1, \dot{m}_2, \ldots)]$ of TM can be interpreted as a position and a velocity. The Lagrangian is a function on TM. The space T^*M is called the cotangent bundle over M. The coordinate of a point (m, p) $[p = (p_1, p_2, \ldots)]$ of T^*M can be interpreted as a position and a momentum so that in physicists' language, T^*M is the phase space. At each m, p belongs to the vector space dual to the vector space of velocities.

Poisson brackets (PB's) can be defined for any cotangent bundle T^*M. In the notation familiar to physicists, they read

$$\begin{aligned}
\{m_i, m_j\} &= \{p_i, p_j\} = 0, \\
\{m_i, p_j\} &= \delta_{ij} \quad .
\end{aligned} \tag{2.1}$$

Now given a Lagrangian L, there exists a map from TM to T^*M defined by

$$(m, \dot{m}) \rightarrow \left(m, \frac{\partial L}{\partial \dot{m}}(m, \dot{m}) \right) \quad . \tag{2.2}$$

If this map is globally one to one and onto, the image of TM is T^*M and we can express any velocity as a function of position and momentum. This is the case in elementary mechanics and leads to the familiar rules for the passage from Lagrangian to Hamiltonian mechanics.

2.2 CONSTRAINT ANALYSIS

It may however happen that the image of TM under the map (2.2) is not all of T^*M. Suppose for instance, that it is a submanifold of T^*M defined by the equations

$$P_j(m, p) = 0 \quad , \qquad j = 1, 2, \ldots \quad . \tag{2.3}$$

Then we are dealing with a theory with constraints. The constraints P_j are said to be *primary*.

The functions P_j do not identically vanish on T^*M. Rather their zeros define a submanifold of T^*M. A reflection of the fact that P_j are not zero functions on $T^\star M$ is that there exist functions g on $T^\star M$ such that $\{g, P_j\}$ do not vanish on the surface $P_j = 0$. These functions g generate canonical transformations which take a point of the surface $P_j = 0$ out of this surface. It follows that it is incorrect to take PB's of arbitrary functions with both sides of the equations $P_j = 0$ and equate them. This fact is emphasized by replacing the "strong" equality signs $=$ of these equations by "weak" equality signs $\approx$, i.e.

$$P_j(m, p) \approx 0 \quad . \tag{2.4}$$

When $P_j(m, p)$ are weakly zero, we can in general set $P_j(m, p)$ equal to zero only after evaluating all PB's.

In the presence of constraints, the Hamiltonian can be shown to be

$$
\begin{aligned}
H &= \dot{m}_j \frac{\partial L}{\partial \dot{m}_j}(m, \dot{m}) - L(m, \dot{m}) + v_j P_j(m, p) \\
&\equiv H_0(m, p) + v_j P_j(m, p) \quad .
\end{aligned}
\tag{2.5}
$$

In obtaining H_0 from the first two terms of the first line, one can freely use the primary constraints. The functions v_j are as yet undetermined Lagrange multipliers. Some of them may get determined later in the analysis while the remaining ones will continue to be unknown functions with even their time dependence arbitrary.

Consistency of dynamics requires that the primary constraints are preserved in time. Thus we require that

$$\{P_m, H\} \approx 0 \quad . \tag{2.6}$$

These equations may determine some of the v_j or they may hold identically when the constraints $P_j \approx 0$ are imposed. Yet another possibility is that they lead to "secondary constraints"

$$P'_m(q, p) \approx 0 \quad . \tag{2.7}$$

The requirement $\{P'_m, H\} \approx 0$ may determine more of the Lagrange multipliers or lead to tertiary constraints (or be identically satisfied). We proceed in this fashion until no more new constraints are generated.

Let us denote all the constraints one obtains in this way by

$$C_k \approx 0 \quad . \tag{2.8}$$

Dirac divides these constraints into *first class* and *second class* constraints. First class constraints $F_\alpha \approx 0$ are those for which

$$\{F_\alpha, C_k\} \approx 0 \quad , \quad \forall k \quad . \tag{2.9}$$

In other words, the Poisson brackets of F_α with C_k vanishes on the surface defined by (2.8)]. The remaining constraints S_a are defined to be second class.

It can be shown that

$$\{F_\alpha, F_\beta\} = C_{\alpha\beta}^\gamma F_\gamma \quad , \tag{2.10}$$

where $C_{\alpha\beta}^\gamma$ $(= -C_{\beta\alpha}^\gamma)$ are functions on T^*M. This may be proved as follows: Eq.(2.9) implies that $\{F_\alpha, F_\beta\} = C_{\alpha\beta}^\gamma F_\gamma + D_{\alpha\beta}^a S_a$. But on using the Jacobi identity

$$\{F_\alpha, \{F_\beta, S_a\}\} + \{F_\beta, \{S_a, F_\alpha\}\} + \{S_a, \{F_\alpha, F_\beta\}\} = 0 \quad ,$$

we find,

$$0 \approx \{F_\alpha, \{F_\beta, S_a\}\} \approx \{S_a, \{F_\alpha, F_\beta\}\} \approx D_{\alpha\beta}^b \{S_a, S_b\} \quad .$$

In obtaining this result, we have used (2.9) which implies that $\{F_\alpha, S_a\}$ is of the form $\sum_k \xi_a^k C_k$. Now as regards S_a, we have the basic property [33-37]

$$\det(\{S_a, S_b\}) \neq 0 \tag{2.11}$$

on the surface $C_k \approx 0$. Thus the matrix $(\{S_a, S_b\})$ is nonsingular on the surface $C_k \approx 0$. It then follows that $D_{\alpha\beta}^b$ vanishes, proving (2.10).

Let $\mathcal{C}$ be the submanifold of T^*M defined by the constraints

$$\mathcal{C} = \{(m,p) \mid C_k(m,p) = 0\} \quad . \tag{2.12}$$

Then since the canonical transformations generated by F_α preserve the constraints, a point of $\mathcal{C}$ is mapped onto another point of $\mathcal{C}$ under the canonical transformations generated by F_α. Since the canonical transformations generated by S_a do not preserve the constraints, such is not the case for S_a.

Second class constraints can be eliminated by introducing the so-called Dirac brackets. They have the basic property that the Dirac bracket of S_a with any function on $T^\star M$ is zero. We will not go into their details having no use for them in this book. Instead, we shall later follow the alternative route of finding all functions $\mathcal{F}$ with zero PB's with S_a. So long as we work with only such functions, we can use the constraints $S_a \approx 0$ as strong constraints $S_a = 0$ and eliminate variables using them even before taking PB's. Assuming no first class constraints, the number N of functionally independent functions $\mathcal{F}$ is dimension of $T^\star M$ - number of S_a, $N = \dim(T^\star M) - s$, s being the range of a. Thus s second class constraints eliminate s variables. Since $(\{S_a, S_b\})$ is nonsingular and antisymmetric, s is even. Since $\dim(T^\star M)$ is even as well, N is even.

2.3 QUANTIZATION PROCEDURE

Let us now instead imagine that there are only first class constraints and that $\mathcal{C}$ is defined exclusively by the zeros of F_α. (If there are second class constraints S_a as well, they can first be eliminated in the manner indicated above.) Dirac's prescription for the implementation of first class constraints in quantum theory is that they be imposed as conditions on the physically allowed states $\mid >$:

$$\hat{F}_\alpha \mid > = 0 \quad . \tag{2.13}$$

Here $\hat{F}_\alpha$ is the quantum operator corresponding to the classical function F_α.

Since the PB's between F's involve only F's, this prescription is consistent (modulo factor ordering problems). That is, both sides of the equation

$$[\hat{F}_\alpha, \hat{F}_\beta] = i C^\gamma_{\alpha\beta} \hat{F}_\gamma \tag{2.14}$$

annihilate the physical states. [A similar argument shows that we can not impose the condition $\hat{S}_a| > 0$ on physical states where $\hat{S}_a$ is the operator corresponding to the function S_a.]

An observable $\hat{\mathcal{O}}$ of the theory must preserve the condition (2.13) on the physical states. Requiring that $\hat{\mathcal{O}}| >$ is physical if $| >$ is, we find, for the set of quantum observables $\hat{\mathcal{A}}$, the condition

$$[\hat{\mathcal{O}}, \hat{F}_\alpha] = id_\alpha^\gamma(\hat{\mathcal{O}})\hat{F}_\gamma \ , \quad \hat{\mathcal{O}} \in \hat{\mathcal{A}} \ . \tag{2.15}$$

For classical observables $\mathcal{O}$ this becomes

$$\{\mathcal{O}, F_\alpha\} = d_\alpha^\gamma(\mathcal{O})F_\gamma \ . \tag{2.16}$$

Since the right hand side is zero on $\mathcal{C}$, we can regard $\mathcal{O}$ as a function on $\mathcal{C}$ which is constant on the orbits generated by F_α. If we regard these orbits as generating an equivalence relation $\sim$ between points of $\mathcal{C}$, then the classical observables are functions on the quotient of $\mathcal{C}$ by $\sim$. This quotient $\mathcal{C}/\sim$ may be regarded as the physical phase space. Note that if there are f first class constraints, then the dimension $\dim[\mathcal{C}/\sim]$ of the physical phase space is $\dim(T^*M) - 2f$, $\mathcal{C}$ having dimension $\dim(T^*M) - f$ and each orbit in $\mathcal{C}$ having dimension f.

An alternative method to deal with F_α consists in directly finding all the classical observables $\mathcal{O}$ and the corresponding classical PB algebra $\mathcal{A}$ of observables. This is the algebra of functions on $\mathcal{C}/\sim$. We then quantize it by replacing $\{.,.\}$ by $-i\,[\cdot,\cdot]$ and thus find $\hat{\mathcal{A}}$, and then look for its representation on a Hilbert space. In this approach, unlike in Dirac's approach we do not first find a vector space V on which $\hat{F}_\alpha$ acts nontrivially and then define physical states by the condition $\hat{F}_\alpha| >= 0$. Rather, we directly look for a suitable irreducible representation of $\hat{\mathcal{A}}$. We will have occasion to use this method later.

In many examples, $C_{\alpha\beta}^\gamma$ are constants so that F_α generate a group. This group is in fact the Hamiltonian version of the group of gauge transformations for the action. Hence one says that first class constraints generate gauge transformations. An important fact one can prove is that the only undetermined

Lagrange multipliers in H at the end of the constraint analysis multiply first class constraints. Since $\{\mathcal{O}, F_\alpha\} \approx 0$ for an observable, it follows that the time evolution of $\mathcal{O}$ does not depend on these arbitrary functions. Thus a well defined Cauchy problem can be posed on $\mathcal{A}$ and the time evolution of $\mathcal{O}$ can be determined uniquely from suitable initial data. The theory is therefore deterministic if we consider only $\mathcal{A}$. This ceases to be the case when nonobservables are also considered since their time evolution is influenced by the unknown Lagrange multipliers v_j.

Finally, we notice that there is an important symmetry structure associated with the first class constrained surfaces in phase space, the so called BRS [38] T [39] symmetry, which is frequently used in the quantization of gauge theories, which are typically theories with a closed algebra of first class constraints. [For a review and further references see e.g. Refs. 40.] It has more recently been developed for handling second class constraints as well [41]. We will not touch upon these considerations since we shall have no compelling reason for using this approach to quantization.

Chapter 3

NONRELATIVISTIC PARTICLES WITH FIXED SPIN

3.1 INTRODUCTION

A classical nonrelativistic particle with fixed spin is an example of an elementary system where the utility of the fibre bundle formalism can be illustrated. The Hamiltonian description of such systems is well known and is recalled below. The construction of a Lagrangian description however is not straightforward. One such construction involves the use of nontrivial fibre bundles as we shall illustrate.

3.2 THE HAMILTONIAN DESCRIPTION

Let $\vec{x} = (x_1, x_2, x_3)$, $\vec{p} = (p_1, p_2, p_3)$ and $\vec{S} = (S_1, S_2, S_3)$ denote the coordinate, the momentum and the spin of the particle. Here we want to describe a particle with a definite spin. We therefore impose the constraint

$$S^2 = S_i S_i = \lambda^2 \quad , \qquad (3.1)$$

where λ is a constant.

The Poisson brackets are

$$\{x_i, x_j\} = \{p_i, p_j\} = 0 \ , \qquad (3.2)$$

$$\{x_i, p_j\} = \delta_{ij} \ , \qquad (3.3)$$

$$\{S_i, S_j\} = \varepsilon_{ijk} S_k \ , \qquad (3.4)$$

where ε_{ijk} is the usual Levi-Civita symbol.

If the particle is free, the Hamiltonian of the system is

$$H_0 = \frac{\vec{p}^2}{2m} \quad , \qquad (3.5)$$

where m is the mass of the particle. If there is an external magnetic field $\vec{B} = (B_1, B_2, B_3)$ present and the particle has a magnetic moment μ, the Hamiltonian has the following form:

$$H = H_0 + \mu \vec{S} \cdot \vec{B} \quad . \qquad (3.6)$$

The equations of motion for the free particle and the interacting particle are, respectively,

$$\dot{x}_i = p_i/m \quad , \qquad (3.7)$$

$$\dot{p}_i = 0 \quad , \qquad (3.8)$$

$$\dot{S}_i = 0 \qquad (3.9)$$

and

$$\dot{x}_i = p_i/m \quad , \qquad (3.10)$$

$$\dot{p}_i = -\mu S_j \partial_i B_j \quad , \quad \partial_i B_j \equiv \frac{\partial B_j}{\partial x_i} \quad , \qquad (3.11)$$

$$\dot{S}_i = \mu \varepsilon_{ijk} B_j S_k \quad . \qquad (3.12)$$

3.3 THE LAGRANGIAN DESCRIPTION

If we know the Hamiltonian description, it is often possible to find the Lagrangian of the system by a Legendre transformation. We can perform the Legendre transformation provided we can find coordinates for the phase space which are canonical. By this we mean the following. Let Q denote the configuration space of the system under consideration. The phase space T^*Q in our case is eight-dimensional. A canonical system of coordinates for this space is by definition of the form

$$T^*Q \equiv \{(Q_1, Q_2, Q_3, Q_4, P_1, P_2, P_3, P_4)\} \quad , \tag{3.13}$$

where

$$\{Q_\alpha, Q_\beta\} = \{P_\alpha, P_\beta\} = 0 \tag{3.14}$$

and

$$\{Q_\alpha, P_\beta\} = \delta_{\alpha\beta} \quad , \tag{3.15}$$

where now α and β run from 1 to 4.

For our system we can, of course, set

$$Q_i = x_i \ , \quad P_i = p_i \ , \quad i \leq 3 \ . \tag{3.16}$$

It remains to find Q_4 and P_4. They will depend on S_i and perhaps on $\vec{x}$ and $\vec{p}$. One can show, however, that there exist no such coordinates Q_4 and P_4 which are smooth functions of $\vec{S}$. From the constraint (3.1), we see that $\vec{S}$ spans a two dimensional sphere which cannot be globally coordinatized by a set of coordinates (Q_4, P_4). Any choice of Q_4 and P_4 will therefore show a singularity for at least one value of $\vec{S}$. This singularity is the analogue of the Dirac string in the theory of magnetic monopoles. We refer to Section 4.5 for a proof of this result.

Thus we cannot find a global Lagrangian by a Legendre transformation when we have a constraint like (3.1). Although it is not possible to find a global Lagrangian by a Legendre transformation, the above system does admit a global Lagrangian description by enlarging the configuration space. We shall now construct it and point out some of its novel features. The Hamiltonian formalism for this Lagrangian is the one discussed above. We will discuss the derivation of this formalism in Chapter 5.

Let $\Gamma = \{s\}$ denote the usual spin $1/2$ representation of the rotation group. Thus we have

$$s^\dagger s = 1 , \quad \det s = 1 . \tag{3.17}$$

Here 1 is the identity matrix.

The configuration space Q for the Lagrangian will be the Cartesian product space $\mathbf{R}^3 \times \Gamma$. The points of $\mathbf{R}^3$ as usual correspond to the position coordinates of the particle while a point $s \in \Gamma$ is related to the spin degrees of freedom S_i through

$$S_i \sigma_i = \lambda s \sigma_3 s^{-1} . \tag{3.18}$$

Here σ_i , $i = 1,2,3$, are the three Pauli matrices. As a consequence of this definition, the constraint (3.1) is fulfilled as an identity. To see this square both sides of (3.18) and take the trace.

The Lagrangian of the free spinning particle is

$$L_{B=0} = \frac{1}{2}m\dot{\vec{x}}^2 + \lambda i \, \mathrm{Tr} \, (\sigma_3 s^{-1}\dot{s}) , \tag{3.19}$$

where the dot indicates differentiation with respect to time. We shall see in detail in Chapter 6 that the second term in $L_{B=0}$ is a Wess-Zumino term [see Chapter 1].

We now verify that $L_{B=0}$ gives the correct equations of motion. Variation of the coordinate $\vec{x}$ leads in a known way to

$$m\ddot{x}_i = 0 , \quad i = 1,2,3 . \tag{3.20}$$

Variation of s can be performed as follows. The most general variations of s can be written in the form

$$\delta s = i \, \varepsilon \cdot \sigma \, s , \tag{3.21}$$

$$\varepsilon \cdot \sigma = \varepsilon_i \sigma_i , \tag{3.22}$$

ε_i being infinitesimal. This is so because $i\varepsilon \cdot \sigma$ is a generic element of the Lie algebra of Γ and the general variation of s is induced by such an element. Equations (3.17) and (3.21) imply that

$$\delta s^{-1} = -i s^{-1} \varepsilon \cdot \sigma \quad . \tag{3.23}$$

Hence for the variation (3.21),

$$\delta L_{B=0} = -\lambda \ \text{Tr} \ (\sigma_3 s^{-1} \dot{\varepsilon} \cdot \sigma s) = -2 S_i \dot{\varepsilon}_i \quad . \tag{3.24}$$

After a trivial partial integration, this yields the required equation of motion

$$\dot{S}_i = 0 \quad . \tag{3.25}$$

If the particle has a magnetic moment μ, the Lagrangian in the presence of an external magnetic field $\vec{B}$ is

$$L_B = L_{B=0} - \frac{\mu}{2} \ \text{Tr}(\mathcal{S}\mathcal{B}) \equiv L_{B=0} - \mu S_i B_i \quad , \tag{3.26}$$

where

$$\mathcal{S} \equiv S_i \sigma_i = \lambda s \sigma_3 s^{-1} \tag{3.27}$$

and

$$\mathcal{B} = B_i \sigma_i \quad . \tag{3.28}$$

In (3.26), during variations, we should regard S_i as a function of s. Now the variation of $\vec{x}$ gives

$$\delta L_B = m \dot{x}_i \delta \dot{x}_i - \mu S_j \partial_i B_j \delta x_i \tag{3.29}$$

where

$$\partial_i B_j \equiv \frac{\partial B_j}{\partial x_i} \quad . \tag{3.30}$$

Hence

$$m \ddot{x}_i = -\mu S_j \partial_i B_j \quad . \tag{3.31}$$

The variation (3.21) of s gives in this case

$$\delta L_B = - \ \text{Tr}(\mathcal{S}\sigma \cdot \dot{\varepsilon}) - \frac{i\mu}{2} \ \text{Tr}([\mathcal{S}, \mathcal{B}]\varepsilon \cdot \sigma) \tag{3.32}$$

where we have used the cyclic property of the trace operation, that is,

$$\text{Tr}(A[B,C]) = \text{Tr}(B[C,A]) \quad . \tag{3.33}$$

Thus

$$\dot{S}_i = \mu\varepsilon_{ijk}B_jS_k \quad . \tag{3.34}$$

Equations (3.31) and (3.34) are the same as those given by the Hamiltonian discussed above.

3.4 GAUGE PROPERTIES OF L_B

The Lagrangian L_B, for arbitrary magnetic field B including zero, exhibits gauge invariance under a gauge group $\mathcal{G}$ which we now discuss in some detail.

Let $U(1) = \{(\exp(i\sigma_3\theta/2)\}$ and consider the transformation

$$s \to s\exp(i\sigma_3\theta/2) \quad , \tag{3.35}$$

where θ in general is time dependent. Under this transformation, L_B changes only by the time derivative of a function, that is,

$$L_B \to L_B + \lambda\dot{\theta} \quad . \tag{3.36}$$

We distinguish this invariance property of a Lagrangian function from the conventional one where the last term in (3.36) is absent by saying that L_B is *weakly invariant* under the gauge transformation (3.35). This weak invariance of L_B clearly suggests that the equations of motion involve only variables invariant under the gauge transformation (3.35). For dynamical variables, "invariance" under the transformation (3.35) of course has the conventional meaning. We may note here that the equations of motion (3.31) and (3.34) in fact only contain the gauge invariant dynamical variables x_i and S_i.

Since L_B changes under the gauge transformation (3.35), it is not possible to write it as it stands in terms of gauge invariant quantities only. We can instead attempt to eliminate the additional gauge degree of freedom in L_B by fixing the gauge. This means the following: We can show that any gauge invariant quantity is a function of S_i (and, of course, of x_i). Gauge fixing means that for each $\vec{S} = (S_1, S_2, S_3)$, we try to find an element $s(\vec{S}) \in SU(2)$ such that

$$S_i \sigma_i = \lambda s(\vec{S}) \sigma_3 s(\vec{S})^{-1} \quad . \tag{3.37}$$

If such an $s(\vec{S})$ existed, we could substitute $s(\vec{S})$ for s in the Lagrangian L_B and thereby eliminate the gauge degree of freedom. We can show, however, that there exists no such choice of $s(\vec{S})$ which is continuous for all $\vec{S}$. The reason for this is as follows. The vectors $\vec{S}$ which satisfy the normalization conditions $S_i S_i = \lambda^2$ span the two sphere S^2. The existence of a smooth $s(\vec{S})$ with the property (3.37) means that

$$SU(2) = S^2 \times U(1) \quad , \tag{3.38}$$

since any point in Γ can then written in the form $s(\vec{S}) \exp(i\sigma_3 \theta/2)$. But $SU(2)$ is simply connected while $U(1)$ on the right hand side of (3.38) is infinitely connected, and so the right hand side of (3.38) is infinitely connected. (Recall that $U(1)$ is topologically identical to the circle S^1.) Hence a smooth $s(\vec{S})$ does not exist.

Thus we have the remarkable situation that a Lagrangian for a non-relativistic system of fixed spin exists only if the space of coordinates and spin variables is nontrivially enlarged to include a $U(1)$ gauge degree of freedom (at least in our approach).

It is often stated in the literature that $U(1)$ gauge invariance implies electromagnetism. But the $U(1)$ gauge invariance of the Lagrangian L seems to have little to do with electromagnetism. Thus the assertions in the literature seem to require qualifications.

3.5 PRINCIPAL FIBRE BUNDLES

The Lagrangians L_B and $L_{B=0}$ are associated with what in differential geometry is called a principal fibre bundle structure [42-45]. We now discuss this bundle structure.

As we have seen above, the configuration space appropriate to the Lagrangian L is the group space $SU(2) = \{s\}$. (More accurately, it is $\mathbf{R}^3 \times SU(2)$. But $\mathbf{R}^3$ being, in this case, the set of positions of the particle under consideration, is not relevant in the present context and will be simply ignored.) On the space $SU(2)$, there is an action

$$s \rightarrow s \exp(i\sigma_3\theta/2) \tag{3.39}$$

of $U(1)$. Under this action, L is weakly invariant for time dependent θ's. If we now define the projection map π by

$$\pi : SU(2) \rightarrow S^2 \quad , \tag{3.40}$$

where

$$\pi : s \rightarrow \lambda s \sigma_3 s^{-1} \equiv S_i \sigma_i \quad , \tag{3.41}$$

weak invariance of L_B implies that the equations of motion depend only on the *base manifold* $S^2 = \{\vec{S}\}$.

Thus we have the following mathematical structure:

1. A manifold $SU(2)$ which topologically is the same as the three sphere S^3,

2. the action of a *structure group* $U(1)$ on the manifold $SU(2)$,

3. the projection map π from $SU(2)$ to the base manifold S^2.

4. Further, the $U(1)$ action is *free*, that is, $sg = s$ for $g \in U(1)$ implies that g equals the identity element e of the structure group $U(1)$.

Note that the projection π maps all the right cosets

$$sU(1) \equiv s\{\exp(i\sigma_3\theta/2)\} \tag{3.42}$$

to a single point on the base space S^2. This right coset is just the orbit of s under the action of the $U(1)$ group. It is also easy to check that distinct orbits have distinct images on S^2 and that the mapping is onto S^2. That is, the space $SU(2)/U(1)$ of the right cosets can be identified with the base space S^2. Thus, if we define an equivalence relation $\sim$ by the statement

$$s_1 \sim s_2 \text{ if } s_1 g = s_2 \text{ for some } g \in U(1) \ , \tag{3.43}$$

then π is just the map from $SU(2)$ to the space of equivalence classes generated by the relation $\sim$, that is, to the space $SU(2)/U(1)$.

The preceding features define a principal fibre bundle with the bundle space S^3 ($\equiv SU(2)$ as a manifold), structure group $U(1)$ and base space S^2. It is a well known structure in mathematics - the Hopf fibration of the two sphere S^2 [see e.g. Ref. 21].

We now give the general definition of a principal fibre bundle. It consists of a bundle space E, a structure group G, a base space B and a projection map π from E onto B. The group $G = \{g\}$ has an action on the bundle space E:

$$E \ni p \to pg \in E \ . \tag{3.44}$$

This action is required to be *free*, that is,

$$\text{for each } p, \ pg = p \text{ implies that } g \text{ equals the identity } e \text{ of } G \ . \tag{3.45}$$

A "fibre" may be defined as a G-orbit in E. The projection π is just the identification of points on a fibre, i.e. points related by the G-action. Thus

$$\pi(p) = \pi(pg) \tag{3.46}$$

while

$$\pi(p) = \pi(q) \tag{3.47}$$

implies that

$$q = pg \tag{3.48}$$

for some $g \in G$. We can think of B as the space of G-orbits in E.

The following point regarding the action $p \to pg$ of G on E may be noted. If it is written in the form $p \to \rho(g^{-1})p \equiv pg$, then $\rho(g_1^{-1})\rho(g_2^{-1}) = \rho([g_2 g_1]^{-1})$. Hence the map $\rho : g \to \rho(g)$ from G to these transformations on E is a homomorphism. It is in fact an isomorphism in view of (3.45). Note that the image of g under ρ acts on E according to $p \to pg^{-1}$ and not according to $p \to pg$. Nevertheless, following the convention in the mathematical literature, we shall often regard the action of g on E as being given by $p \to pg$.

A global section is a map

$$\varphi : B \to E \qquad (3.49)$$

such that

$$\pi \circ \varphi = \text{ identity map on } B . \qquad (3.50)$$

Thus for $b \in B$, $\varphi(b)$ is in E and

$$\pi(\varphi(b)) = b \text{ for all } b \text{ in } B . \qquad (3.51)$$

A local section is defined analogously by restricting the domain of definition of the map $B \to E$ to an open set in B. For suitable open sets in B, a local section always exists. In fact, there is always a covering $\{V_\alpha\}$ of B by open sets V_α where $\underset{\alpha}{\cup} V_\alpha = B$, such that each V_α admits a (local) section.

The principal fibre bundle E is said to be trivial if $E = B \times G$. A principal fibre bundle is trivial *if and only if* it admits a global section. Note that a point p in a trivial bundle is of the form $p = (b, g)$ where $b \in B$ and $g \in G$ while the group acts on E as follows:

$$(b, g) \to (b, gg') , \; g' \in G \quad . \qquad (3.52)$$

Thus the projection map is just

$$\pi(b, g) = b \quad . \qquad (3.53)$$

A (smooth) bijective map $\phi : E \to E$ such that $\phi(pg) = \phi(p)g$ is called a (bundle) automorphism. The gauge group $\mathcal{G}$ is the group of fibre preserving, base invariant automorphisms of E. If e.g. E is a trivial bundle, then $\mathcal{G}$ is the group of (smooth) maps from B to G.

3.6 GAUGE FIXING

In the conventional treatment of gauge theories there is a procedure called *gauge fixing*. It may be explained in the following way. Suppose the space of fields on which the action depends is $\{\xi\}$. Here ξ can be a multi-component, as well as a space-time dependent field. Let the gauge group $\mathcal{G}$ be described by the set $\{g\}$. $\mathcal{G}$ is a time (and possibly space) dependent group, and has the action

$$\xi \to \xi g \qquad\qquad (3.54)$$

on $\{\xi\}$. Fixing the gauge consists of picking exactly one point from each orbit $\{\xi g\}$. This is accomplished by imposing a condition of the form

$$\chi(\xi) = 0 \qquad\qquad (3.55)$$

on ξ. [Here χ of course can be multi-component, $\chi = (\chi_1, \chi_2, \ldots, \chi_n)$. Thus (3.55) can actually be many conditions.] This equation defines a surface M. From the previous remarks, it is clear that the surface M must be such that each orbit cuts M once and exactly once.

If the action (3.54) is free, the previous discussion shows that M is a global section in a principal fibre bundle. In this case, M exists if and only if $\{\xi\}$ is a trivial bundle. Global gauge fixing is possible only in such a case.

In general, the action of the gauge group $\mathcal{G}$ on $\{\xi\}$ can be quite involved. Thus: a) The action of $\mathcal{G}$ may not be free. Then the orbit $\xi\mathcal{G}$ is not diffeomorphic to $\mathcal{G}$ since some elements of $\mathcal{G}$ leave ξ fixed, that is, some degrees of freedom of $\mathcal{G}$ disappear in the map

$$g \to \xi g \quad . \qquad\qquad (3.56)$$

b) The little group (the stability group, the isotropy group) $\mathcal{G}_\xi$ of ξ is the set

$$\mathcal{G}_\xi = \{g \in \mathcal{G} | g = \xi g\} \ . \tag{3.57}$$

It may happen that two distinct points ξ and ξ' have little groups $\mathcal{G}_\xi$ and $\mathcal{G}_{\xi'}$ which are not conjugate in $\mathcal{G}$. That is, there exist no element $\bar{g} \in \mathcal{G}$, such that

$$\bar{g}\mathcal{G}_\xi\bar{g}^{-1} \equiv \{\bar{g}g\bar{g}^{-1} | g \in \mathcal{G}_\xi\} = \mathcal{G}_{\xi'} \ . \tag{3.58}$$

In fact, $\mathcal{G}_\xi$ and $\mathcal{G}_{\xi'}$ may not even be isomorphic. An example is the action of the connected Lorentz group $L_+^\uparrow$ on the Minkowski space M^4. In this case, if for instance $x \in M^4$ is timelike, the little group is $SO(3)$, while if x is spacelike, the little group is the connected $2+1$ Lorentz group. In case b), the different orbits are not diffeomorphic.

If the orbits $\xi\mathcal{G}$ for different ξ are diffeomorphic, we have a *fibration* of the space $\{\xi\}$ by the group $\mathcal{G}$. If there are topologically distinct orbits, we have a *singular fibration* of the space $\{\xi\}$ by the group $\mathcal{G}$.

In Yang-Mills field theory, there are some results which show the nonexistence of a global gauge condition, that is, of a global surface M with the properties discussed above [46]. These results are usually proved either when the Euclidean space-time is compactified to the four sphere S^4 or its time slices are compactified to three spheres S^3.

It may be noted that in principle, it is unnecessary to fix a gauge. The orbits of $\mathcal{G}$ in $\{\xi\}$ are well defined. We can work on the space of these orbits. That is, $\mathcal{G}$ defines an equivalence relation $\sim$ on $\{\xi\}$, ξ and ξ' being equivalent if they are connected by the $\mathcal{G}$-action that is,

$$\xi \sim \xi' \Longleftrightarrow \xi' = \xi g \text{ for some } g \in \mathcal{G} \ . \tag{3.59}$$

The space of orbits is just the space $\{\xi\}$ with $\mathcal{G}$-equivalent points identified, that is, $\{\xi\}/\sim$. Thus for the spinning system discussed above, it is unnecessary to fix a gauge. In fact, a global gauge does not exist for this system since the bundle is non-trivial. For each fixed time, the space $\{\xi\}$ in this case is the three sphere S^3, the group G which is gauged is $U(1)$ and the space of orbits

$S^3/\sim$ is S^2. This example also shows that even if a global gauge does not exist, the space of orbits, or the space of gauge invariant variables, can still be well defined. This space, however, may not be a manifold.

However, the sort of systems like the spinning particle we have discussed are rather exceptional. Here we can readily identify the space of gauge invariant variables in a concrete way. In field theoretical problems, this usually turns out to be difficult to do. The practice in these problems is to fix the gauge by some convenient procedure. We have seen that a global gauge fixing is not always possible. Such a circumstance can cause difficulties in such problems during quantization.

The Lagrangians (3.19) and (3.26) can easily be extended to relativistic motion as well as to coupling to an external gravitational field [3,4] and to an arbitrary number of space-time dimensions [8,9]. They can also be generalized to describe supersymmetric spinning particles [47]. There exist alternative descriptions of spinning particles in terms of anticommuting variables [48-52], but the latter do not exhibit non-trivial fibre bundle structures.

3.7 CONNECTIONS IN A PRINCIPAL FIBRE BUNDLE

In this section we briefly introduce the concept of connections and connection one-forms in a principal fibre bundle. (For more details see e.g. [45].) It is then applied to the example of the Hopf bundle of a two-sphere, and we show that the action associated with the Wess-Zumino term in equation (3.19) is proportional to the integral of the connection one-form. In the discussion which follows and in some later discussions we assume that the reader has a minimal familiarity with vector fields and differential forms.

Let E be the bundle space of a principal fibre bundle. Then for each point $p \in E$ we can define the space $T_p E$ consisting of all tangent vectors at p. Any vector $v \in T_p E$ can be decomposed into *vertical* and *horizontal* components, by which we mean the following: A tangent vector v^v at point p in E is said to be *vertical* if its projection $\pi(v^v)$ to the base manifold B is zero. This is ensured if v^v points in the directions of the fibre passing through p. Let $V_p E$ denote the subspace of $T_p E$ of vertical vectors and $H_p E$ be a complementary subspace of *horizontal* vectors, so that any vector $v \in T_p E$ can be written in the form

$$v = v^v + v^h \quad , \quad v^v \in V_p E \quad , \quad v^h \in H_p E \quad . \tag{3.60}$$

A *connection* is defined to be a *smooth* assignment of $H_p E \subset T_p E$ for all points p in E. In addition, the structure group G of the principal fibre bundle is required to have the following action on $H_p E$:

$$g \circ H_p E = H_{pg} E \tag{3.61}$$

for any $g \in G$.

Let us now apply the above discussion to the Hopf fibration of the two-sphere. There $E = SU(2)$, $G = U(1)$ and the base manifold $B = S^2$. The projection map π was given in equations (3.40) and (3.41). Let us parametrize the group element s by a set of three angles (e.g. Euler angles) $\xi = (\xi_1, \xi_2, \xi_3)$. Thus $s = s(\xi)$. Let $\{v_\alpha , \alpha = 1, 2, 3\}$ denote a basis of tangent vectors at p, i.e. of $T_p E$. They can be written according to

$$v_\alpha(\xi) = M_{\beta\alpha}(\xi) \frac{\partial}{\partial \xi_\beta} \quad , \tag{3.62}$$

where M is a 3×3 matrix which we determine in what follows.

Define v_3 to be a *vertical* tangent vector. It should generate a change in the value of $s(\xi)$ along the direction of the fibre. Thus for specificity we consider $v_3 s(\xi) = s(\xi) \frac{i}{2} \sigma_3$. Let $s[f(\theta)]$, where $f(\theta) = (f_1(\theta), f_2(\theta), f_3(\theta))$, correspond to the fibre parametrized by θ passing through the point $s(\xi) \equiv s[f(0)]$. Then by (3.39) we can take

$$s[f(\theta)] = s(\xi) \exp(i\sigma_3 \theta/2) \quad . \tag{3.63}$$

Upon differentiating (3.63) with respect to θ and setting $\theta = 0$, we find

$$\frac{i}{2}s(\xi)\sigma_3 = \frac{\partial s(\xi)}{\partial \xi_\beta}M_\beta(\xi) \quad ,$$

$$M_\beta(\xi) = \frac{\partial f_\beta(\theta)}{\partial \theta}\Big|_{\theta=0} \quad . \tag{3.64}$$

When multiplied by an infinitesimal quantity, the left hand side of (3.64) represents an infinitesimal change in the value of $s(\xi)$ along the direction of the fibre. From equation (3.62), we can then identify $M_{\beta 3}(\xi) = M_\beta(\xi)$.

Equation (3.63) can be generalized to accomodate arbitrary right transformations on $SU(2)$. Then (3.64) takes the form

$$\frac{i}{2}s(\xi)\sigma_\alpha = \frac{\partial s(\xi)}{\partial \xi_\beta}M_{\beta\alpha}(\xi) \quad ,$$

$$M_{\beta\alpha}(\xi) = \frac{\partial f_\beta(\theta)}{\partial \theta_\alpha}\Big|_{\theta=0} \quad , \tag{3.65}$$

where now $\theta = (\theta_1, \theta_2, \theta_3)$ parametrizes an arbitrary right transformation on $SU(2)$. By definition, horizontal vectors generate changes in $s(\xi)$ in directions transverse to the fibre. From (3.62) and (3.65), one such infinitesimal change is given by

$$\varepsilon_a v_a s(\xi) = s(\xi)(\frac{i}{2}\varepsilon_a\sigma_a) \quad , \tag{3.66}$$

where ε_a are infinitesimal and a runs over 1 and 2. Hence v_1 and v_2 define a basis for a horizontal vector space $H_p E$. [The subspace $H_p E$, and hence the connection, are of course not unique. Another possible choice of $H_p E$ would for instance be a subspace spanned by $v_1 + v_3$ and v_2.]

The matrix M in (3.65) is nonsingular. For if not, there exist χ_α, not all zero, such that $M_{\beta\alpha}\chi_\alpha = 0$. But by (3.65) this would imply that $s(\xi)\sigma_\alpha\chi_\alpha = 0$, and hence that $\sigma_\alpha\chi_\alpha = 0$. But this contradicts the linear independence of the σ's. From (3.65) the inverse of M is explicitly given by

$$M_{\alpha\beta}^{-1}(\xi) = -i \operatorname{Tr}(\sigma_\alpha s(\xi)^{-1}\frac{\partial s(\xi)}{\partial \xi_\beta}) \quad . \tag{3.67}$$

Under the action of the $U(1)$ structure group [Cf. Eq. (3.39)],

$$M_{\alpha\beta}^{-1} \rightarrow R_{\alpha\gamma}(\theta)M_{\gamma\beta}^{-1} \quad , \quad R(\theta) = \begin{bmatrix} \cos\theta & -\sin\theta & 0 \\ \sin\theta & \cos\theta & 0 \\ 0 & 0 & 1 \end{bmatrix} . \tag{3.68}$$

Hence the vertical vector v_3 is invariant and v_1 and v_2 rotate, while moving up and down the fibre.

The connection one-form ω_p at the point $p \in E$ is defined such that it maps vertical tangent vectors at p to elements of the Lie algebra associated with structure group G in a smooth way. It does so in such a manner that the infinitesimal generators of the action of G evaluated at p are mapped into the corresponding elements in the Lie algebra. On the other hand, ω_p maps horizontal vectors to zero. Under the action of G, the connection one-form transforms under the adjoint action of the group.

Note that we can associate a connection one-form ω to any connection (as defined by H_pE) and vice versa. Thus the vertical subspace V_pE is spanned by the infinitesimal generators for the action of the structure group evaluated at p. The one-form ω can hence be defined by requiring that ω_p [the evaluation of ω at p] vanishes on H_pE and maps the infinitesimal generators for the action of the structure group into the corresponding elements in its Lie algebra. Conversely, given ω we can reconstruct H_pE as it is the subspace on which ω_p vanishes. In this correspondence between H_pE and ω_p, Eq. (3.61) is equivalent to the transformation property of ω.

For the case of the Hopf bundle of the two-sphere, the connection one-form ω_s defined at $s \in SU(2)$ is invariant under the action of the $U(1)$ structure group. It can be expressed as

$$\omega_s = \Omega_\alpha(\xi)d\xi_\alpha \quad , \quad s \in E \quad , \tag{3.69}$$

where $d\xi_\alpha$ is the one-form which maps $\partial/\partial\xi_\beta$ to $d\xi_\alpha(\partial/\partial\xi_\beta) = \delta_{\alpha\beta}$. Then (3.69) maps tangent vector v_α [Cf. Eq. (3.62)] to $\omega_s(v_\alpha) = \Omega_\beta(\xi)d\xi_\beta(M_{\gamma\alpha}(\xi)\partial/\partial\xi_\gamma) = M_{\beta\alpha}(\xi)\Omega_\beta(\xi)$. We must require the latter to vanish for $\alpha = 1, 2$, if we choose

v_1 and v_2 to be horizontal vectors. This is possible for the choice

$$\Omega_\alpha(\xi) = M_{3\alpha}^{-1}(\xi) = -i \operatorname{Tr}(\sigma_3 s^{-1} \frac{\partial s}{\partial \xi_\alpha}) \ . \tag{3.70}$$

Consequently,

$$\omega_s = -i \ Tr(\sigma_3 s^{-1} ds) \ . \tag{3.71}$$

Upon comparing with equation (3.19), we conclude that the action associated with the Wess-Zumino term is proportional to the integral of the connection one-form ω_s.

Gauge potentials are often utilized in physics for the description of fibre bundles. The former are (local) functions on the base manifold B and take values in the Lie algebra associated with the structure group G. They can be obtained from the connection one-form ω after specifying (local) sections in E. Let V_a be an open set in B. A local section φ_a maps V_a to a submanifold in the bundle space E in such a way that $\pi(\varphi_a(b)) = b$ for all b in V_a. Then any $p \in E$ such that $\pi(p) \in V_a$ can be written in the form

$$p = \varphi_a(b)g \ , \tag{3.72}$$

where $b \in V_a$ and g is an element of the structure group G. The "pull-back" map φ_a^* is defined in such a way that it maps a function f defined on E to a function defined on V_a , i.e. if $b \in V_a$ then $\varphi_a^* f(b) = f(\varphi_a(b))$. In a similar manner one can define the "pull-back" of a one-form ω, defined on E, to be a one-form $\varphi_a^* \omega$ defined on V_a. [It is done as follows. Let us define a mapping φ_{a*} from $T_b V_a$ to $T_{\varphi_a(b)} E$ for each b in V_a by pulling f back by φ_a^*, i.e. by setting $(\varphi_{a*} X)(f) = X(\varphi_a^* f)$ for any vector field X on V_a. The "pull-back" $\varphi_a^* \omega$ is then defined by $((\varphi_a^* \omega)(X))(b) = \omega(\varphi_{a*} X)(\varphi_a(b))$.] Gauge potentials $A_i^{(a)}$ correspond to the components of the one-form $\varphi_a^* \omega$. They are easily obtained for the example of the Hopf bundle using Eq. (3.71) as we now show.

For the Hopf bundle, V_a are open sets in $S^2 = \{\vec{S}|S_i S_i = \lambda^2\}$. For example, we can choose to define the open set V_N to cover all of S^2 except for one point, say the south pole $[\vec{S} = (0,0,-\lambda)]$. For this consider the stereographic projection of S^2 to $\mathbf{R}^2$, the latter being parametrized by (b_1, b_2), i.e.

$$S_1 = \lambda \frac{2b_1}{1 + (b_1)^2 + (b_2)^2} \ , \quad S_2 = \lambda \frac{2b_2}{1 + (b_1)^2 + (b_2)^2} \ \text{and}$$

$$S_3 = \lambda \frac{1 - (b_1)^2 - (b_2)^2}{1 + (b_1)^2 + (b_2)^2} \ . \tag{3.73}$$

From this mapping we note that the south pole corresponds to the "points at infinity" $[(b_1)^2 + (b_2)^2 \to \infty]$ of $\mathbf{R}^2$. We thus exclude the latter point in defining V_N.

Let φ_N be a section of the Hopf bundle which maps V_N to $SU(2)$. It is required to satisfy $\lambda \varphi_N(b) \sigma_3 \varphi_N^{-1}(b) = \sigma_i S_i$. Any s such that $\pi(s)$ is in V_N can be written in the form

$$s = \varphi_N(b) \exp(\frac{i\theta\sigma_3}{2}) \ . \tag{3.74}$$

The representative of ω on $V_N \times U(1)$ is obtained by substituting (3.74) into (3.71), the result being

$$\omega_s = -i\text{Tr}(\sigma_3 \varphi_N^{-1} d\varphi_N) + d\theta \tag{3.75}$$

The first term in Eq. (3.75) is a one-form on V_N and can be identified with $\varphi_N^* \omega$. Then the associated potential is given by

$$A_i^{(N)} = -i\text{Tr}(\sigma_3 \varphi_N^{-1}(b) \frac{\partial \varphi_N(b)}{\partial b_i}) \ . \tag{3.76}$$

We now make the following choice for the section φ_N on V_N:

$$\varphi_N(b) = \frac{1}{2}\{\alpha \mathbf{1} - \frac{1}{\lambda\alpha}[\sigma_3, S_i\sigma_i]\} \ , \quad \alpha = [2(1 + S_3/\lambda)]^{1/2} \ . \tag{3.77}$$

Here $\vec{S}$ are functions of b as given in Eq. (3.73). It is easy to verify that $\varphi_N(b)$ is a unimodular matrix and fulfills $\lambda \varphi_N(b) \sigma_3 \varphi_N^{-1}(b) = \sigma_i S_i$. Upon substituting (3.77) in Eq. (3.76) we obtain

$$\begin{aligned} A_l^{(N)} &= -\frac{1}{\lambda} \frac{\varepsilon_{3ji} S_i}{(\lambda + S_3)} \frac{\partial S_j}{\partial b_l} \\ &= \frac{1}{\lambda}\varepsilon_{3il} S_i \ . \end{aligned} \tag{3.78}$$

We have used Eq. (3.73) to derive the second line in Eq. (3.78).

Now let φ_1 and φ_2 be two local sections on V_1 and V_2, respectively, where $V_1 \cap V_2 \neq \emptyset$. From Eq. (3.72), the two sections are related by a "gauge transformation" $\varphi_2(b) = \varphi_1(b)g_{12}(b)$ for b in the intersecting region $V_1 \cap V_2$. g_{12} takes values in G.

For the case of the Hopf bundle of the two-sphere the structure group G is $U(1)$ and $g_{12} = \exp(i\Lambda(b)\sigma_3/2)$. Upon substituting into Eq. (3.76), we obtain the usual gauge transformation rule for $U(1)$ potentials:

$$A_i^{(2)} = A_i^{(1)} + \frac{\partial \Lambda}{\partial b_i} \quad . \tag{3.79}$$

[For a generalization of (3.79) to non-Abelian structure groups G, see for example Ref. 45.]

To illustrate the above gauge transformation for the Hopf bundle of S^2, identify V_1 and φ_1 with V_N (defined previously) and φ_N, respectively. Let V_2 be the open set V_S which covers all of S^2 except the north pole $\vec{S} = (0, 0, \lambda)$. The union of V_S and V_N covers all of S^2, while $V_N \cap V_S$ corresponds to S^2 with both poles removed. In analogy with Eq. (3.73) we can parametrize V_S with $(c_1, c_2) \in \mathbf{R}^2$ using a stereographic projection, i.e.

$$S_1 = \lambda \frac{2c_1}{1 + (c_1)^2 + (c_2)^2} \quad , \quad S_2 = \lambda \frac{2c_2}{1 + (c_1)^2 + (c_2)^2} \quad \text{and}$$

$$S_3 = \lambda \frac{(c_1)^2 + (c_2)^2 - 1}{1 + (c_1)^2 + (c_2)^2} \quad . \tag{3.80}$$

Eq. (3.80) associates the north pole to the points at infinity on $\mathbf{R}^2$. The latter points are excluded from V_S. We identify the section φ_2 with φ_S, where

$$\varphi_S(c) = \frac{-i}{2}\{\beta\mathbf{1} + \frac{1}{\lambda\beta}[\sigma_3, S_i\sigma_i]\}\sigma_2 \quad , \quad \beta = [2(1 - S_3/\lambda)]^{1/2} \quad . \tag{3.81}$$

Here S_i are functions of c_1 and c_2 as per Eq. (3.80). Eq. (3.81) satisfies $\lambda\varphi_S(b)\sigma_3\varphi_S^{-1}(b) = \sigma_i S_i$. Away from the north and south poles φ_S is related to φ_N by $\varphi_S = \varphi_N g_{NS}$ where

$$g_{NS} = \frac{2}{\lambda\alpha\beta}(S_1 - iS_2\sigma_3) = \exp(i\Lambda\sigma_3/2) \ ,$$
$$\Lambda = -2\tan^{-1}(S_2/S_1) \ . \tag{3.82}$$

In view of Eq. (3.79) we can define the two-form F, with components $F_{ij} = F_{ij}^{(a)} \equiv \partial_i A_j^{(a)} - \partial_j A_i^{(a)}$ which is independent of the local section φ_a of the Hopf bundle. (There is a similar global two-form for any U(1) bundle.) [For the generalization to non-Abelian structure groups G, see e.g. Ref. 45.] Furthermore, we can define the total flux passing through S^2. For this, let $\{V_a\}$ be a set of nonoverlapping patches of S^2 which covers all of the two-sphere. For example this set can consist of the northern and southern hemispheres, S_N^2 and S_S^2, respectively, with the corresponding local sections φ_N and φ_S. Using Stokes theorem and Eq. (3.78) the flux through S_N^2 is found to be

$$\int_{S_N^2} F = \int_{S_N^2} F_{12} d^2b = \oint_{b_1^2 + b_2^2 = 1} A_i^{(N)} db_i = 2\pi \ . \tag{3.83}$$

The same result applies for the flux passing through S_S^2, giving a total flux of 4π.

3.8 HORIZONTAL LIFTS IN A PRINCIPAL FIBRE BUNDLE

Given a connection one-form, we can define the notion of a *horizontal lift*. Let $C = \{C(t)|0 \leq t \leq 1\}$ be a curve in the base manifold B. The horizontal lift $\hat{C} = \{\hat{C}(t)|0 \leq t \leq 1\}$ of C is defined to be the curve in the bundle manifold E for which $\pi(\hat{C}(t)) = C(t)$ and all vectors tangent to $\hat{C}$ at t are horizontal, that is they are elements of $H_{\hat{C}(t)}E$. (The horizontal lift $\hat{C}$ to E is unique provided the point $p = \hat{C}(0)$ is specified. This is so because it is determined by an ordinary differential equation.) Utilizing the horizontal lift, we can map tangent vectors on the base manifold to horizontal vectors in the bundle space. Thus if $T_b B$ be the space of tangent vectors at the point $b \in B$, we can define the (smooth) linear mapping

$$\sigma_p : T_b B \to H_p E \quad , \quad b = \pi(p) \quad , \tag{3.84}$$

such that $\pi \circ \sigma_p$ is the identity map on $T_b B$. It is constructed as follows. If $w \in T_b B$, let C be any curve in B with tangent vector w at b. Let $\hat{C}$ be its horizontal lift passing through p. The tangent to $\hat{C}$ at p is then defined to be $\sigma_p(w)$. Note that because of the transformation property of $H_p E$ given by Eq. (3.61), the structure group G acts on σ_p according to

$$g \circ \sigma_p = \sigma_{pg} \quad , \quad g \in G \quad . \tag{3.85}$$

Conversely, from the uniqueness property of horizontal lifts, we can use σ_p, as defined above, to reconstruct the horizontal vector space $H_p E$ starting from $T_b B$. σ_p therefore provides yet another means for determining a connection on a principal fibre bundle.

Let us apply the above to the case of the Hopf bundle with $C = \{\vec{S}(t) | 0 \leq t \leq 1\}$ corresponding to a curve on S^2 and $\hat{C} = \{s(t)\}$ denoting its horizontal lift in $SU(2)$. Let d/dt be a vector tangent to $\hat{C}(t)$. It is required to be horizontal. If we choose $H_p E$ to be spanned by v_1 and v_2, then from Eq. (3.66) it follows that

$$\text{Tr}(\sigma_3 s^{-1}(t) \frac{d}{dt} s(t)) = 0 \quad . \tag{3.86}$$

On the open set V_N we can define a local section $\varphi_N(b)$ as in Eq. (3.77). Then $s(t)$ can be expressed in terms of $b = b(t)$ and $\theta = \theta(t)$ using Eq. (3.74). Now write

$$\frac{d}{dt} = \frac{db_i}{dt} \frac{\partial}{\partial b_i} + \frac{d\theta}{dt} \frac{\partial}{\partial \theta} \quad . \tag{3.87}$$

and substitute (3.86) to obtain the following first order differential equation for $\theta(t)$

$$i \frac{d\theta}{dt} + \frac{db_i}{dt} \text{Tr}(\sigma_3 \varphi_N^{-1} \frac{\partial \varphi_N}{\partial b_i}) = 0 \quad , \tag{3.88}$$

or from Eq. (3.76), we find

$$\frac{d\theta}{dt} = A_i^{(N)} \frac{db_i}{dt} \quad . \tag{3.89}$$

So given the curve $b_i(t)$ on B, $\theta(t = 0)$ and the gauge potential $A_i^{(N)}$, we can uniquely determine $\theta(t)$ and hence we can construct the associated horizontal lift $\hat{C}$ of C. It is given by

$$s(t) = \varphi_N(b(t)) \exp\left\{\frac{i\sigma_3}{2}\left[\int_{b(0)}^{b(t)} A_i^{(N)} db_i + \theta(0)\right]\right\} \quad . \tag{3.90}$$

Furthermore, we can use Eq. (3.89) to define the mapping $\sigma_p : T_pB \to H_pE$. Eq. (3.89) gives the tangent to C while $db_i/dt \cdot \partial/\partial b_i$ gives the corresponding tangent vector to $\hat{C}$. Hence the mapping σ_p associated with $A_i^{(N)}$ can be given by

$$\frac{\partial}{\partial b_i} \to \frac{\partial}{\partial b_i} + A_i^{(N)}\frac{\partial}{\partial \theta} \quad . \tag{3.91}$$

Chapter 4

MAGNETIC MONOPOLES

4.1 INTRODUCTION

In this Chapter, we discuss the classical formalism for the description of a nonrelativistic (NR) charged particle in the field of a Dirac magnetic monopole [19-21; for a review of Dirac as well as nonabelian monopoles see e.g. Ref. 22]. This system as well illustrates the utility of the fibre bundle formalism in an elementary context.

4.2 EQUATIONS OF MOTION

Let $\vec{x} = (x_1, x_2, x_3)$ denote the relative coordinates and m the reduced mass of the system. We assume that the magnetic field is Coulomb-like. Then the conventional Lorentz force equation for the relative coordinates reads

$$m\ddot{x}_i = n\frac{1}{r^3}\varepsilon_{ijk}x_j\dot{x}_k \quad . \tag{4.1}$$

Here r is the radial coordinate, ε_{ijk} is the Levi-Civita symbol, $4\pi n$ is the product eg of the electric and magnetic charges e and g, and dots denote time differentiation.

The equation (4.1) reveals a remarkable structure when written in terms of radial and angular variables. Let

$$x_i = r\hat{x}_i \ , \quad \sum_i (\hat{x}_i)^2 = 1 \ . \tag{4.2}$$

Then (4.1) is equivalent to

$$\ddot{r} = r \sum_i (\dot{\hat{x}}_i)^2 \ , \tag{4.3}$$

$$\frac{d}{dt}[\varepsilon_{ijk}x_j(m\dot{x}_k) + n\hat{x}_i] = 0 \ . \tag{4.4}$$

The radial equation (4.3) has the same form as for a NR free particle. But from (4.4), the conserved angular momentum

$$J_i = \varepsilon_{ijk}x_j(m\dot{x}_k) + n\hat{x}_i \tag{4.5}$$

has an additional piece $n\hat{x}_i$ as compared to that of the free particle. It can be interpreted as contributing a helicity

$$\hat{x}_i J_i = n \tag{4.6}$$

along the line joining the particle and the monopole.

4.3 THE HAMILTONIAN FORMALISM

It is much easier to write down a Hamiltonian description of this system than it is to write a Lagrangian description. We describe the former in this Section.

Let $B = \{\vec{x} \in \mathbf{R} | \vec{x} \neq 0\}$ denote the configuration space. Note that we have excluded the origin $\vec{x} = 0$ from B. Thus the charge and the monopole are forbidden to occupy the same spacetime point. The phase space $T^{\star}B$ can be chosen to have coordinates $(\vec{x}, \vec{v})$ where $\vec{v} = (v_1, v_2, v_3)$ denotes the relative velocity of the charge and the monopole.

The equation of motion (4.1) is readily verified to be produced by the Hamiltonian

$$
\begin{aligned}
H & = \frac{1}{2}m\left(\sum_i v_i^2\right) \\
& \equiv \frac{1}{2}mv^2 \quad ,
\end{aligned}
\tag{4.7}
$$

provided the Poisson brackets (PB's) are chosen as follows:

$$\{x_i, x_j\} = 0 \quad , \tag{4.8}$$

$$\{x_i, v_j\} = \delta_{ij}/m \quad , \tag{4.9}$$

$$\{v_i, v_j\} = -\frac{n}{m^2}\varepsilon_{ijk}\frac{x_k}{r^3} \quad . \tag{4.10}$$

Note that the right hand side of (4.10) is proportional to the magnetic field.

As was the case for the spinning NR particle, a global Lagrangian can be found if a canonical system of coordinates (Q, P) for $T^{\star}B$ can be found. It may again be shown, however, that no such global system of coordinates exists. Thus, it is not possible to construct a global Lagrangian by application of a simple Legendre transformation.

4.4 THE LAGRANGIAN FORMALISM

The global Lagrangian can be constructed by enlarging the configuration space B appropriate to the Hamiltonian to a $U(1)$ bundle E over B [21,23,24]. This Lagrangian exhibits weak gauge invariance under time dependent $U(1)$ transformations. As a consequence, the equations of motion are defined entirely on B. The structure of the Lagrangian formalism bears a strong resemblance to the one for the nonrelativistic spinning system, although there are important points of difference as well.

Let $\{s\}$ denote the set of all two by two unitary unimodular matrices, that is, the elements of the $SU(2)$ group in the defining representation. The space E is

$$E = R^1 \times SU(2) \equiv \{(r,s)\} \quad . \tag{4.11}$$

Here r is the radial variable with the restriction $r > 0$. So the charge and the monopole are again forbidden to occupy the same spacetime point. The relation of s to the relative coordinates x_i is given by

$$
\begin{aligned}
\hat{X} &= \sigma_i \hat{x}_i \\
&= s\sigma_3 s^{-1} \quad .
\end{aligned}
\tag{4.12}
$$

In the Lagrangian below, the basic dynamical variables are r and s. So, wherever x_i occurs, it is to be regarded as written in terms of r and s.

The Lagrangian is

$$
\begin{aligned}
L &= \frac{1}{2}m \sum_i \dot{x}_i^2 + ni \, \mathrm{Tr} \, \sigma_3 s^{-1}\dot{s} \\
&= \frac{1}{2}m\dot{r}^2 + \frac{1}{4}mr^2 \, \mathrm{Tr} \, \dot{\hat{X}}^2 + ni \, \mathrm{Tr} \, \sigma_3 s^{-1}\dot{s} \quad ,
\end{aligned}
\tag{4.13}
\tag{4.14}
$$

where the last term is a Wess-Zumino term [see Chapter 1], as we shall see in Chapter 6. In writing (4.14) the identity $\mathrm{Tr} \, \hat{X}\dot{\hat{X}} = 0$ has been used. Variation of r in (4.14) leads directly to Eq.(4.3). The most general variation of s is

$$\delta s = i\varepsilon_i \sigma_i s \quad , \quad \varepsilon_i \text{ real.} \tag{4.15}$$

Hence

$$\delta\hat{X} = i[\varepsilon \cdot \sigma, \hat{X}] \quad , \quad \varepsilon \cdot \sigma = \varepsilon_i \sigma_i \; ; \tag{4.16}$$

$$\delta \, \mathrm{Tr} \, \sigma_3 s^{-1}\dot{s} = i \, \mathrm{Tr} \, \dot{\varepsilon} \cdot \sigma \hat{X} \quad . \tag{4.17}$$

The variation of s in (4.14) thus leads to

$$\mathrm{Tr} \, \varepsilon \cdot \sigma \frac{d}{dt}\{-\frac{1}{2}[\hat{X}, mr^2\dot{\hat{X}}] + n\hat{X}\} = 0 \tag{4.18}$$

where we have used the identity (3.33). The bracketed expression in (4.18) is a linear combination of Pauli matrices and ε_i is arbitrary. Therefore,

$$\frac{d}{dt}\{-\frac{i}{2}[\hat{X}, mr\dot{\hat{X}}] + n\hat{X}\} = 0 \tag{4.19}$$

which is equivalent to

$$\frac{dJ_i}{dt} = 0 \quad , \tag{4.20}$$

that is, to Eq. (4.4).

Thus L leads to both the equations of motion (4.3) and (4.4).

4.5 GAUGE PROPERTIES OF L

The Lagrangian L shows a weak gauge invariance under gauge transformations associated with the $U(1)$ group

$$U(1) = \{g = e^{i\frac{\sigma_3}{2}\theta}\} \quad . \tag{4.21}$$

This is similar to the weak gauge invariance of the Lagrangian for the spinning systems. Under the gauge transformation

$$s \to s e^{i\frac{\sigma_3}{2}\theta} \quad , \quad \theta = \theta(t) \quad , \tag{4.22}$$

we have the weak gauge invariance

$$L \to L - n\dot{\theta} \quad . \tag{4.23}$$

As for the spinning system, associated with L, there is the fibre bundle structure

$$U(1) \to S^3 \to S^2 \quad . \tag{4.24}$$

Again, it is impossible to fix a gauge globally so as to eliminate the $U(1)$ gauge degree of freedom. This is because L is only weakly gauge invariant, and $S^3 \neq S^2 \times U(1)$. Thus there does not exist an $s(\hat{X}) \in SU(2)$ which is continuous for all $\hat{X}$ such that

$$\hat{X} = s(\hat{X})\sigma_3 s(\hat{X})^{-1} \quad . \tag{4.25}$$

As in Eq. (3.75) it is of course possible to find an $s(\hat{X})$ which fails to be continuous only at one point, say the south pole [$\hat{x} = (0, 0, -1)$]. Such an $s(\hat{X})$ is

$$s(\hat{X}) = \frac{1}{2}\{\alpha \mathbf{1} - \frac{1}{\alpha}[\sigma_3, \hat{X}]\} \quad ,$$

$$\alpha = [2(1 + \hat{x}_3)]^{1/2} \quad . \tag{4.26}$$

$s(\hat{X})$ appearing in Eq. (4.26) is a unimodular unitary matrix and fulfills (4.25). $s(\hat{X})$ is also not differentiable at the south pole. Substitution of (4.26) into the interaction term appearing in Eq.(4.13) yields

$$ni \, \text{Tr} \, \sigma_3 s(\hat{X})^{-1}\dot{s}(\hat{X}) = n\varepsilon_{3ij}\hat{x}_i\dot{\hat{x}}_j/(1 + \hat{x}_3) \tag{4.27}$$

which is a conventional form of the interaction Lagrangian with a string singularity along the $-x_3$ axis.

Alternatively, we can cover the two sphere $S^2 = \{\hat{X}\}$ by two coordinate patches U_1 and U_2 and find group elements $s_i(\hat{X})$ which are defined and continuous in U_i and which fulfill (4.25). Substitution of $s_i(\hat{X})$ for s in (4.13) leads to Lagrangians L_i defined on U_i. In the intersection region $U_1 \cap U_2$, in view of (4.25),

$$[s_1(\hat{X})^{-1}s_2(\hat{X})]\sigma_3[s_1(\hat{X})^{-1}s_2(\hat{X})]^{-1} = \sigma_3 \quad . \tag{4.28}$$

This means that s_i differ from each other in the overlapping region by a gauge transformation,

$$s_1(\hat{X}) = s_2(\hat{X})e^{i\frac{\sigma_3}{2}\theta} \tag{4.29}$$

for some $\theta = \theta(t)$. Hence L_1 and L_2 differ by the total time derivative of a function in $U_1 \cap U_2$:

$$L_1 = L_2 - n\dot{\theta} \quad . \tag{4.30}$$

Such a (singularity free) formulation which works with two local Lagrangians is the nonrelativistic analogue of the work of Wu and Yang [20].

Chapter 5

THE CANONICAL FORMALISM AND QUANTIZATION

5.1 INTRODUCTION

In this Chapter we carry out the canonical quantization for the systems discussed in the previous two Chapters. All the Lagrangians presented there are singular, that is, there exist constraints amongst the corresponding phase space variables. We will rely on the Dirac-Bergmann quantization procedure, discussed in Chapter 2, to handle these constraints.

A common feature of the systems which we have presented (and certain ones in later Chapters) is that elements of a group G appear as dynamical variables. In the discussion which follows, we describe a method of treating group elements in setting up the canonical formalism [6,53].

Let $s \in G$ be parametrized by a set of variables (local coordinates) $\xi = (\xi_1, \xi_2, \ldots, \xi_n)$ so that $s = s(\xi)$, n being the dimension of G. (The functional form of $s(\xi)$ will not be important for us.) We can then regard the Lagrangian as a function of ξ and $\dot{\xi}$ (as well as of any other configuration space variables present in the system and of their velocities).

We first note a preliminary identity. Let us define a set of functions $f(\varepsilon) = (f_1(\varepsilon), f_2(\varepsilon), \ldots, f_n(\varepsilon))$, $\varepsilon = (\varepsilon_1, \varepsilon_2, \ldots, \varepsilon_n)$, by

$$e^{i\, T(\alpha)\, \varepsilon_\alpha} s(\xi) = s[f(\varepsilon)], \quad f(0) = \xi \quad , \tag{5.1}$$

where $T(\alpha)$'s form a basis for the Lie algebra of the group with

$$[T(\alpha), T(\beta)] = i\, c_{\alpha\beta\gamma} T(\gamma) \quad . \tag{5.2}$$

Differentiating (5.1) with respect to ε_α and setting $\varepsilon = 0$, we find

$$iT(\alpha)s(\xi) = \frac{\partial s(\xi)}{\partial \xi_\beta} N_{\beta\alpha}(\xi) \quad , \tag{5.3}$$

$$N_{\beta\alpha}(\xi) = \frac{\partial f_\beta(\varepsilon)}{\partial \varepsilon_\alpha}\Big|_{\varepsilon=0} \quad . \tag{5.4}$$

Here $\det N \neq 0$. The proof of this result is the the same as in Section 3.7. If $\det N = 0$, there exist χ_α, not all zero, such that $N_{\rho\sigma}\chi_\sigma = 0$. By (5.3), $\chi_\sigma T(\sigma)s(\xi) = 0$, and hence $\chi_\sigma T(\sigma) = 0$. But this contradicts the linear independence of the $T(\alpha)$'s.

Now the coordinates ξ_α and their conjugate momenta π_α fulfill the Poisson bracket (PB) relations

$$\begin{aligned}
\{\xi_\alpha, \xi_\beta\} &= \{\pi_\alpha, \pi_\beta\} = 0 \quad , \\
\{\xi_\alpha, \pi_\beta\} &= \delta_{\alpha\beta} \quad .
\end{aligned} \tag{5.5}$$

Since N is nonsingular we can replace the phase space variables π_α by t_α where

$$t_\alpha = -\pi_\beta N_{\beta\alpha} \quad . \tag{5.6}$$

From (5.3) and (5.5) it follows that

$$\{t_\alpha, s\} = i\, T(\alpha)s \quad , \tag{5.7}$$

$$\{t_\alpha, s^{-1}\} = -i\, s^{-1}T(\alpha) \quad , \tag{5.8}$$

$$\{t_\alpha, t_\beta\} = c_{\alpha\beta\gamma}\, t_\gamma \quad . \tag{5.9}$$

To prove (5.9) note that from the Jacobi identity and (5.7), we have,

$$\{\{t_\alpha, t_\beta\}, s\} = -\{\{t_\beta, s\}, t_\alpha\} - \{\{s, t_\alpha\}, t_\beta\}$$
$$= i\, c_{\alpha\beta\gamma} T(\gamma) s \quad . \tag{5.10}$$

Thus

$$\{t_\alpha, t_\beta\} = c_{\alpha\beta\gamma} t_\gamma + F \quad , \tag{5.11}$$

where $\{F, s(\xi)\} = 0$. Consequently F is independent of the π's. Substituting $\pi_\alpha = 0$ in (5.11), we find $F = 0$. This proves (5.9). It also follows from a direct calculation using (5.3) and (5.6).

The PB's (5.7) (5.8) (5.9) involving t_α and s are simple and do not require a particular parametrization for $s(\xi)$. We therefore find it convenient to use these variables in canonically quantizing the systems below.

5.2 NONRELATIVISTIC SPINNING PARTICLES

Here we show how the Hamiltonian description for a spinning particle (Eqs. (3.1) - (3.5)) is obtained from the Lagrangian (3.19). [A similar method works for the Lagrangian (3.26).] Now G is $SU(2) = \{s\}$, $T(i) = \sigma_i/2$ and $c_{ijk} = \epsilon_{ijk}$. The phase space coordinates are x_i, p_i, s and t_i, where p_i is canonically conjugate to x_i. From (3.19), we obtain the following primary constraint:

$$P_i = t_i - S_i \approx 0 \quad , \tag{5.12}$$

where S_i is defined in (3.18). From (5.7), (5.8), (5.9),

$$\{P_i, P_j\} = \epsilon_{ijk}(P_k - S_k) \quad . \tag{5.13}$$

Applying Dirac's procedure, the following Hamiltonian is obtained from the Lagrangian (3.19):

$$H = \frac{p^2}{2m} + v_j P_j \quad , \tag{5.14}$$

v_j are Lagrange multipliers. From the requirement that

$$\{P_i, H\} \approx 0 \quad , \tag{5.15}$$

we find that there exist no secondary constraints. Instead, we obtain conditions on v_j:

$$\varepsilon_{ijk} v_j t_k = 0 \quad . \tag{5.16}$$

Since those $v(= v^{(t)})$ in a direction parallel to t are left arbitrary by (5.16), only those variables which have weakly zero PB's with $v_i^{(t)} P_i$ have a well defined time evolution. Only such variables are of physical interest. We will call them observables. Of course, x_i and p_i are observables. In addition, so are t_i and S_i. This follows from

$$\{t_i, P_j\} \approx \varepsilon_{ijk} P_k \quad , \tag{5.17}$$

$$\{s, v_i^{(t)} P_i\} \approx \frac{i}{2} v_i^{(t)} \sigma_i s \sim \frac{i}{2} s \sigma_3 \quad . \tag{5.18}$$

Eq. (5.18) corresponds to an infinitesimal version of the $U(1)$ gauge transformation discussed in Chapter 3. Hence only those functions of s which are invariant under the gauge transformation (3.35) are observables. But these are precisely S_i (or functions thereof). However, we can eliminate S_i by applying the constraints. Thus a complete set of observables describing the reduced phase space are

$$x_i, p_i \text{ and } t_i \tag{5.19}$$

since S_i can be eliminated via the constraints. In so doing note that

$$t_i t_i = \lambda^2 \quad . \tag{5.20}$$

It is easily verified that there is one (independent) first class constraint and two (independent) second class constraints. They can be written in terms of a unit vector $\hat{v}^{(t)}$ parallel to t and two orthogonal unit vectors $\hat{v}^{(a)}$ ($a = 1, 2$) perpendicular to t. [Thus $\hat{v}^{(a)} \cdot \hat{v}^{(t)} = 0$, $\hat{v}^{(a)} \cdot \hat{v}^{(b)} = \delta_{ab}$.] The first class constraint is $\hat{v}_i^{(t)} P_i$ and the second class constraints are $\hat{v}_i^{(a)} P_i$. Our discussion in Chapter 2 shows that four variables must be eliminated by these constraints so that the space of observables must be eight dimensional. The space of observables we have constructed above, given by (5.19) and (5.20), is indeed eight dimensional.

In the quantum theory we elevate x_i, p_i and t_i to operators $\hat{x}_i, \hat{p}_i$ and $\hat{t}_i$ acting on some Hilbert space. Since $\hat{t}_i$ fulfill angular momentum commutation relations when we quantize the system, and also fulfill the constraint (5.20), the quantum mechanics of this system is that of a particle of fixed spin. From elementary quantum mechanics, we know that $t_i t_i = \lambda^2$ can take on only one of the values $0, 1/2(1/2 + 1), 1(1 + 1), \ldots$. Thus quantization is possible in this approach to quantization only if λ^2 has one of these values. In particular, the only integer value λ can assume in this approach to quantization is 0.

Actually, as we shall illustrate below, quantization of the above classical system is not unique. This becomes plausible if we view condition (5.20) as a classical limit of a quantum relation, rather than as a condition which has the precise quantum analogue

$$\hat{t}_i \hat{t}_i = \lambda^2 \quad . \tag{5.21}$$

In the classical limit, λ takes on a large value (compared to $\hbar$, which we have set to one). Thus, unlike in Eq. (5.21), the λ appearing in (5.20) (and hence in the classical Lagrangian) is to be thought of as "large". Hence we obtain an identical classical limit starting from quantum theories whose operators $\hat{t}_i$ obey

$$\hat{t}_i \hat{t}_i = \lambda(\lambda + c) \quad , \tag{5.22}$$

where c is an arbitrary constant. The value of c can in general depend on the quantization procedure employed. In the Dirac approach $c = 0$. We next illustrate another quantization approach, the "Gupta-Bleuler" approach, where $c = 1$ and λ is integral or half-odd integral.

5.3 THE "GUPTA-BLEULER" APPROACH TO QUANTIZATION

There is an alternative method to quantize the Lagrangian (3.19) which leads to the quantization condition

$$\lambda \in \{0, \pm 1/2, \pm 1 \ldots\} \tag{5.23}$$

rather than to the one above. In this method, $|\lambda|$ acquires the meaning of the spin of the state. It is inequivalent to the method described above although in the classical limit spin $\rightarrow \infty$, the ratio of the two values of $|\lambda|$ for a given spin approaches 1.

The difference between the two methods of quantization comes about as follows. In the method of quantization described above, we found a complete set of variables t_i having zero PB's with all the constraints and subject to the requirement $t_i t_i = \lambda^2$. Quantum theory was then constructed by quantizing t_i. In the second method of quantization, on the other hand, the constraints (5.12) are effectively treated as conditions on the physical states using a variant of the Gupta-Bleuler method, whereas the quadratic condition (5.20) is no longer imposed on t_i. This difference in the treatment of the constraints is the source of the inequivalence between these approaches to quantization.

In order to describe the second approach to quantization, it is convenient to first rewrite the constraints (5.12) in an alternative but equivalent way. For this purpose, note first that in view of (5.7), t_α are generators of the $SU(2)$ group which acts from the *left* on s. There is also an $SU(2)$ group acting from the *right* on s. Its generators $\tilde{t}_\alpha$ can be constructed as follows. Let $s \rightarrow R(s)$ define the standard two to one homomorphism from $SU(2)$ to $SO(3)$ defined by

$$s\sigma_\alpha s^{-1} = \sigma_\beta R_{\beta\alpha}(s) \quad . \tag{5.24}$$

A simple computation shows that

$$\{t_\beta R_{\beta\alpha}(s), s\} = is\frac{\sigma_\alpha}{2} \quad , \tag{5.25}$$

so that we can set

$$\tilde{t}_\alpha = -t_\beta R_{\beta\alpha}(s) \quad . \tag{5.26}$$

It follows from this, as before, that

$$\{\tilde{t}_\alpha, \tilde{t}_\beta\} = \varepsilon_{\alpha\beta\gamma}\tilde{t}_\gamma \quad . \tag{5.27}$$

Now if λ is positive in (3.19), we can set $s = s'e^{i\pi\sigma_1/2}$ and get a new Lagrangian with a negative λ and with s' instead of s. We therefore assume without loss of generality that

$$\lambda \leq 0 \quad . \tag{5.28}$$

The constraints (5.12) look very simple in terms of $\tilde{t}_\alpha$:

$$\tilde{t}_\alpha \approx -\lambda\delta_{\alpha 3} = |\lambda|\delta_{\alpha 3} \quad . \tag{5.29}$$

Classically, these are equivalent to one real and one complex constraint:

$$\tilde{t}_3 \approx |\lambda| \quad , \quad \tilde{t}_+ \equiv \tilde{t}_1 + i\tilde{t}_2 \approx 0 \quad . \tag{5.30}$$

We therefore replace (5.20) by the constraints (5.30). It is advantageous to do so because (5.30) form a first class set:

$$\{\tilde{t}_3, \tilde{t}_+\} = -i\tilde{t}_+ \quad . \tag{5.31}$$

Here we do not include $\tilde{t}_+^*$ in the set of constraints (* denotes complex conjugation).

In quantum theory, we will be imposing the constraints (5.30) as conditions on the physical states. For consistency it is then necessary that the observables of the system have (weakly) zero PB's with the constraints (5.30). As t_α and these constraints act on the opposite sides of s, it is evident that

$$\{t_\alpha, \tilde{t}_3\} = \{t_\alpha, \tilde{t}_+\} = 0 \quad , \tag{5.32}$$

so that we can as before regard t_α and functions thereof as observables when we pass to quantum theory. It is legitimate to do since classically the constraints (5.30) implies (5.20).

It may be reiterated here that in contrast to the method of quantization we have already described, we shall not retain the constraint (5.20) when we pass to quantum theory. It is legitimate to do so since classically the constraints (5.30) imply (5.20).

The wave functions of quantum theory can be constructed by starting with the space $\mathcal{F}$ of complex valued functions on $SU(2)$. [The dependence on $\vec{x}$ of wave functions is trivial to take into account and will therefore be ignored here.] The physical wave functions ψ then span the subspace of $\mathcal{F}$ subject to the conditions

$$\tilde{t}_3\psi = |\lambda|\psi \quad ,$$
$$\tilde{t}_+\psi = 0 \quad , \tag{5.33}$$

where now $\tilde{t}_\alpha$ and $\tilde{t}_+$ are quantum operators. These equations have a simple interpretation. Since $\tilde{t}_3$ and $\tilde{t}_+$ are, respectively, the third generator and raising operator of the group $SU(2)$ acting from the right on s, Eqs. (5.31) mean that ψ transforms as a highest weight state under this action. In other words, ψ transforms as a state of spin $|\lambda|$ and third component of spin $|\lambda|$ under this action.

Since $|\lambda|$ is the third component of spin, the quantization condition on λ is

$$|\lambda| \in \{0, 1/2, 1, \ldots\} \quad . \tag{5.34}$$

Let us next display the physical states explicitly. For this purpose, we introduce the scalar product $(\cdot, \cdot)$ on functions in $\mathcal{F}$ by the formula

$$(\alpha, \beta) = \int_{SU(2)} d\mu(s)\alpha(s)^*\beta(s) \quad , \tag{5.35}$$

$d\mu(s)$ being the invariant measure on $SU(2)$. By the Peter-Weyl theorem, any L^2 function f for the norm defined by (5.35) has the expansion [54]

$$f(s) = \sum_{j,\rho,\sigma} \alpha_{\rho\sigma}^j D_{\rho\sigma}^j(s) \quad , \quad \alpha_{\rho,\sigma}^j = \text{constants} \quad , \tag{5.36}$$

D^j being spin j rotation matrices in the conventional basis where the third component of angular momentum is diagonal. Since the physical wave functions are subject to the conditions (5.33), it follows that they are linear combinations of the form

$$\sum_{\rho=-|\lambda|}^{|\lambda|} C_\rho D_{\rho|\lambda|}^{(|\lambda|)}(s) \quad , \quad C_\rho = \text{constants} \quad . \tag{5.37}$$

Their scalar product is also given by (5.35).

The observables $\hat{t}_\alpha$ (the quantum operators associated with t_α) act on the left of s. Under this action, $D_{\rho|\lambda|}^{(|\lambda|)}(s)$ transform as a basis for the carrier space of the $(2|\lambda|+1)$ dimensional unitary irreducible representations (UIR) of $SU(2)$. The states of the system therefore describe a particle of spin $|\lambda|$.

This completes our description of the "Gupta-Bleuler" approach to the quantization of (3.19). There is a generalization of the Lagrangian (3.19) which describes particles whose internal states transform by any UIR of a compact semisimple Lie group. Both methods of quantization described in this Chapter can be adapted to this Lagrangian as well. For details, see Refs. [6,55,53].

We remark finally that the second method of quantization is closely related to the Borel-Weil-Bott approach to the representation theory of compact semisimple Lie groups [56].

5.4 MAGNETIC MONOPOLES

The canonical quantization for the magnetic monopole theory proceeds in a similar fashion to the preceding Section. The essential difference is due to Eq. (4.12). Because of it, the independent phase space coordinates consist of r, p_r, s and t_i, where p_r is canonically conjugate to r. From Eq.(4.14), we find only one primary constraint,

$$P = \hat{x}_i t_i - n \approx 0 \quad , \tag{5.38}$$

where x_i is defined in Eq. (4.12). Computing the Hamiltonian, we have,

$$H = \frac{p_r^2}{2m} + \frac{1}{2mr^2}(t_i t_i - n^2) + vP \quad . \tag{5.39}$$

Here v is a Lagrange multiplier. The constraint (5.38) is rotationally invariant: $\{P, H\} \approx 0$. Hence the requirement that $\{P, H\} \approx 0$ on the reduced phase space leads to no secondary constraints. Since there is only one constraint, it is first class.

As before, observables are those variables which have zero PB's with P. Among them are

$$r, p_r, t_i \text{ and } \hat{x}_i \quad . \tag{5.40}$$

The latter follows from

$$\{P, s\} = \frac{i}{2} s \sigma_3 \tag{5.41}$$

which is analogous to (5.18). As before, (5.41) corresponds to a $U(1)$ gauge transformation, and only those functions of s which are invariant under such transformations are observables. But these are precisely x_i (or functions thereof), so (5.40) corresponds to a complete set of observables subject to the constraint (5.38).

A representation for the quantum theory can be constructed as follows. Let us regard the wave functions as functions of r and s:

$$\psi = \psi(r, s) \quad . \tag{5.42}$$

The position coordinates are diagonal in this representation in view of (5.14). The momentum p_r acts as the usual differential operator on ψ. The operators t_i are the differential operators which represent the elements $\sigma_i/2$ in the left regular representation of $SU(2)$, that is,

$$[\exp(i\theta_i t_i)\psi](r, s) = \psi(r, \exp(-i\theta_i \sigma_i/2)s) \quad . \tag{5.43}$$

The constraint (5.38) is taken into account by imposing the condition

$$\hat{x}_i t_i \psi = n\psi \tag{5.44}$$

on the wave functions. In view of (5.43) and (4.12), this means

$$\psi(r, s \exp(-i\theta \sigma_3/2)) = \psi(r, s) \exp(i\theta n) \quad . \tag{5.45}$$

The scalar product on wave functions is

$$(\psi, \chi) = \int_0^\infty dr r^2 \int_{SU(2)} d\mu(s) \psi^\star(r,s) \chi(r,s) \tag{5.46}$$

where $d\mu$ is the invariant Haar measure on $SU(2)$.

Let $\{D^j(s)\}$ again define the representation of $SU(2)$ with angular momentum j. Wave functions ψ with finite norm have the expansion

$$\psi(r,s) = \sum_j \sum_{\rho,\sigma} \alpha^j_{\rho\sigma}(r) D^j_{\rho\sigma}(s) \quad , \tag{5.47}$$

$D^j_{\rho\sigma}$ being the matrix elements of D^j in the conventional basis with the third component of angular momentum diagonal.

The constraint (5.45) means that in (5.47), only those $\alpha^j_{\rho\sigma}$ with $\sigma = -n$ are nonzero. Thus

$$\{D^j_{\rho,-n}\} \quad , \quad \text{fixed } n \tag{5.48}$$

is a basis for expansions of the form (5.47). Since n is necessarily integral or half integral, we have the Dirac quantization condition

$$2n = \text{ integer} \quad . \tag{5.49}$$

In (5.48), j and ρ are half odd-integral if $2n$ is odd and integral if $2n$ is even.

The quantum mechanics outlined here is essentially equivalent to conventional treatments.

Chapter 6

THE WESS-ZUMINO TERM AND THE PATH SPACE

6.1 INTRODUCTION

In the previous Chapters, we have found global Lagrangians and actions for topologically nontrivial systems like the charge-monopole system. These Lagrangians were functions on $T\hat{Q}$ where $\hat{Q}$ were suitable $U(1)$ bundles over Q. There is an alternative formulation of global actions which works instead with the so called path space $\mathcal{P}Q$ of Q rather than with Q. This formulation is more common in the theoretical particle physics literature. We shall now describe it for the charge-monopole system.

6.2 THE CHARGE-MONOPOLE SYSTEM

We consider a particle of electric charge e which moves in the magnetic field of a monopole of strength g located at the origin. The equation of motion of the charge is thus

$$m\ddot{x}_i = \frac{eg}{4\pi}\varepsilon_{ijk}\frac{x_j\dot{x}_k}{r^3} , \ r^2 = x_\ell x_\ell \ . \tag{6.1}$$

We shall also forbid the charged particle from going through the origin. Thus the configuration space Q is $\mathbf{R}^3 \setminus \{0\}$.

As we know, the Coulomb magnetic field does not admit a globally well defined vector potential. A vector potential A_i which fulfills the equation

$$F_{jk} = \partial_j A_k - \partial_k A_j = -\frac{g}{4\pi} \varepsilon_{jk\ell} \frac{x_\ell}{r^3} \tag{6.2}$$

is necessarily singular. The singularity set can be taken to be a line through the origin, this singular line being the Dirac string. Since the standard action for a charged particle in an electromagnetic field is expressed in terms of the vector potential, it will necessarily be singular for the charge-monopole system. Is there a way of writing an action for this system free of such a singularity? We have seen that the answer is in the affirmative and that a smooth Lagrangian exists if Q is enlarged to a $U(1)$ bundle $\hat{Q}$ over Q. A smooth action for the charge-monopole system can also be written if the configuration space Q is enlarged to the so called path space $\mathcal{P}Q$ over Q. This result was first shown by Zaccaria et al. [57]. We now discuss the construction of this action.

The path space $\mathcal{P}Q$ is defined as follows. Let x_0 be a fixed reference point of Q. (This point may be chosen at will.) Then a point of $\mathcal{P}Q$ is a path Γ from x_0 to a point x. If the path is parametrized by σ, we have thus

$$\Gamma = \{\vec{\Gamma}(\sigma)|0 \leq \sigma \leq 1 \ , \ \vec{\Gamma}(0) = \vec{x}_0, \vec{\Gamma}(1) = \vec{x}\} \ . \tag{6.3}$$

Such a path is time independent. For a time dependent path we use the notation $\vec{\Gamma}(\sigma, t)$ instead of $\vec{\Gamma}(\sigma)$.

The action which describes the interaction of the charge and the monopole is described on the space of these paths. It reads

$$S_{WZ} = e \int_{t_1}^{t_2} dt \int_0^1 d\sigma F_{ij}[\Gamma] \frac{\partial \Gamma_i(\sigma, t)}{\partial \sigma} \frac{\partial \Gamma_j(\sigma, t)}{\partial t} \ ,$$

$$F_{ij}[\Gamma] \equiv -\frac{g}{4\pi} \varepsilon_{ijk} \frac{\Gamma_k(\sigma, t)}{|\vec{\Gamma}(\sigma, t)|^3} \ ,$$

$$|\vec{\Gamma}(\sigma, t)| \equiv [\Gamma_k(\sigma, t)\Gamma_k(\sigma, t)]^{1/2} \ . \tag{6.4}$$

This action may also be called as the Wess-Zumino term [1]. It is manifestly singularity free. We can show that on variation, it gives the correct equation of motion as follows. A general variation appropriate to the action principle is $\delta\Gamma_i$ which vanishes at the initial and final times t_1 and t_2:

$$\delta\Gamma_i(\sigma, t_1) = \delta\Gamma_i(\sigma, t_2) = 0 \quad . \tag{6.5}$$

Further, of course,

$$\delta\Gamma_i(0, t) = 0 \quad . \tag{6.6}$$

Now

$$\delta S_{WZ} = e \int_{t_1}^{t_2} dt \int_0^1 d\sigma [\partial_k F_{ij} \delta\Gamma_k \partial_\sigma \Gamma_i \partial_t \Gamma_j +$$

$$+ F_{ij} \partial_\sigma \delta\Gamma_i \partial_t \Gamma_j + F_{ij} \partial_\sigma \Gamma_i \partial_t \delta\Gamma_i] \quad . \tag{6.7}$$

Since the divergence of the magnetic field is zero away from the origin, we have the Bianchi identity

$$\partial_k F_{ij} + \partial_i F_{jk} + \partial_j F_{ki} = 0 \quad . \tag{6.8}$$

It follows from this identity and the boundary conditions on the variations that

$$\delta S_{WZ} = e \int_{t_1}^{t_2} dt \int_0^1 d\sigma [(-\partial_i F_{jk} - \partial_j F_{ki}) \delta\Gamma_k \partial_\sigma \Gamma_i \partial_t \Gamma_j$$

$$- \partial_\sigma (F_{ij} \partial_t \Gamma_j) - \delta\Gamma_i - \partial_t (F_{ij} \partial_\sigma \Gamma_i) \delta\Gamma_j]$$

$$+ e \int_{t_1}^{t_2} dt [F_{ij} \partial_t \Gamma_j \delta\Gamma_i]_{\sigma=1}$$

$$= e \int_{t_1}^{t_2} dt [F_{ij} \partial_t \Gamma_j \delta\Gamma_i]_{\sigma=1} \quad . \tag{6.9}$$

With the identification of $\Gamma_i(1, t)$ with the trajectory $x_i(t)$ of the charged particle, we see that the variation δS_{WZ} is of the required form.

The integral in S_{WZ} runs over a triangle Δ in Q. The boundary of Δ is

$$\begin{aligned}
\partial\Delta &= \partial\Delta_1 \cup \partial\Delta_1 \cup \partial\Delta_3 \ , \\
\partial\Delta_1 &= \{\Gamma(\sigma, t_1)\} \ , \\
\partial\Delta_2 &= \{\Gamma(\sigma, t_2)\} \ , \\
\partial\Delta_3 &= \{\Gamma(1, t)\} \ .
\end{aligned} \tag{6.10}$$

The variables associated with the boundaries $\partial\Delta_1$ and $\partial\Delta_2$, and the interior Int Δ, of Δ are unphysical since only the variables $\Gamma(1, t) = x(t)$ are identified with the positions of the particle. The variational principle should give no equations of motion for the unphysical variables. No such equations can arise for $\Gamma(\sigma, t_1)$ and $\Gamma(\sigma, t_2)$ since they are not varied in S_{WZ}. The variation of Γ in Int Δ gives no equation because of Bianchi identities. Thus suppose we deform Δ in the interior a little bit forming a surface Δ'. The surfaces Δ and Δ' together form a closed surface $D = \Delta \cup \Delta'$ and the variation of S_{WZ} is

$$\delta S_{WZ} = \frac{e}{2} \int_D F_{ij} d\Gamma_i \wedge d\Gamma_j \ , \tag{6.11}$$

where we have written the integral using differential forms. We can convert δS_{WZ} into an integral over the volume Int$D$ enclosed by D using Stokes' theorem:

$$\delta S_{WZ} = \frac{e}{2} \int_{\text{Int}D} d[F_{ij} d\Gamma_i \wedge d\Gamma_j] \ . \tag{6.12}$$

Since the Bianchi identities are equivalent to the statement

$$d[F_{ij} d\Gamma_i \wedge d\Gamma_j] = 0 \ , \tag{6.13}$$

we see that this δS_{WZ} is identically zero and leads to no equation of motion.

For small variations of Δ, D is a closed surface which does not enclose the origin where the monopole is located. Now given the boundary $\partial\Delta$, there is another surface Δ'' with the same boundary $\partial\Delta$ such that $\Delta \cup \Delta''$ is (diffeomorphic to) a two sphere S^2 and encloses the monopole. In this case, Δ'' can not be deformed to Δ since the origin is not a part of the configuration space Q. Quantum mechanically, we do not want any physical effect which distinguishes our choice of Δ or Δ'' in defining S_{WZ}, such a choice having no physical meaning in the dynamics of the charge-monopole system. This requirement can be enforced at the semiclassical level by demanding that

$$\exp\left[\frac{ie}{2}\int_{\Delta} F_{ij}d\Gamma_i \wedge d\Gamma_j\right] = \exp\left[-\frac{ie}{2}\int_{\Delta''} F_{ij}d\Gamma_i \wedge d\Gamma_j\right] \quad , \tag{6.14}$$

where the minus sign on the right hand side is due to requiring that Δ and Δ'' should be oriented in the same way. This consistency condition can be motivated for instance by the functional integral: when it is fulfilled, the functional integral will have the same value for either of the surfaces Δ and Δ''. Since

$$\frac{e}{2}\int_{\Delta \cup \Delta''} F_{ij}d\Gamma_i \wedge d\Gamma_j = -\frac{eg}{2} \quad , \tag{6.15}$$

we see that the consistency condition is the Dirac quantization condition:

$$eg = 4\pi n, \ n = 0, \ \pm 1, \ , \pm 2, \ldots \quad . \tag{6.16}$$

Summarizing, the two principal features of S_{WZ} are the following:

i) It is the integral of a *closed* differential form:

$$S_{WZ} = \frac{e}{2}\int_{\Delta} F_{ij}d\Gamma_i \wedge \Gamma_j \quad , $$

$$d[F_{ij}d\Gamma_i \wedge d\Gamma_j] = 0 \quad . \tag{6.17}$$

ii) It fulfills the quantization condition

$$\frac{e}{2}\int_{S^2} F_{ij}d\Gamma_i \wedge d\Gamma_j = 2\pi \times \text{ integer} \quad , \tag{6.18}$$

where the integral is over any two sphere in Q.

The relation between the topological action for the charge-monopole system we wrote in Chapter 4,

$$\int dt \ ni \ \text{Tr} \ \sigma_3 s(t)^{-1}\dot{s}(t)$$

$$\equiv \int ni \ \text{Tr} \ \sigma_3 s(t)^{-1}ds(t) \quad , \quad n = \frac{eg}{4\pi} \tag{6.19}$$

and the action S_{WZ} can be explained as follows. First recall that under the gauge transformations

$$s(t) \to s(t) \exp(i\frac{\sigma_3}{2}\theta(t)) \ , \tag{6.20}$$

we have

$$ni \ \text{Tr} \ \sigma_3 s(t)^{-1} ds(t) \to ni \ \text{Tr} \ \sigma_3 s(t)^{-1} ds(t) - n d\theta(t) \ . \tag{6.21}$$

Now let us extend s and θ from functions of one coordinate t to functions $s(\sigma, t)$ and $\theta(\sigma, t)$ of two coordinates σ, t. The range of σ is taken to be $[0, 1]$ and $s(1, t)$ and $\theta(1, t)$ are identified with $s(t)$ and $\theta(t)$. The equation (6.21) is still valid for the gauge transformation (6.20) with s and θ being regarded as functions of σ and t. The differentiation implied by d is now with regard to both σ and t. Since d^2 is zero, it follows that

$$\omega = ni \ d\left[\text{Tr} \ \sigma_3 s(\sigma, t)^{-1} ds(\sigma, t)\right] \to ni \ d\left[\text{Tr} \ \sigma_3 s(\sigma, t)^{-1} ds(\sigma, t)\right] \tag{6.22}$$

under the gauge transformation $s(\sigma, t) \to s(\sigma, t) e^{i\sigma_3 \theta(\sigma, t)/2}$. Since ω is thus gauge invariant, it can be expressed in terms of $\hat{x}_i(\sigma, t)$ defined by $\sigma_i \hat{x}_i(\sigma, t) = s(\sigma, t)\sigma_3 s(\sigma, t)^{-1}$. It is a matter of algebra to verify that the resultant expression is precisely the integrand of S_{WZ} with $\hat{x}(\sigma, t) = \vec{\Gamma}(\sigma, t)/|\Gamma(\sigma, t)|$.

The integral

$$S_{WZ} = \int_\Delta \omega \tag{6.23}$$

can now be expressed using Stokes' theorem as an integral over the boundary:

$$\begin{aligned} S_{WZ} = \ &ni \ \int_0^1 d\sigma \left[\text{Tr} \ \sigma_3 s(\sigma, t_1)^{-1}\frac{\partial s}{\partial \sigma}(\sigma, t_1)\right.\\ &\left. - \text{Tr} \ \sigma_3 s(\sigma, t_2)^{-1}\frac{\partial s}{\partial \sigma}(\sigma, t_2)\right] +\\ &+ ni \int_{t_1}^{t_2} dt \left[\text{Tr} \ \sigma_3 s(t)^{-1}\frac{ds}{dt}(t)\right] \ . \end{aligned} \tag{6.24}$$

The first two terms which involve initial and final times are not varied when the equations of motion are obtained from varying the action. Thus S_{WZ} and the action of Chapter 4 lead to the same equations of motion. We have of course verified this fact directly earlier.

It is worth emphasizing that it is not possible to write the integrand of S_{WZ} globally as $d\psi$ where ψ involves only $\hat{x}(\sigma, t) \in S^2$. Attempts to do so will lead to Dirac string singularities. However it is possible to write it as $ni\ d[\text{Tr}\ \sigma_3 s(\sigma, t)^{-1} ds(\sigma, t)]$, where $ni\ \text{Tr}\ \sigma_3 s(\sigma, t)^{-1} ds(\sigma, t)$ involves $s(\sigma, t)$ which is a point of a $U(1)$ bundle over S^2.

The construction of the Wess-Zumino action described here for the charge-monopole system is readily generalized to a number of other examples. This is illustrated in the next Section and in Chapter 15. Here, for the charge-monopole system, we have also explained the relation between the topological (Wess-Zumino) Lagrangian involving the $U(1)$ bundle $\hat{Q}$ over Q, the Wess-Zumino action S_{WZ} and the two form ω which is its integrand. Although we will not go into the details here, such relationships are generic.

There exists a general method for the construction of the bundle $\hat{Q}$ and of wave functions in the presence of a Wess-Zumino term. The ideas behind this construction are elementary and go back to the earliest work of Dirac on monopoles. We shall describe this construction in the remaining part of this Chapter.

6.3 WAVE FUNCTIONS AND THE BUNDLE $\hat{Q}$ FOR A WESS-ZUMINO TERM

We shall begin this discussion by recapitulating the properties of the Wess-Zumino action once more.

In the mechanics of point particles with configuration space Q, the action involves the "fields" x^i of one variable τ, $(x^1(\tau), x^2(\tau), \ldots)$ being the location of the particle in the configuration space when the parameter labelling its trajectory has the value τ. This mechanics can be regarded as a one dimensional field theory. As we discussed earlier, if it admits a Wess-Zumino term, this term will be the two dimensional integral of a closed two form ω. Written in detail, it will read

$$\int_\Delta \omega \equiv \int_\Delta \omega_{ij}(x)dx^i \wedge dx^i \ , \ x = (x^1, x^2, \ldots) \ , \qquad (6.25)$$

where

$$d\omega = 0 \text{ or } \partial_i \omega_{jk} + \partial_j \omega_{ki} + \partial_k \omega_{ij} = 0 \ , \ \partial_i \equiv \frac{\partial}{\partial x^i} \ , \qquad (6.26)$$

Δ being a two dimensional surface. The integrand of the Wess-Zumino term thus defines a closed two form on the configuration space Q.

The structure of the Wess-Zumino term for general systems is similar to its structure for point particles. The integrand of this term when suitably interpreted always defines a closed two form ω on the configuration space Q. Let us illustrate this explicitly for the charge-monopole system (once more) and for the Skyrmion. The remarkable consequences of this term for the Skyrmion will be discussed later.

For the charge-monopole system, let $\vec{x}$ be the relative coordinate and eg the product of electric and magnetic charges. The space Q is the space of relative coordinates and the Wess-Zumino action is the interaction term

$$S_{WZ} = -\frac{eg}{8\pi} \int_\Delta \varepsilon_{ijk} r^{-3} x^i dx^j \wedge dx^k \ , \ r^2 = \sum_i x_i^2 \ , \qquad (6.27)$$

where Δ is a two surface in Q. The integrand of S_{WZ} is a closed two form ω:

$$\omega = \frac{e}{2} F_{jk}(x)dx^j \wedge dx^k \ ,$$

$$F_{jk} \equiv -\frac{g}{4\pi} \varepsilon_{ijk} \frac{x^i}{r^3} \ ,$$

$$d\omega = 0 \text{ or } \partial_i F_{jk} + \partial_j F_{ki} + \partial_k F_{ij} = 0 \ . \qquad (6.28)$$

In the chiral model [[58], also see Part II and III], for N_f flavours, the dynamical variables are fields U which map the three space $R^3 = \{\vec{x}\}$ to the manifold of the group $SU(N_f)$. We shall think of $SU(N_f)$ concretely in terms of $N_f \times N_f$ unitary matrices of determinant 1. The fields U obey the boundary condition $U(\vec{x}) \to 1$ as $|\vec{x}| \to \infty$ so that they can be regarded as maps of the three sphere S^3 to $SU(N_f)$. We now want to display a closed two form ω on this infinite dimensional space Q using the integrand of the Wess-Zumino term. This amounts to doing the following: Consider a two surface Δ in Q; a point of this two surface is a field parametrized by two variables s and t and can be written as $U_{s,t}$. (We can take the boundary of this surface to be $\partial\Delta = \{U_{1,t}\}$). What is to be done then is to give an integral formula which associates a number to Δ thereby defining a two form ω on Q, further this number must not change under small deformations of the interior of Δ in order that ω is closed as well. Let us denote the value of the field $U_{s,t}$ at $\vec{x}$ by $U(\vec{x}, s, t)$. Then such a formula is, for $N_f \geq 3$,

$$\frac{-iN_c}{240\pi^2} \int_{(s,t)} \int_{S^3} \text{Tr}\, [U(\vec{x}, s, t)^{-1} dU(\vec{x}, s, t)]^5 \ , \tag{6.29}$$

where the constant N_c denotes the number of colours in the underlying theory of strong interactions [59]. In this equation, we have adopted the convention that the kth power of a differential form A denotes its k-fold wedge product: $A^k = A \wedge A \wedge \ldots \wedge A$. The integral over S^3 is the same as the integral over $\vec{x}$ while the remaining integral is the integral over s and t. The value of (6.29) does not change when the interior of Δ is slightly deformed because as a simple calculation shows, $\text{Tr}\,(U^{-1}dU)^5$ is closed:

$$d\text{Tr}\,(U^{-1}dU)^5 = 0 \ . \tag{6.30}$$

Thus the formula defines a closed two form on Q. Note that the integrand of (6.29) is identically zero for $N_f = 2$.

The expression (6.29) is the Wess-Zumino action S_{WZ} for the chiral model for three or more flavours. The fields on Δ can be interpreted using the path space approach as discussed for the charge-monopole system. [The factors in (6.29) have been inserted in order to fulfill the (generalized) Dirac quantization condition. This condition can be derived following our derivation of (6.16). See below as well.]

There is a more formal interpretation of the two form defined by (6.29) which we now indicate. Let V denote the elements of $SU(N_f)$. Then

$$\Omega = \frac{-iN_c}{240\pi^2} \operatorname{Tr} (V^{-1}dV)^5 \qquad (6.31)$$

is a closed five form on $SU(N_f)$. Let θ be the evaluation map:

$$\theta: \quad S^3 \times Q \to SU(N_f) \,,$$
$$(\vec{x}, U) \to U(\vec{x}) \,. \qquad (6.32)$$

Then Ω can be pulled back to $S^3 \times Q$ by the map θ to give the closed five form $\theta^\star\Omega$. Its integral over S^3 gives the closed two form ω on Q.

In Dirac's [19] approach and subsequent developments, the wave functions are to be regarded as special sorts of functions on the path space $\mathcal{P}Q$ of Q rather than as functions on Q. Let us first recall the definition of the path space $\mathcal{P}Q$. Suppose first that Q is connected. Let P_0 be any point of Q which once chosen is held fixed in all subsequent considerations. Then $\mathcal{P}Q$ is the collection of all paths from P_0 to any point P of Q. If Q is not connected, it is the union of several connected components Q_n. For each such Q_n, we can construct the path space $\mathcal{P}Q_n$ by choosing a point P_n in Q_n and considering paths which radiate from P_n. The path space $\mathcal{P}Q$ of Q in such a case is the union of the $\mathcal{P}Q_n$.

Let Γ_P denote a path from P_0 to P so that $\Gamma_P \in \mathcal{P}Q$. Then we assume that the wave function assigns a complex number to Γ_P, $\psi(\Gamma_P) \in \mathcal{C}$. [This excludes the possibility that $\psi(\Gamma_P)$ is an element of a Hilbert space of more than one dimension.] It is not an arbitrary function of this sort however, but is subject to further constraints. We explain these constraints under the assumption that Q is simply connected (or rather that the first homology group $H_1(Q, \mathbf{Z})$ of Q is zero). (Knowledge of homology theory is not required to follow this discussion. Comments involving this theory are being inserted only for completeness. For the discussion of the case where $H_1(Q, \mathbf{Z}) \neq 0$, see Ref. [60].) With such an assumption, if Γ_P and $\tilde{\Gamma}_P$ are two paths from P_0 to P, then their union encloses a two surface (more precisely a two chain) D, its boundary ∂D being this union $\Gamma_P \cup \tilde{\Gamma}_P$. This boundary and hence D is given an orientation by saying that ∂D is to be traversed by first going from P_0 to P along Γ_P and then returning to P_0 along $\tilde{\Gamma}_P$. We require that the values of the wave function at Γ_P and $\tilde{\Gamma}_P$ are related by the formula

$$\psi(\tilde{\Gamma}_P) = \left[\exp i \int_D \omega\right] \psi(\Gamma_P) \quad . \tag{6.33}$$

This guarantees that the probability density $|\psi(\Gamma_P)|^2$ depends only on the point P and not on the path Γ_P.

A semiclassical analysis of systems with Wess-Zumino terms provides a justification for this hypothesis about the nature of the wave functions. We shall adopt this hypothesis in what follows and refer to the literature for this justification [27]. We shall also assume hereafter that Q is connected and simply connected.

The following point about the formula (6.33) is noteworthy. Since the wave function does not depend on the interior Int D enclosed by D, the equation (6.33) makes sense only if the phase

$$\exp i \int_D \omega$$

is unchanged when D is replaced by another surface D' with the same boundary ∂D. For small deformations of Int D, this phase is in fact unchanged because ω is closed. However if D' is not homotopic (or rather, homologous) to D, it is unchanged only if

$$\int_{D \cup D'} \omega = 2\pi \times \text{ integer } . \tag{6.34}$$

We are therefore obliged to impose on ω the condition that its integral over any two cycle is $2\pi \times$ integer. This is the usual quantization condition on the Wess-Zumino term. For the charge-monopole system, if S^2 is the two sphere surrounding the origin $\vec{x} = 0$, it reads

$$\frac{eg}{2} = 2\pi \times \text{ integer } , \tag{6.35}$$

and leads as we know to the result $eg = 4\pi \times$ integer of Dirac. [It may be noted here that the Wess-Zumino term for the nonrelativistic spinning particle did not fulfill (6.34) in one method of quantization [Cf. Chapter 5.2].]

The path space $\mathcal{P}Q$ is an enormously large space. We do not need such a large space to define the wave functions properly because they are not completely general functions but are required to satisfy (6.33). Thus if Γ_P and $\tilde{\Gamma}_P$ are two paths on which all wave functions have the same value, we can identify these paths in $\mathcal{P}Q$ to get a smaller space $\hat{Q}$. Let us therefore define an equivalence relation $\sim$ on paths as follows:

$$\Gamma_P \sim \tilde{\Gamma}_P \quad \text{if} \quad \int_D \omega = 2\pi \times \text{integer} \quad . \tag{6.36}$$

Then

$$\hat{Q} = Q/\sim \tag{6.37}$$

and the wave functions are well defined on $\hat{Q}$.

A point of $\hat{Q}$ is thus an equivalence class of paths ending at the same point P such that any two equivalent paths enclose the flux $2\pi \times$ integer. Let us denote the equivalence class containing the path Γ_P by $[\Gamma_P]$. How many equivalence classes are there among paths having the same end point? Let $[\Gamma_P]$ and $[\tilde{\Gamma}_P]$ be two such distinct equivalence classes. Now $\tilde{\Gamma}_P$ and $\tilde{\Gamma}'_P$ both belong to $[\tilde{\Gamma}_P]$ if and only if $\tilde{\Gamma}_P \cup \tilde{\Gamma}'_P$ encloses the flux $2\pi \times$ integer. An implication is that if $\Gamma_P \cup \tilde{\Gamma}_P$ encloses the flux ϕ, then the flux enclosed by $\Gamma_P \cup \tilde{\Gamma}'_P$ for any member $\tilde{\Gamma}'_P$ of $[\tilde{\Gamma}_P]$ is ϕ mod $2\pi \times$ integer. This is because the two surface D with the boundary $\Gamma_P \cup \tilde{\Gamma}'_P$ can be taken to be the union of the two surfaces D_1 and D_2 with boundaries $\Gamma_P \cup \tilde{\Gamma}_P$ and $\tilde{\Gamma}_P \cup \tilde{\Gamma}'_P$. Furthermore, if $\tilde{\Gamma}''_P$ is any curve such that $\Gamma_P \cup \tilde{\Gamma}''_P$ encloses the flux ϕ mod $2\pi \times$ integer, then $\tilde{\Gamma}''_P \in [\tilde{\Gamma}_P]$. Thus each such equivalence class can be labelled by a point $e^{i\phi}$ of a circle S^1. Thus, assuming that ω and hence ϕ are not zero, there is an S^1 worth of points in $\hat{Q}$ which come from paths having the same end point.

Let us define a projection map π from $\hat{Q}$ to Q by the rule

$$\pi : [\Gamma_P] \to P \quad . \tag{6.38}$$

As we just saw, the inverse image of any point by this map is a circle S^1. Thus $\hat{Q}$ is a circle bundle over Q. It will become a principal $U(1)$ bundle if we can define an action of $U(1)$ on $\hat{Q}$ which will move a point $\hat{q}$ of $\hat{Q}$ along the circle on which it sits (and which acts freely). [The latter means that only the identity element of $U(1)$ maps any point $\hat{q}$ to the same point $\hat{q}$.] This is easily done. If $e^{i\phi} \in U(1)$ we define its action on $[\Gamma_P]$ by

$$e^{i\phi}[\Gamma_P] = [\tilde{\Gamma}_P] \quad , \tag{6.39}$$

where the equivalence class $[\tilde{\Gamma}_P]$ is defined by

$$\int_D \omega = \phi \bmod 2\pi \times \text{ integer} . \tag{6.40}$$

Here D is the two surface defined by a member each of $[\Gamma_P]$ and $[\tilde{\Gamma}_P]$. Since $e^{i\psi}[\Gamma_P] = [\Gamma_P]$ only if $\psi = 2\pi\times$ integer and $e^{i\psi}$ is the identity, this action of $U(1)$ on $\hat{Q}$ by definition is free. With this action, we can thus regard $\hat{Q}$ as a principal bundle over Q with structure group $U(1)$. When all points of $\hat{Q}$ which are carried into each other by this $U(1)$ action are identified, we get back Q. It is precisely this identification which is used in the construction of the projection map π.

We refer the reader to the literature [see e.g. Refs. [57,61]] for further discussion of this bundle and for a detailed study of its application to the charge-monopole system.

Chapter 7

QUANTUM SYMMETRIES AND THE WESS-ZUMINO TERM

7.1 INTRODUCTION

In the presence of a topologically nontrivial Wess-Zumino term, the wave functions are not functions on the configuration space Q of the system; instead as we have seen they are functions on a $U(1)$ bundle $\hat{Q}$ over this configuration space [57,61]. If a group G operates on Q and is a symmetry of the action, then it can happen that G does not operate on $\hat{Q}$ and on wave functions. Instead, the analogous "quantum mechanical" symmetry group G_{QM} which does act on $\hat{Q}$ and thereby on wave functions, is in general a central extension of G. G_{QM} may not even contain G as a subgroup. The central extensions of a group will be defined later.

In the discussion which follows, we shall assume that G is connected.

We have seen that wave functions are functions on $\hat{Q}$ and not on Q. The group G need not act on $\hat{Q}$ which is a different space from Q. We thus have to find a group $\hat{G}$ which does act on $\hat{Q}$ and which *in some appropriate sense* is a quantum theoretic generalization of G. [The group G_{QM} referred to earlier is a subgroup of $\hat{G}$. The precise definition of the former is often motivated by physical considerations. Section 7.3 illustrates that $\hat{G}$ can be different from the group G_{QM}.] In order that it deserves being considered as such a generalization, it clearly has to fulfill the property that its action on $\hat{Q}$ when projected to the base Q using projection operator π (defined in Chapter 3) is the action of G on Q. This requirement in detail means the following: There must exist a homomorphism ρ from $\hat{G}$ onto G,

$$\hat{G} \ni \hat{g} \xrightarrow{\rho} \rho(\hat{g}) \equiv g \in G \tag{7.1}$$

which is such that

$$\pi(\hat{g}\hat{q}) = \rho(\hat{g})\pi(\hat{q})$$
$$\equiv gq \ . \tag{7.2}$$

Here q is $\pi(\hat{q})$ and the transforms of $\hat{q}$ and q by $\hat{g}$ and g, respectively, have been denoted by $\hat{g}\,\hat{q}$ and $g\,q$.

There is one more condition we have to impose on G whose necessity is perhaps a little less obvious. It is that the kernel of the homomorphism ρ must be contained in $U(1)$. We now explain the need for this requirement. If $\hat{g}$ is in the kernel of ρ, then $\rho(\hat{g})q = q$. This means that

$$\pi(\hat{g}\hat{q}) = q = \pi(\hat{q}) \tag{7.3}$$

or that $\hat{g}$ moves $\hat{q}$ along the fibre on which it sits. Its action on $\hat{q}$ thus corresponds to the action of an element $e^{i\phi(\hat{q})}$ of the structure group $U(1)$ on $\hat{q}$. It follows that the kernel of ρ is in general a subset of the group of maps from $\hat{Q}$ to $U(1)$. As we will show, however, we cannot tolerate such a potential $\hat{q}$ dependance of $e^{i\phi(\hat{q})}$. This is first of all because G acts not only on q, but also on the variables conjugate to q and it is reasonable to regard $\hat{g}$ as belonging to the kernel only if these variables as well are unaltered under this action. For reasons of simplicity, we did not state this point in the preceding discussion. With this refined definition of the kernel of ρ, $\hat{g}$ is in this kernel only if the associated phase ϕ is independent of $\hat{q}$. We can prove this result as follows.

The point $\hat{q}$ in the bundle is some equivalence class $[\Gamma_P]$ (as discussed in Chapter 6) so that we may write $\phi(\hat{q})$ as $\phi([\Gamma_P])$. Under the action of an element in the kernel of ρ, the equivalence class $[\Gamma_P]$ becomes

$$e^{i\phi([\Gamma_P])}\, [\Gamma_P] \ ,$$

where this expression is defined by (6.39) and (6.40). Thus under the action of this element, a wave function ψ becomes a wave function χ where

$$\chi(\Gamma_P) = e^{-i\phi([\Gamma_P])}\, \psi(\Gamma_P) \ .$$

Since both ψ and χ must fulfill (6.33), it follows that $\phi([\Gamma_P])$, that is $\phi(\hat{q})$, cannot change as we move along the fibre on which $\hat{q}$ sits, i.e. we must require that $e^{-i\phi([\Gamma_P])} = e^{-i\phi([\tilde{\Gamma}_P])}$, for any two equivalence classes Γ_P and $\tilde{\Gamma}_P$. A consequence is that we can regard ϕ as a function on the configuration space Q, more precisely ϕ projects down to a function $\phi_\star$ on Q defined by $\phi_\star[\pi(\hat{q})] = \phi(\hat{q})$.

Thus under the action of an element in the kernel of ρ, the values of the wave functions at $\hat{q}$ are multiplied by

$$e^{-i\phi_\star[\pi(\hat{q})]} \ .$$

If the variables p of the classical theory are realized as operators p_{OP} in quantum theory, they are therefore transformed under the action of this element to

$$e^{-i\phi_\star} p_{OP} e^{i\phi_\star} \ .$$

Consequently we require that

$$e^{-i\phi_\star} p_{OP} e^{i\phi_\star} = p_{OP}$$

for all operators p_{OP}. The left hand side of this expression can be computed using the values of the multiple commutators $[\phi_\star, [\phi_\star, \ldots [\phi_\star, p_{OP}]\ldots]]$ and is equal to p_{OP} provided all such multiple commutators vanish. Now the values of these commutators are uniquely given by the algebra of observables. In particular if we are quantizing a classical theory, they are uniquely given by the values of the Poisson brackets $\{\phi_\star, \{\phi_\star, \ldots \phi_\star, p\} \ldots\}$ and the rules of quantization. If these Poisson brackets vanish for all p, we know that $\phi_\star$ must be a constant function. We can thus conclude that $\phi_\star$ and hence ϕ are constant functions.

When ϕ is a constant function, $e^{i\phi(\hat{q})}$ is a fixed element $e^{i\alpha}$ of $U(1)$ independent of $\hat{q}$. Since the kernel $\hat{K}$ of ρ acts on Q by such elements $e^{i\alpha}$ of $U(1)$, it is contained in $U(1)$. In the construction below, $\hat{K}$ will in fact be in the centre of $\hat{G}$, that is the elements of $\hat{K}$ will commute with all elements of $\hat{G}$. Thus $\hat{K}$ is in the centre of $\hat{G}$ and $\hat{G}/\hat{K} = G$. As these properties of $\hat{G}$ characterize a central extension of G by $\hat{K}$, $\hat{G}$ is a central extension of G by $\hat{K}$.

Below, we describe an explicit and elementary construction of $\hat{G}$. The construction is then applied to the charge-monopole system and used to explain why the states of this system may be spinorial in nature even though the action contains only tensorial (integral spin) variables.

7.2 THE GROUP $\hat{G}$

We now proceed to construct the group $\hat{G}$. For this purpose, we start with paths γ_g in G from the identity e to any point g. Let us write such a path as

$$\{\gamma_g(t)|0 \le t \le 1; \quad \gamma_g(0) = e, \quad \gamma_g(1) = g\} \quad . \tag{7.4}$$

Suppose $[\Gamma_P]$ is the equivalence class containing the path Γ_P in Q. We can write Γ_P explicitly as

$$\{\Gamma_P(t)|0 \le t \le 1; \quad \Gamma_P(0) = P_0, \quad \Gamma_P(1) = P\} \quad . \tag{7.5}$$

Then we define the transform of $[\Gamma_P]$ by γ_g to be the equivalence class $[\gamma_g \Gamma_P]$ containing the path

$$\gamma_g \Gamma_P = \{\gamma_g(t)P_0|0 \le t \le 1\} \cup \{g\Gamma_P(t)|0 \le t \le 1\} \quad . \tag{7.6}$$

Thus we first act with the path $\{\gamma_g(t)\}$ on the preferred point P_0 to get the path $\{\gamma_g(t)P_0\}$ from P_0 to $\gamma_g(1)P_0 \equiv gP_0$. Then we slide the path $\{\Gamma_P(t)\}$ by g to get the path $\{g\Gamma_P(t)\}$ from gP_0 to gP. This path is then attached to the path $\{\gamma_g(t)P_0\}$ to get the continuous path $\gamma_g\Gamma_P$.

Such paths as they stand do not form a group. Since we have not required that if a path is followed from e to g and then retraced back to e, the result is to be identified with the trivial path consisting of e alone, inverses cannot be defined on this space of paths. We have thus to introduce identifications in this space before we can hope to get a group $\hat{G}$.

Such identifications are naturally provided by the properties of the wave functions. Suppose that there are two paths γ_g and $\tilde{\gamma}_g$ with the same end point. Then from (6.33),

$$\psi[\tilde{\gamma}_g\Gamma_P] = e^{i\int_D \omega}\, \psi[\gamma_g\Gamma_P] \ , \tag{7.7}$$

D being the two surface with the boundary formed by the paths $\{\gamma_g(t)P_0\}$ and $\{\tilde{\gamma}_g(t)P_0\}$, where $\{\tilde{\gamma}_g(t)|0 \leq t \leq 1\}$ is the path $\tilde{\gamma}_g$. Thus if

$$\int_D \omega = 2\pi \times \text{ integer} \ , \tag{7.8}$$

then $\psi(\tilde{\gamma}_g\Gamma_P) = \psi(\gamma_g\Gamma_P)$. We are in this way led to introduce the equivalence relation $\sim$ on paths in G where γ_g and $\tilde{\gamma}_g$ are equivalent if the flux enclosed by the union of the paths $\gamma_g\Gamma_P$ and $\tilde{\gamma}_g\Gamma_P$ is $2\pi \times$ integer:

$$\gamma_g \sim \tilde{\gamma}_g \Leftrightarrow \int_D \omega = 2\pi \times \text{ integer} \ . \tag{7.9}$$

If $\mathcal{P}G$ is the path space of G whose elements are paths from e to any point g, then we define $\hat{G}$ to be the quotient of $\mathcal{P}G$ by the equivalence relation $\sim$:

$$\hat{G} = \mathcal{P}G/\sim \tag{7.10}$$

We now show that an appropriate composition law for elements in $\hat{G}$ can be defined so that it becomes a group.

The composition law in $\hat{G}$ can be inferred from the action of γ_g on Γ_P. If $\gamma_g = \{\gamma_g(t)\}$ and $\gamma_{g'} = \{\gamma_{g'}(t)\}$ are two paths with end points $g = \gamma_g(1)$ and

$g' = \gamma_{g'}(1)$, then

$$
\begin{aligned}
\gamma_{g'}(\gamma_g \Gamma_P) &= \gamma_{g'}[\{\gamma_g(t)P_0\} \cup \{g\Gamma_P(t)\}] \\
&= \{\gamma_{g'}(t)P_0 \cup g'[\{\gamma_g(t)P_0\} \cup \{g\Gamma_P(t)\}] \\
&= \{\gamma_{g'}(t)P_0\} \cup \{g'\gamma_g(t)P_0\} \cup \{g'g\Gamma_P(t)\} \quad .
\end{aligned}
\tag{7.11}
$$

The right hand side is equal to the action of the path $\gamma_{g'}\gamma_g$ on Γ_P where in $\gamma_{g'}\gamma_g$, we first trace the path $\{\gamma_{g'}(t)\}$ and then attach the path $\{g'\gamma_g(t)\}$ to its end point. Thus if $[\gamma_g]$ denotes the equivalence class in $\hat{G}$ containing γ_g, the group composition law in $\hat{G}$ is

$$
\begin{aligned}
[\gamma_{g'}][\gamma_g] &= [\gamma_{g'}\gamma_g] \quad , \\
\gamma_{g'}\gamma_g &\equiv \{\gamma_{g'}(t)\} \cup \{g'\gamma_g(t)\} \quad .
\end{aligned}
\tag{7.12}
$$

It may be verified that this composition law has all the properties required to define a group provided ω is invariant under the action of G. Thus $\hat{G}$ is a group. [See later for a discussion of the role of G-invariance of ω in this construction of $\hat{G}$.]

Note that the inverse of the equivalence class containing the path $\gamma_g = \{\gamma_g(t)|0 \le t \le 1\}$ is the equivalence class containing the "inverse" path $\gamma_g^{-1} = \{\gamma_g(1)^{-1}\gamma_g(1 - t)\}$. Further, if $\{\gamma_g(t)\}$ is a path and $\{\tilde{\gamma}(t) \equiv \gamma_g[\beta(t)]\}$ is the path obtained after a reparametrization $t \to \beta(t)$, then $\{\gamma_g(t)\}$ and $\{\tilde{\gamma}(t)\}$ belong to the same equivalent class, that is equivalence classes are reparametrization invariant.

It is worth noting that the group $\hat{G}$ has been constructed in such a way that it acts on $\hat{Q}$. This action is defined by

$$
\begin{aligned}
[\gamma_g] : [\Gamma_P] &\to [\gamma_g][\Gamma_P] \\
&= [\{\gamma_g(t)P_0\} \cup \{g\Gamma_P(t)\}] \quad .
\end{aligned}
\tag{7.13}
$$

[Note that this action depends on the choice of P_0.] Further $\hat{G}$ acts on wave functions in the standard way:

$$
\begin{aligned}
[\gamma_g]^{-1} : \psi &\to [\gamma_g]^{-1}\psi \quad , \\
([\gamma_g]^{-1}\psi)(\Gamma_P) &\equiv \psi(\gamma_g\Gamma_P) \quad .
\end{aligned}
\tag{7.14}
$$

This rule makes sense because in view of (6.33) and (6.34), the right hand side does not depend on the element γ_g of the equivalence class $[\gamma_g]$ we pick to write it.

The projection map $\pi : \hat{Q} \to Q$ sends the equivalence class $[\Gamma_P]$ of paths with end point P to point $P : \pi([\Gamma_P]) = P$. The equivalence class $[\gamma_g \Gamma_P]$ consists of paths with end point gP, under π they are mapped to gP. Thus the homomorphism ρ from $\hat{G}$ to G is defined by

$$\rho([\gamma_g]) = g \quad . \tag{7.15}$$

The kernel of this homomorphism consists of equivalence classes $[\gamma_e]$ containing closed loops γ_e. Now γ_e and γ_e' are in the same equivalence class $[\gamma_e]$ if and only if the fluxes enclosed by $\gamma_e P_0$ and $\gamma_e' P_0$ differ by $2\pi \times$ integer. Thus the equivalence class $[\gamma_e]$ is uniquely labelled by the element $e^{i\phi}$ of $U(1)$ where ϕ is the flux enclosed by the curve $\gamma_e P_0$:

$$\phi = \int_D \omega, \quad \partial D = \gamma_e P_0 \quad . \tag{7.16}$$

∂D denotes the boundary of D. The kernel $\hat{K}$ of the homomorphism ρ is therefore contained in $U(1)$ and we have the result that $\hat{G}$ is an extension of G by some subgroup of $U(1)$:

$$\hat{G}/\hat{K} = G \quad . \tag{7.17}$$

It is easy to show that $\hat{G}$ is in fact a central extension of G by $\hat{K}$, that is to say the elements of $\hat{K}$ commute with all elements of $\hat{G}$. This is so since an element of $\hat{K}$ is an equivalence class $[\gamma_e]$ containing a closed loop γ_e. Let ϕ be the flux through the loop $\gamma_e P_0$. Now the flux through the loop $\gamma_g \gamma_e \gamma_g^{-1} P_0$ is equal to the flux through the loop $g\gamma_e P_0$ based at gP_0. The latter is the loop obtained by translation of $\gamma_e P_0$ by g, that is by multiplying each point of $\gamma_e P_0$ by g. Since ω is invariant under the action of g, the flux through $g\gamma_e P_0$ is also ϕ. Thus $[\gamma_g][\gamma_e][\gamma_g]^{-1} = [\gamma_e]$, $\hat{K}$ is in the center of $\hat{G}$ and $\hat{G}$ is a central extension of G by $\hat{K}$.

The role of the invariance of ω under the action of G in this construction merits special emphasis. There are inconsistencies in the action of the group $\hat{G}$ on wave functions if ω is not G-invariant. To see this, consider a path Γ_P and another path $\tilde{\Gamma}_P$ which is the union of Γ_P and a loop C based at P_0, $\tilde{\Gamma}_P = \Gamma_P \cup C$. Let C be the boundary of a two surface D. Then if ψ is any wave function, by (6.33),

$$\psi(\tilde{\Gamma}_P) = e^{i\int_D \omega} \, \psi(\Gamma_P) \ . \tag{7.18}$$

This equation is true for every wave function. Writing it out for the transform of ψ by $[\gamma_g]^{-1}$,

$$([\gamma_g]^{-1}\psi)(\tilde{\Gamma}_P) = e^{i\int_D \omega} \, ([\gamma_g]^{-1}\psi)(\Gamma_P) \ , \tag{7.19}$$

and using (7.14), we find

$$\psi(\gamma_g \tilde{\Gamma}_P) = e^{i\int_D \omega} \, \psi(\gamma_g \Gamma_P) \ . \tag{7.20}$$

But $\gamma_g \tilde{\Gamma}_P$ is the curve $\gamma_g \Gamma_P$ to which is attached the loop gC, gC being the loop obtained by rigidly translating C by g. The loop gC bounds the two surfaces gD. Thus the left hand side of (7.20) is

$$e^{i\int_{gD} \omega} \, \psi(\gamma_g \Gamma_P) \ . \tag{7.21}$$

It follows that we must have

$$\int_{gD} \omega = \int_D \omega + 2\pi n \ , \tag{7.22}$$

where n is an integer. This integer must be zero. For being an integer, it cannot change under continuous changes of C or γ_g. By letting C or γ_g shrink continuously to a point, we see that $n = 0$. With $n = 0$, (7.22) states that ω is G-invariant. Thus there are inconsistencies in the way we transform wave functions unless ω is G-invariant.

We shall illustrate this construction of $\hat{G}$ for the charge-monopole system below.

7.3 THE CHARGE-MONOPOLE SYSTEM
REVISITED

For the charge-monopole-system, the two form ω is given by (6.28). It is invariant under the group $G = SO(3)$ of spatial rotations. The group $\hat{G}$ which acts on wave functions is thus well defined. It is a central extension of $SO(3)$ by some group $\hat{K}$ where the group $\hat{K}$ is actually $U(1)$. Let us show this result. For any choice of P_0, we can clearly find a tiny loop γ_e in $SO(3)$ such that $\gamma_e P_0$ encloses a flux $\varphi \neq 0$. By shrinking γ_e to a point, we see that there are loops in $SO(3)$ which are associated with fluxes between zero and φ. Multiple circuits around these loops in G generate fluxes of all magnitudes and hence all elements $e^{i\varphi}$ of $U(1)$. Therefore $\hat{K}$ is $U(1)$.

There are precisely two central extensions of $SO(3)$ by $U(1)$, namely $SO(3) \times U(1)$ and $U(2) = [SU(2) \times U(1)]/Z_2$. Using our formalism, we want to show the known result that if eg is an integer, the former occurs and states have integral spin. On the other hand, if eg is one half of an odd integer the latter occurs and states change sign under 2π rotation.

For such a demonstration, it is first desirable to identify a subgroup G_{QM} in $\hat{G}$ which can be regarded as representing rotations and which is thus an $SO(3)$ or $SU(2)$ subgroup of $\hat{G}$. Since $\hat{G}$ is given to us in terms of equivalence classes of paths where it is not obvious which elements represent pure rotations and which correspond to $U(1)$, the identification of G_{QM} in $\hat{G}$ is not immediately apparent.

In order to resolve this problem, let us first recall the definition and some properties of commutator subgroups. If a and b are elements of $\hat{G}$, the commutator of a and b is the element $aba^{-1}b^{-1}$. The group generated by taking products of such elements in $\hat{G}$ is the commutator subgroup $[\hat{G}, \hat{G}]$ of $\hat{G}$. If $\{T(\alpha)\}$ is a basis for the Lie algebra of $\hat{G}$, the Lie algebra of $[\hat{G}, \hat{G}]$ is spanned by all the commutators $[T(\alpha), T(\beta)]$. The $U(1)$ generators in the Lie algebras of $SO(3) \times U(1)$ or $U(2)$ commute with all $T(\alpha)$ and do not appear in these commutators which therefore span an $SO(3)$ or $SU(2)$ Lie algebra. Thus it is the commutator subgroup $[\hat{G}, \hat{G}]$ which is $SO(3)$ or $SU(2)$ and which must be regarded as the "quantum mechanical rotation group" G_{QM}.

There is a simple criterion involving commutators which distinguishes $SU(2)$ from $SO(3)$ and which is useful in the present context. Let R_1 and R_3 denote rotations by π around the x- and z- axes in $SO(3)$. In terms of 3×3 matrices, they are

$$R_1 = \begin{bmatrix} 1 & & 0 \\ & -1 & \\ 0 & & -1 \end{bmatrix} \quad , \quad R_3 = \begin{bmatrix} -1 & & 0 \\ & -1 & \\ 0 & & 1 \end{bmatrix} \tag{7.23}$$

so that

$$R_1 R_3 R_1^{-1} R_3^{-1} = \mathbf{1} \ . \tag{7.24}$$

In $SU(2)$ realized by 2×2 matrices, R_1 can be represented by either of the elements

$$\eta_1 \exp\left[i\frac{\tau_1}{2}\pi\right] = \eta_1 i\tau_1 \ , \quad \eta_1 = 1 \text{ or } -1 \ .$$

Similarly R_3 can be represented by either of the elements $\eta_3 i\tau_3$, $\eta_3 = 1$ or -1. Now $\eta_1(i\tau_1)\eta_3(i\tau_3)[\eta_1(i\tau_1)]^{-1}[\eta_3(i\tau_3)]^{-1}$ is always $-\mathbf{1}$ despite the ambiguities in the values of η_α. Thus the commutator formed out of π rotations around the x and z axes distinguishes $SU(2)$ from $SO(3)$.

We can use this fact in our problem as follows. Let us take the inverse images of R_1 and R_3 in $\hat{G}$ to be the equivalence classes containing the curves

$$\gamma_1 \equiv \{e^{i\pi t J_1}\}$$

and

$$\gamma_3 \equiv \{e^{i\pi t J_3}\}$$

$(0 \le t \le 1)$ where J_1 and J_3 are the x and z-components of spin 1 angular momenta. These inverse images are ambiguous by $U(1)$ elements of $\hat{G}$, but we have seen that this ambiguity is immaterial in computing the commutator. Thus in $\hat{G}$, $R_1 R_3 R_1^{-1} R_3^{-1}$ becomes

$$[\gamma_1][\gamma_3][\gamma_1]^{-1}[\gamma_3]^{-1} \equiv [\gamma_1 \gamma_3 \gamma_1^{-1} \gamma_3^{-1}] \ , \tag{7.25}$$

where

$$\gamma_{1,3}^{-1} = \{e^{i\pi J_{1,3}} e^{i\pi(1-t)J_{1,3}} = e^{-i\pi t J_{1,3}}\} \ .$$

Now $SO(3)$ is a subgroup of $SO(3) \times U(1)$ but not of $U(2)$, while $SU(2)$ is a subgroup of $U(2)$ but not of $SO(3) \times U(1)$. Therefore if this commutator is $+1$ on wave functions, $\hat{G}$ is $SO(3) \times U(1)$ and the states have integral spins, if it is -1, $\hat{G}$ is $U(2)$ and the states have half odd integral spins.

For the charge-monopole system, it is convenient to take P_0 to be on the x-axis:

$$P_0 = (1, 0, 0) \ . \tag{7.26}$$

Consider $\gamma_1 \gamma_3 \gamma_1^{-1} \gamma_3^{-1} \Gamma_P$. It differs from Γ_P by a closed loop ∂D based at P_0. The loop ∂D is obtained by successively tracing the curves

1) $\gamma_1 P_0 = \{P_0\}$,

2) $e^{i\pi J_1} \gamma_3 P_0 = \{(\cos \pi t, \sin \pi t, 0) | 0 \le t \le 1\}$,

3) $e^{i\pi J_1} e^{i\pi J_3} \gamma_1^{-1} P_0 = \{(-1, 0, 0)\}$,

4) $e^{i\pi J_1} e^{i\pi J_3} e^{-i\pi J_1} \gamma_3^{-1} P_0 = \{(-\cos \pi t, -\sin \pi t, 0) | 0 \le t \le 1\}.$

$$\tag{7.27}$$

In this computation, we have used the formulae

$$\gamma_1 = \left\{ \begin{bmatrix} 1 & 0 & 0 \\ 0 & \cos \pi t & \sin \pi t \\ 0 & -\sin \pi t & \cos \pi t \end{bmatrix} \right\} , \tag{7.28}$$

$$\gamma_3 = \left\{ \begin{bmatrix} \cos \pi t & \sin \pi t & 0 \\ -\sin \pi t & \cos \pi t & 0 \\ 0 & 0 & 1 \end{bmatrix} \right\} . \tag{7.29}$$

If the third axis points towards the north pole, this loop ∂D is the equator through P_0.

It follows that under the action of $[\gamma_1][\gamma_3][\gamma_1]^{-1}[\gamma_3]^{-1}$, every wave function acquires the phase $e^{i \int_D \omega}$ where D may be taken to be the upper half of the two sphere with the equator ∂D as the boundary. The integral of ω over the full two sphere is $-4\pi eg$, thus this phase is $e^{-i2\pi eg}$. It is $+1$ and $\hat{G} = SO(3) \times U(1)$

if eg is an integer, and -1 and $\hat{G} = U(2)$ if eg is half an odd integer. We have thus proved the result we were after.

Chapter 8

QUANTUM THEORY FOR MULTIPLY CONNECTED CONFIGURATION SPACES

8.1 INTRODUCTION

In this Chapter, we shall study the novel features which can arise in quantum theory when the configuration space is multiply connected.

Multiply connected configuration spaces play an important role in field theory and particle physics. Interesting examples where such spaces occur are provided by nonabelian gauge theories like QCD [59]. The "θ vacua" in any such gauge theory are known to be consequences of the infinite connectivity of the space of gauge invariant observables. The θ vacua will be discussed in detail in Chapter 20. Other examples with multiply connected configuration space are certain $2 + 1$ dimensional nonlinear models and the two flavour chiral model. Nonlinear models will be the subject of Chapter 10, while the chiral model will be covered in Chapters 12 to 18.

As a prelude to the discussion of multiply connected configuration spaces, we shall first generalize our remarks of Chapter 1 on the nature of wave functions in quantum theory. The argument there was that since only observables like $\psi^{\star}\psi$ are required to be functions on Q, it is permissible to consider wave functions ψ which are functions on a $U(1)$ bundle $\hat{Q}$ over Q provided all wave functions fulfill the property $\psi(qe^{i\theta}) = \psi(q)e^{in\theta}$. We shall now show that we can meet this requirement on observables even with vector valued wave functions $\psi = (\psi_1, \ldots, \psi_N)$ which are functions on an H bundle $\overline{Q}$ over Q, the group H not being necessarily $U(1)$. Recall from Chapter 3.5 that in such an H bundle $\overline{Q} = \{\overline{q}\}$, there is a free action $\overline{q} \rightarrow \overline{q}\, h$ of the structure group H on the bundle space $\overline{Q}$, and that if we identify points of $\overline{Q}$ related by this H action, then we get the base space Q. It follows that any function ρ on $\overline{Q}$ which is invariant under the H action $[\rho(\overline{q}h) = \rho(\overline{q})]$ can be regarded as a function on Q. Let $h \rightarrow D(h)$ define a representation Γ of H by unitary matrices. Let us demand of our wave functions that they transform by Γ under the action of H:

$$\psi_i(\overline{q}h) = \psi_j(\overline{q})D_{ji}(h) \ . \tag{8.1}$$

Then for any two wave functions ψ and ψ', the expression

$$< \psi, \psi' > (\overline{q}) \equiv \psi_i^{\star}(\overline{q})\psi_i'(\overline{q}) \tag{8.2}$$

is invariant under H and $< \psi, \psi' >$ may be thought of as a function on Q. If we define the scalar product (ψ, ψ') on wave functions by appropriately integrating $< \psi, \psi' >$ over Q, then it is clear that there is no obvious conceptual problem in working with wave functions of this sort.

We shall see that such vector valued wave functions with $N \geq 2$ will occur in the general theory of multiply connected configuration spaces if H is nonabelian. When that happens, as Sorkin [62, see also Refs. 61 and 63] has proved, the space of wave functions we have described above is too large when the dimension of Γ exceeds 1, even when Γ is irreducible. The reduction of this space to its proper size will also be described following Sorkin and will be seen to lead to interesting consequences.

A result of particular importance we shall see later and which merits emphasis is that the quantum theory of systems with multiply connected configuration spaces is ambiguous, there being as many inequivalent ways of quantizing the system as there are distinct unitary irreducible representations (UIR's) of $\pi_1(Q)$. The angle θ which labels the vacua in QCD, for example, can be thought as the label of the distinct UIR's of $\mathbf{Z}$, $\mathbf{Z}$ being $\pi_1(Q)$ for such a theory. As is well known, the quantum theories associated with different $e^{i\theta}$ are inequivalent.

8.2 THE UNIVERSAL COVERING SPACE AND THE FUNDAMENTAL GROUP

Given any manifold such as a configuration space Q, it is possible to associate another manifold $\overline{Q}$ to Q which is simply connected. The space $\overline{Q}$ is known as the universal covering space of Q. The group $\pi_1(Q) = H$ acts freely on $\overline{Q}$ and the quotient of $\overline{Q}$ by this action is Q. Thus $\overline{Q}$ is a principal fibre bundle over Q with structure group H. The space $\overline{Q}$ plays an important role in the construction of possible quantum theories associated with Q. In this Section, we shall describe the construction of $\overline{Q}$. [The similarity of this construction to the construction of the $U(1)$ bundle $\hat{Q}$ in Chapter 6 should be noted.] We shall also explain the concept of the fundamental group $\pi_1(Q)$ of Q and its action on $\overline{Q}$.

We shall assume in what follows that Q is path-connected. That is, if q_0, q_1 are any two points of Q, we can find a continuous curve $Q(t) \in Q$ with $q(0) = q_0$, $q(1) = q_1$.

The first step in the construction of $\overline{Q}$ is the construction of the path space $\mathcal{P}Q$ associated with Q. Let us recall its definition from Chapter 6.3. Let q_0 be any point of Q which once chosen is held fixed in all subsequent considerations. Then $\mathcal{P}Q$ is the collection of all paths which starts at q_0 and end at any point q of Q. We shall denote the paths ending at q by $\Gamma_q, \tilde{\Gamma}_q, \Gamma'_q$ etc. It is to be noted that these paths Γ_q are *oriented* and *unparametrized*.

The former means that they are to be regarded as *starting* at the base point q_0 and *ending* at q. Each of these paths has thus an arrow attached pointing from q_0 to q. The implication of the statement that Γ_q is "unparametrized" is that (besides its orientation) only its geographical location in Q matters. If we introduce a parameter s to label points of Γ_q and write the associated parametrized path as

$$\gamma_q = \{\gamma_q(s)|\ \gamma_q(0) = q_0,\ \ \gamma_q(1) = q\}\ , \tag{8.3}$$

then Γ_q is the equivalence class of all such parametrized paths (with parameters compatible with the orientation of Γ_q) with the same location in Q.

We next introduce an equivalence relation $\sim$ on the paths known as homotopy equivalence. We say that two paths Γ_q and $\tilde{\Gamma}_q$ with the same end point q are *homotopic* and write

$$\Gamma_q \sim \tilde{\Gamma}_q \tag{8.4}$$

if Γ_q can be continuously deformed to $\tilde{\Gamma}_q$ while holding q (and of course q_0) fixed.

A more formal definition of homotopy equivalence is the following: If there exists a continuous family of paths $\Gamma_q(t)$ $[0 \leq t \leq 1]$ in Q (all from q_0 to q) such that

$$\Gamma_q(0) = \Gamma_q\ ,\ \ \Gamma_q(1) = \tilde{\Gamma}_q\ , \tag{8.5}$$

then $\Gamma_q \sim \tilde{\Gamma}_q$.

Let $[\Gamma_q]$ denote the equivalence class of all paths ending at q which are homotopic to Γ_q. The *universal covering space* $\overline{Q}$ of Q is just the collection of all these equivalence classes:

$$\overline{Q} = \{[\Gamma_q]\}\ . \tag{8.6}$$

It can be shown that $\overline{Q}$ is simply connected.

Of particular interest to us are the equivalence classes $[\Gamma_{q_0}]$ of all loops Υ_{q_0} starting and ending at q_0. These equivalence classes have a natural group structure. The group product is defined by

$$[\Gamma_{q0}][\tilde{\Gamma}_{q0}] = [\Gamma_{q0} \cup \tilde{\Gamma}_{q0}] \quad , \qquad (8.7)$$

where in the loop $\Gamma_{q0} \cup \tilde{\Gamma}_{q0}$, we first traverse Γ_{q0} and then traverse $\tilde{\Gamma}_{q0}$. The inverse is defined by

$$[\Gamma_{q0}]^{-1} = [\Gamma_{q0}^{-1}] \quad , \qquad (8.8)$$

where the loop Γ_{q0}^{-1} has the same geographical location in Q as Γ_{q0}, but has the opposite orientation. The identity e is the equivalence class of the loop consisting of the single point q_0. It is clear that

$$[\Gamma_{q0}][\Gamma_{q0}^{-1}] = [\Gamma_{q0}^{-1}][\Gamma_{q0}] = e \quad . \qquad (8.9)$$

The group $\pi_1(Q)$ with elements $[\Gamma_{q0}]$ and the group structure defined above is known as the *fundamental group* of Q. If $\pi_1(Q)$ is nontrivial $[\pi_1(Q) \neq \{e\}]$, the space Q is said to be *multiply connected*. We shall see examples of multiply connected spaces in Section 3. They will show in particular that $\pi_1(Q)$ can be abelian or nonabelian. In any case, it is always discrete.

The group $\pi_1(Q)$ has a free action on $\overline{Q}$. It is defined by

$$[\Gamma_{q0}] : [\Gamma_q] \to [\Gamma_{q0}][\Gamma_q] = [\Gamma_{q0} \cup \Gamma_q] \quad , \qquad (8.10)$$

where in $\Gamma_{q0} \cup \Gamma_q$, we first traverse Γ_{q0} and then traverse Γ_q. It is a simple exercise to show that this action is free.

We now claim that the quotient of $\overline{Q}$ by this action is Q, the associated projection map $\pi : \overline{Q} \to Q$ being defined by

$$\pi : [\Gamma_q] \to \pi([\Gamma_q]) = q \quad . \qquad (8.11)$$

This means the following: a) All the points $[\Gamma_q]$, $[\tilde{\Gamma}_q]$, ... with the same image q under π are related by $\pi_1(Q)$ action, and b) these are the only points related by $\pi_1(Q)$ action. To show a) let $\tilde{\Gamma}_q \cup \Gamma_q^{-1}$ be the loop based at q_0 where we first go along $\tilde{\Gamma}_q$ from q_0 to q and then return to q_0 along Γ_q (in a sense opposite to the orientation of Γ_q). It is clear that

$$[\tilde{\Gamma}_q] = [\tilde{\Gamma}_q \cup \Gamma_q^{-1}][\Gamma_q] \quad , \quad [\tilde{\Gamma}_q \cup \Gamma_q^{-1}] \in \pi_1(Q) \quad . \qquad (8.12)$$

This proves a). As regards b), elements of $\pi_1(Q)$ act by attaching loops at the starting point q_0 of Γ_q and hence map $[\Gamma_q]$ to some $[\tilde{\Gamma}_q]$. Both $[\Gamma_q]$ and $[\tilde{\Gamma}_q]$ project under π to the same point q of Q. This proves b).

We have now proved that $\overline{Q}$ is a principal fibre bundle over Q with structure group $\pi_1(Q)$.

8.3 EXAMPLES OF MULTIPLY CONNECTED CONFIGURATION SPACES

It is appropriate at this point to give some examples of multiply connected spaces. We will be discussing nonlinear models and gauge and gravity theories in some detail in later chapters so that we shall here give examples which are not obviously related to such theories. There are several such relevant examples and we shall pick three.

1. Let $x_1, x_2, \ldots, x_N$ be N distinct points in the plane $\mathbf{R}^2$ and let Q be the complement of the set $\{x_1, x_2, \ldots, x_N\}$ in $\mathbf{R}^2$:

$$Q = \mathbf{R}^2 \backslash \{x_1, x_2, \ldots, x_N\} \ . \tag{8.13}$$

Thus Q is the plane with N holes $x_1, x_2, \ldots, x_N$. The fundamental group $\pi_1(Q)$ of this Q is of infinite order. It is nonabelian for $N \geq 2$. The generators of this group are constructed as follows: Let q_0 be any fixed point of Q and let C_M be any closed curve from q_0 to q_0 which encloses x_M and none of the remaining holes. It is understood that C_M winds around x_M exactly once with a particular orientation. Let C_M^{-1} be the curve with orientation opposite to C_M, but otherwise the same as C_M. Let $[C_M]$ and $[C_M^{-1}] = [C_M]^{-1}$ be the homotopy classes of C_M and C_M^{-1}. Then $\pi_1(Q)$ consists of all possible products like $[C_M][C_{M'}][C_{M''}]^{-1} \ldots$ and is the free group with generators $[C_M]$. The products of homotopy classes are defined here as in the last Section. For example, $[C_M][C_{M'}] = [C_M \cup C_{M'}]$ where $C_M \cup C_{M'}$ is the curve where we first trace C_M and then trace $C_{M'}$. For $N = 1$, the group $\pi_1(Q)$ has one generator and is **Z**. The relevance of this Q for the treatment of the Aharonov-Bohm effect [64] should be evident.

2. In the collective model of nuclei, one considers nuclei with asymmetric shapes with three distinct moments of inertia I_i along the three principal axes [65]. There are also polyatomic molecules such as the ethylene molecule C_2H_4 which can be described as such asymmetric rotors [65]. The configuration space Q in these cases is the space of orientations of the nucleus or the molecule. These orientations can be described by a real symmetric 3×3 matrix T (the moment of inertia tensor) with three distinct but fixed eigenvalues I_i. We now show that this Q has a nonabelian fundamental group.

Any $T \in Q$ can be written in the form

$$T \equiv \mathcal{R}T_0\mathcal{R}^{-1} \quad ,$$

$$T_0 = \begin{bmatrix} I_1 & & 0 \\ & I_2 & \\ 0 & & I_3 \end{bmatrix} \quad , \tag{8.14}$$

where $\mathcal{R}$ being in $SO(3)$ is regarded as a real orthogonal matrix of determinant 1. Hence Q is the orbit of T_0 under the action of $SO(3)$ given by (8.14). If $\mathcal{R}_i(\pi)$ is the rotation by π around the ith axis,

$$\mathcal{R}_1(\pi) = \begin{bmatrix} 1 & & 0 \\ & -1 & \\ 0 & & -1 \end{bmatrix}, \quad \mathcal{R}_2(\pi) = \begin{bmatrix} -1 & & 0 \\ & 1 & \\ 0 & & -1 \end{bmatrix} \quad ,$$

$$\mathcal{R}_3(\pi) = \begin{bmatrix} -1 & & 0 \\ & -1 & \\ 0 & & 1 \end{bmatrix} \quad , \tag{8.15}$$

then T is invariant under the substitution $\mathcal{R} \to \mathcal{R}\mathcal{R}_i(\pi)$. So Q is the space of cosets of $SO(3)$ with respect to the four element subgroup $\{\mathbf{1}, \mathcal{R}_1(\pi), \mathcal{R}_2(\pi), \mathcal{R}_3(\pi)\}$.

It is convenient to view this coset space as the coset space $SU(2)/H$ of $SU(2)$ with regard to an appropriate subgroup H. For this purpose let us introduce the standard homomorphism $R : SU(2) \to SO(3)$. The definition of R is

$$s\tau_i s^{-1} = \tau_j R_{ji}(s) \ , \quad s \in SU(2) \quad , \tag{8.16}$$

τ_i being Pauli matrices. [Here we think of $SU(2)$ concretely as the group of 2×2 unitary matrices of determinant 1.] Then we can write any T in the form

$$T = R(s)T_0 R(s^{-1}) \tag{8.17}$$

and hence view Q as the orbit of T_0 under $SU(2)$. Since by (8.16),

$$R(-s) = R(s),$$
$$R(\pm si\tau_i) = R(\pm se^{i\pi\tau_i/2}) = R(s)\mathcal{R}_i(\pi) \quad , \tag{8.18}$$

the stability group H of T_0 is the quaternion (or binary dihedral) group $D_8^\star$:

$$H = D_8^\star = \{\pm\mathbf{1}, \pm i\tau_1, \pm i\tau_2, \pm i\tau_3\} \quad . \tag{8.19}$$

Thus

$$Q = SU(2)/D_8^\star \quad . \tag{8.20}$$

It is well known that $SU(2)$ is simply connected $[\pi_1(SU(2)) = \{e\}]$. A consequence of this fact is that

$$\pi_1(Q) = D_8^\star \quad . \tag{8.21}$$

The loops in Q associated with the elements of $D_8^\star$ can be constructed as follows. Consider a curve $\{s(t)\}$ in $SU(2)$ from identity to $h \in D_8^\star$:

$$s(t) \in SU(2) \ , \quad s(0) = \mathbf{1} \ , \quad s(1) = h \quad . \tag{8.22}$$

The image of this curve in Q is $\{T(t)\}$ where

$$T(t) = R[s(t)]T_0 \ R[s(t)^{-1}] \quad . \tag{8.23}$$

Since $T(0) = T(1) = T_0$, this is a loop in Q based at T_0. Two loops $T(t)$ and $T'(t)$ with different $s(1) \in D_8^\star$ are not homotopic, whereas loops $T(t)$ and $T'(t)$

with the same $s(1) \in D_8^\star$ are homotopic and form a homotopy class. Such homotopy classes can be thought of as the elements h of $\pi_1(Q)$.

The relation (8.20) shows that Q is the quotient of $SU(2)$ by the free action

$$s \to sh \ , \quad s \in SU(2) \ , \quad h \in D_8^\star \tag{8.24}$$

of $D_8^\star$. Furthermore $\pi_1(Q) = D_8^\star$. Therefore in this example, $SU(2)$ as a manifold is the universal covering space of Q.

3. The last example we shall give is relevant for discussing possible statistics of particles in k spatial dimensions. Consider N *identical* spinless particles in $\mathbf{R}^k$ [for $N \geq 2$] and assume first that $k \geq 3$. The configuration of these particles is given by the *unordered* set $[x_1, x_2, \ldots, x_N]$ where $x_j \in \mathbf{R}^k$. The set must be regarded as unordered (so that for example $[x_1, x_2, \ldots, x_N] = [x_2, x_1, \ldots, x_N]$) because of the assumed indistinguishability of the particles. Let us also assume that no two particles can occupy the same position so that $x_i \neq x_j$ if $i \neq j$. The resultant space of these sets can be regarded as the configuration space Q of this system. It can be shown that $\pi_1(Q)$ is identical to the permutation group S_N. The closed curves in Q associated with the transpositions $s_{ij} \in S_N$ of two particles can be constructed as follows. Choose the base point q_0 to be $[x_1^0, x_2^0, \ldots, x_N^0]$. Let $\{\gamma_{ij}(t) \ ; \ 0 \leq t \leq 1\}$ be the loop in Q defined by

$$
\begin{aligned}
\gamma_{ij} &= [x_1^0, x_2^0, \ldots, x_{i-1}^0, x_i(t), x_{i+1}^0, \ldots, x_{j-1}^0, x_j(t), x_{j+1}^0, \ldots, x_N^0] \ , \\
x_i(0) &= x_i^0 \ , \quad x_i(1) = x_j^0 \ , \\
x_j(0) &= x_j^0 \ , \quad x_j(1) = x_i^0
\end{aligned}
\tag{8.25}
$$

$\{\gamma_{ij}(t)\}$ is a loop since the set $[x_1^0, x_2^0, \ldots, x_N^0]$ is unordered. The homotopy class of this loop can be identified with s_{ij}.

The distinct quantum theories of this system are labelled by the UIR's of S_N and are associated with parastatistics. Special cases of these theories describe bosons and fermions.

We can describe the configuration space of N identical particles for $k = 2$

as well in a similar way. The fundamental group $\pi_1(Q)$ for $k = 2$ however is not S_N, but a very different (infinite) group known as the braid group B_N [65]. It is because $\pi_1(Q) = B_N$ for $k = 2$ that remarkable possibilities for statistics (such as fractional statistics) arise in two spatial dimensions [66].

8.4 QUANTIZATION ON MULTIPLY CONNECTED CONFIGURATION SPACES

We shall now describe the general approach to quantization when the configuration space Q is multiply connected.

As indicated previously, this quantization can be carried out by introducing a Hilbert space $\mathcal{H}$ of complex functions on $\overline{Q}$ with a suitable scalar product and realizing the classical observables as quantum operators on this space. Since the classical configuration space is Q and not $\overline{Q}$, classical observables are functions of $q \in Q$ and of their conjugate momenta. Let us concentrate on functions of q. In this Section, we will consider the quantization of a classical system characterized by a multiply connected configuration space Q and the realization of the associated quantum operators. Let $\alpha(q)$ define a function of q and let $\hat{\alpha}$ be the corresponding quantum operator. The definition of $\hat{\alpha}$ consists in specifying the transformed function $\hat{\alpha} f$ for a generic function $f \in \mathcal{H}$. Thus given the function f, we have to specify the value of $\hat{\alpha}f$ at every $\overline{q}$. This is done by the rule

$$(\hat{\alpha}f)(\overline{q}) = \alpha[\pi(\overline{q})]f(\overline{q}) \ . \tag{8.26}$$

The group $\pi_1(Q)$ acts on $\mathcal{H}$. Let t denote a generic element of $\pi_1(Q)$. If $\hat{t}$ is the operator which represents t on $\mathcal{H}$, and $\hat{t}f$ is the transform of a function $f \in \mathcal{H}$ by $\hat{t}$, $\hat{t}$ is defined by specifying the function $\hat{t}f$ as follows:

$$(\hat{t}f)(\overline{q}) \equiv f(\overline{q}t) \ . \tag{8.27}$$

Now $\hat{\alpha}$ commutes with $\hat{t}$:

$$
\begin{aligned}
(\hat{\alpha}\hat{t}f)(\overline{q}) &= \alpha[\pi(\overline{q})](\hat{t}f)(\overline{q}) \\
&= \alpha[\pi(\overline{q})]f(\overline{q}t) \ , \\
(\hat{t}\hat{\alpha}f)(\overline{q}) &= (\hat{\alpha}f)(\overline{q}t) \\
&= \alpha[\pi(\overline{q}t)]f(\overline{q}t) \\
&= \alpha[\pi(\overline{q})]f(\overline{q}t) \\
&= (\hat{\alpha}\hat{t}f)(\overline{q}) \ .
\end{aligned}
\tag{8.28}
$$

Here we have used the fact that $\pi(\overline{q}t) = \pi(\overline{q})$. [See (8.11) and the remarks which follow.]

Since the operators $\hat{t}$ are not all multiples of the identity operator, Schur's lemma tells us that this representation of the observables $\hat{\alpha}$ on $\mathcal{H}$ is not irreducible. We can proceed in the following way to reduce it to its irreducible components. Let $\Gamma_1, \Gamma_2, \ldots$ denote the distinct irreducible representations of $\pi_1(Q)$. Let $\mathcal{H}_\beta^\ell$ ($\beta = 1, 2, \ldots$) be the subspaces of $\mathcal{H}$ which transform by Γ_ℓ, β being an index to account for multiple occurrences of Γ_ℓ in the reduction. Let us also define

$$
\mathcal{H}^{(\ell)} = \bigoplus_\beta \mathcal{H}_\beta^{(\ell)} \ .
\tag{8.29}
$$

Then

$$
\mathcal{H} = \bigoplus_\ell \mathcal{H}^{(\ell)} \ .
\tag{8.30}
$$

Since $\hat{\alpha}$ commutes with $\hat{t}$, it can not map a vector transforming Γ_ℓ to one transforming by Γ_m ($m \neq \ell$) since Γ_ℓ and Γ_m are inequivalent. Thus

$$
\hat{\alpha}\mathcal{H}^{(\ell)} \subset \mathcal{H}^{(\ell)} \ .
\tag{8.31}
$$

In other words, we can realize our observables on any one subspace $\mathcal{H}^{(\ell)}$ and ignore the remaining subspaces. Quantization on the subspaces $\mathcal{H}^{(\ell)}$ and $\mathcal{H}^{(m)}$ are known to be inequivalent when $\ell \neq m$. Thus there are at least as many distinct ways to quantize the system as the number of inequivalent irreducible representations of $\pi_1(Q)$. It may also be shown that the representation of the algebra of observables on any one $\mathcal{H}^{(\ell)}$ is irreducible if $\pi_1(Q)$ is abelian, while some additional reduction is possible if it is nonabelian as shown by Sorkin [62] and as we shall see below.

Here we have not discussed how the momentum variables conjugate to the coordinates are realized on $\mathcal{H}^{(\ell)}$. It can be shown that for the problems at hand, these momentum variables can also be consistently realized.

8.5 NONABELIAN FUNDAMENTAL GROUPS

Let us now consider nonabelian $\pi_1(Q)$ in more detail. Let γ_ℓ $(\ell = 1, 2, \ldots)$ denote its distinct one dimensional representations and let $\overline{\gamma}_\alpha$ $(\alpha = 1, \ldots)$ denote its distinct UIR's of dimension greater than 1. The subspaces of $\mathcal{H}$ which carry γ_ℓ will be called $h_k^{(\ell)}$ and the subspaces which carry $\overline{\gamma}_\alpha$ will be called $\overline{h}_\sigma^{(\alpha)}$, k and σ being indices to account for multiple occurrences of a given UIR in the reduction of $\mathcal{H}$. If we set

$$h^{(\ell)} = \bigoplus_k h_k^{(\ell)} \quad , \tag{8.32}$$

then as in the abelian case the algebra of observables is represented irreducibly on $h^{(\ell)}$, and the representations on different $h^{(\ell)}$ are inequivalent. The novelty is associated with the representation on

$$\overline{h}^{(\alpha)} = \bigoplus_\sigma \overline{h}_\sigma^{(\alpha)} \quad . \tag{8.33}$$

They are inequivalent for different α, but they are not irreducible. We now show this fact.

Let $e_\sigma(j)(j = 1, 2, \ldots, n > 1)$ be a basis for $\overline{h}_\sigma^{(\alpha)}$ chosen so that they transform in the same way under $\pi_1(Q)$ for different σ:

$$\hat{t} e_\sigma(j) = e_\sigma(k) D(t)_{kj} \quad . \tag{8.34}$$

Here $t \to D(t)$ defines the representation $\overline{\gamma}_\alpha$. [Since α can be held fixed in the ensuing discussion, an index α has not been put on the vectors $e_\sigma(j)$ or on the matrices $D(t)$.] Now if $\hat{L}$ is any linear operator such that $\hat{L} e_\sigma(j)$ transforms in the same way as $e_\sigma(j)$,

$$\hat{t} \hat{L} e_\sigma(j) = [\hat{L} e_\sigma(k)] D_{kj}(t) \quad , \tag{8.35}$$

that is if $[\hat{L}, \hat{t}] = 0$, then by Schur's lemma $\hat{L}$ acts only on the index σ:

$$\hat{L}e_\sigma(j) = e_\lambda(j)\mathcal{D}_{\lambda\sigma}(\hat{L}) \ . \tag{8.36}$$

Furthermore, again by Schur's lemma, $\mathcal{D}(\hat{L})$ is independent of j. Since $\hat{\alpha}$ in (8.17) shares the preceeding property of $\hat{L}$, it follows that

$$\hat{\alpha}e_\sigma(j) = e_\lambda(j)\mathcal{D}_{\lambda\sigma}(\hat{\alpha}) \ , \tag{8.37}$$

It can be shown that there is a similar formula for momentum observables as well. Thus the subspace spanned by the vectors $e_\sigma(j)[\sigma = 1, 2, \ldots]$ for any fixed j is invariant under the action of observables. Also, since $\mathcal{D}(\hat{\alpha})$ is independent of j, the representation of the algebra of observables on the subspaces associated with different j are equivalent. It is thus sufficient to retain just one such subspace, the remaining ones may be discarded. When we do so, we also obtain an irreducible representation of the algebra of observables.

Further insight into the nature of this representation is gained by working with a "basis" for $\mathcal{H}$ consisting of states localized at points of Q. These are analogous to the states $|\vec{x}>$ which are localized at positions $\vec{x}$ in the standard nonrelativistic quantum mechanics of spinless particles. But while there is only one such linearly independent state for a given $\vec{x}$, we have $\dim \pi_1(Q)$ [$\equiv$ dimension of $\pi_1(Q)$] worth of such linearly independent states $\{|\overline{q}t>\}$ localized at q, because under π, $\overline{q}t$ projects to q independently of t. [Here $\overline{q}$ is any conveniently chosen point of $\overline{Q}$ with $\pi(\overline{q}) = q$.] The group $\pi_1(Q)$ acts on these states according to

$$\hat{s}|\overline{q}t> = |\overline{q}ts^{-1}> \ , \quad s \in \pi_1(Q) \ . \tag{8.38}$$

Clearly this representation of $\pi_1(Q)$ on the subspace spanned by $\{|t\overline{q}>\}$ (for fixed $\overline{q}$) is isomorphic to the regular representation of $\pi_1(Q)$. As is well known, when this representation is fully reduced, each UIR occurs as often as its dimension. Thus each γ_ℓ occurs once and is carried by a one dimensional vector space with basis $F^{(\ell)}$ say, while each $\overline{\gamma}_\alpha$ occurs $\dim \overline{\gamma}_\alpha$ times and is carried by a vector space with basis $E_\sigma^{(\alpha)}(j)[j, \sigma = 1, 2 \ldots \dim \overline{\gamma}_\alpha]$ say. The transformation law of $E_\sigma^{(\alpha)}(j)$ under $\pi_1(Q)$ is

$$\hat{t}E_\sigma^{(\alpha)}(j) = E_\sigma^{(\alpha)}(k)D_{kj}(t) \quad . \tag{8.39}$$

According to our previous argument, the reduction of the representation of the algebra of observables is achieved by retaining only the subspace $V_j(q)$ spanned by the vectors $E_\sigma^{(\alpha)}(j)$ for a fixed j [and a fixed α].

Now every nonzero vector in $V_j(q)$ is localized at q. Thus even after this reduction, there are $\dim \overline{\gamma}_\alpha$ linearly independent vectors localized at q. In nonrelativistic quantum mechanics, if the system has internal symmetry (or quantum numbers like intrinsic spin), the linearly independent states localized at $\vec{x}$ are of the form $|\vec{x}, m >$ $(m = 1, 2, \ldots, k)$ where the index m carries the representation of internal symmetry. In this case, there are k linearly independent vectors localized at $\vec{x}$. The situation we are finding when $\pi_1(q)$ is nonabelian has points of resemblance to this familiar quantum mechanical situation in the sense that here as well there are many states localized at q.

It is of interest to know the physical observables $\hat{O}$ which mix the indices σ of the basis $E_\sigma^{(\alpha)}(j)$. That is, it is of interest to find the observables $\hat{O}$ with the property

$$\hat{O}E_\sigma^{(\alpha)}(j) = E_\lambda^{(\alpha)}(j)\mathcal{D}_{\lambda\sigma}(\hat{O}) \tag{8.40}$$

such that their representation on $V_j(q)$ is irreducible. There is an elegant, but local, geometrical construction for a family of such operators which we now describe. Consider loops from q to q, they can be divided into homotopy classes $[C_t(q)][t \in \pi_1(Q)]$ labelled by elements of $\pi_1(Q)$. The class $[C_t(q)]$ consists of closed loops which are homotopic to each other. The labels can be so chosen that $[C_s(q)][C_t(q)] = [C_{st}(q)]$ where the multiplication of homotopy classes has been described in Section 8.2. [Note however that the loops $C_t(q)$ are based at q and not at the base point q_0 of Section 8.2.] Pick one closed curve $C_t(q)$ from $[C_t(q)]$ and consider the operator which parallel transports the wave function around $C_t(q)$. It can be shown that the change of the wave function as a result of parallel transporting it around a loop in $C_t(q) \in [C_t(q)]$ is independent of the choice of the loop in the class $[C_t(q)]$. Thus the parallel transport operator depends only on the homotopy class $[C_t(q)]$ and not on the choice of the closed curve in $[C_t(q)]$. It can hence be denoted by $\hat{O}_t$. These operators $\hat{O}_t$ can serve as the observables we are seeking.

The above description of the operators $\hat{O}_t$ is rather loose however since $\hat{O}_t$ is defined only if the transform $\hat{O}_t\psi$ of a wave function ψ is defined and this involves specifying $(\hat{O}_t\psi)(\bar{q})$ for *all* $\bar{q}$. Hence we must associate a homotopy class $[C_t(q)]$ to each $t \in \pi_1(Q)$ and *all* q. This association must be smooth in q and fulfill the property $[C_s(q)][C_t(q)] = [C_{st}(q)]$. Consider what happens if we smoothly change $[C_t(q)]$ as q is taken around a closed loop in the homotopy class $[C_s(q)]$, $s \in \pi_1(Q)$. It is then easy to convince oneself that $[C_t(q)]$ evolves into the homotopy class $[C_{sts^{-1}}(q)]$. When $\pi_1(Q)$ is nonabelian, $[C_t(q)]$ will not be equal to $[C_{sts^{-1}}(q)]$ for all t and s. A consequence is that the operators $\hat{O}_t$ are not all well defined when the UIR's of $\pi_1(Q)$ defining the quantum theory is nonabelian. [Nonetheless, the representation of the algebra of observables we have described can be shown to be irreducible.] The obstruction in defining all the operators $\hat{O}_t$ here is similar to the obstruction in defining the colour group in the presence of nonabelian monopoles [68] or the helicity group for massless particles in higher dimensions [8,9].

It is remarkable that when $\pi_1(Q)$ is nonabelian, quantization can lead to a multiplicity of states all localized at the same point. The consequences of this multiplicity have not yet been sufficiently explored in the literature. As we shall see later, one of the interesting physical contexts where this phenomenon occurs is provided by the gravitational geons of Friedman and Sorkin [69].

8.6 THE CASE OF THE ASYMMETRIC ROTOR

We shall now briefly illustrate these ideas by the example of the asymmetric rotor described in Section 8.3. The treatment given here is equivalent for example to the standard treatment of nuclei with three distinct moments of inertia in the collective model approach to nuclei [65].

Let $\overline{Q}$ be the manifold of the group $SU(2)$ and let s denote a point of $\overline{Q}$. We regard s as a 2×2 unitary matrix of determinant 1. Let $D_8^\star$ be the quaternion subgroup of $SU(2)$:

$$D_8^\star = \{\pm 1, \pm i\tau_i \ (i = 1, 2, 3)\} \ . \tag{8.41}$$

It has the free action

$$s \to sh \ , \quad h \in D_8^\star \tag{8.42}$$

on $\overline{Q}$. If we identify all the eight points which are taken into each other by this action, we get a space Q which as we saw in Section 8.3 is the configuration space of the asymmetric rotor.

The group $D_8^\star$ has five inequivalent UIR's. Four of these are abelian and may be described as follows. In one, the trivial one, all elements of H are represented by the unit operator. In one of the remaining three, ± 1 and $\pm i\tau_1$ are represented by $+1$ while $\pm i\tau_2$ and $\pm i\tau_3$ are represented by -1. The two remaining one dimensional UIR's are constructed similarly, $\pm i\tau_2$ and ± 1 being represented by $+1$ in one and $\pm i\tau_3$ and ± 1 being represented by $+1$ in the other. As regards the two dimensional UIR, it is the defining representation (8.41) involving Pauli matrices.

There are thus five ways of quantizing this system. We now concentrate on the quantization method involving the two dimensional nonabelian UIR of $D_8^\star$.

A basis for all functions on $SU(2)$ are the matrix elements $D_{\rho\sigma}^j(s)[s \in SU(2)]$ of the rotation matrices. The group $D_8^\star = \{h\}$ acts by operators $\hat{h}$ on these functions according to the rule

$$(\hat{h}D_{\rho\sigma}^j)(s) = D_{\rho\sigma}^j(sh) \ . \tag{8.43}$$

Since

$$D_{\rho\sigma}^j(sh) = D_{\rho\lambda}^j(s)D_{\lambda\sigma}^j(h) \tag{8.44}$$

and since for integer j, $h \to D^j(h)$ for $h \in D_8^\star$ defines an abelian representation of $D_8^\star$, we can and shall restrict j to half odd integer values. The next step is to reduce the representation $h \to D^j(h)$ into its irreducible components. It then splits into a direct sum of the two dimensional UIR's (8.41). [Only the two dimensional UIR's occur in this reduction. This is because the image of $(i\tau_i)^2$ being a 2π rotation is represented by -1, j being half an odd integer.] The basis vectors for the vector spaces which carry such UIR's are of the form $e^j_{\rho,m,a}$, $m = 1, 2, \ldots, N$; $a = 1, 2$ where $2N$ equals $2j + 1$. Under the transformation $s \to sh$, their behavior is given by

$$e^j_{\rho,m,a}(sh) = e^j_{\rho,m,b}(s)h_{ba} \quad . \tag{8.45}$$

The vector space which carries the algebra of observables irreducibly is spanned by e^j_{ρ,m,a_0} with one fixed value a_0 for a and with j, ρ, and m taking on all allowed values. The vectors $e^j_{\rho,m,a'}$ with the remaining values a' for a are to be discarded. When the asymmetric rotor model is used to describe nuclei, m can be interpreted in terms of the third component of angular momentum in the body fixed frame.

We have not discussed a scalar product for this vector space. A suitable scalar product may be

$$(\alpha, \beta) = \int_{SU(2)} d\mu(s) \alpha^\star(s) \beta(s) \quad . \tag{8.46}$$

Here we have regarded the elements of our vector space as functions on $SU(2)$ and $d\mu(s)$ is the invariant measure on $SU(2)$.

In the preceding discussion, we have not referred to a Lagrangian or a Hamiltonian. They are of course important from a dynamical point of view. They do not however play a critical role in the construction of the vector space for wave functions that we have outlined because this construction is valid for a large class of Lagrangians and Hamiltonians.

PART II

TOPOLOGICAL SOLITONS

AND

NONLINEAR MODELS

Chapter 9

TOPOLOGICAL SOLITONS IN ONE AND TWO DIMENSIONS

9.1 INTRODUCTION

In this Chapter, we shall introduce the subject of topological solitons through a discussion of examples in one and two spatial dimensions. There are many excellent reviews [see e.g. Refs. [70-72]] on this subject, which we shall occasionally rely upon in this Chapter. The main topics of interest for us in soliton physics are the Skyrmions and gravitational geons (which are both three dimensional solitons). They will be treated in Parts III and IV, respectively. The examples in one and two dimensions exhibit many of the features of these solitons in a simpler context and will serve as a good introduction to the latter. In two dimensions we study solitons in two different field theories, the linear and nonlinear σ-models. Solitons in the latter model can be considered as the two dimensional analogue of Skyrmions. In the literature they are sometimes referred to as baby Skyrmions.

We shall first give a loose definition of topological solitons in this Section and then take up the examples in one and two dimensions in Sections 2 and 3, respectively.

Basic features of Solitons

Suppose that $F_\phi = 0$ is a differential equation or a system of differential equations involving a field or a set of fields denoted by ϕ which are functions of D space coordinates $\vec{x}$ and a time coordinate t. We shall assume that $F_\phi = 0$ is the equation of motion which results from some Lagrangian field theory. We also assume that the system has a corresponding "energy" $\mathcal{E}_\phi$ and an "energy density" $E_\phi(\vec{x}, t)$,

$$\mathcal{E}_\phi = \int d^D x \; E_\phi(\vec{x}, t) \quad , \tag{9.1}$$

where for all allowed field configurations ϕ, E_ϕ is greater than or equal to zero. If E_ϕ is zero for all $\vec{x}$, we call ϕ a ground state or a vacuum solution, and denote it by ϕ_{vac}. The vacuum solution need not be unique.

Now suppose that $\phi = \phi_{c\ell}$ is a non-vacuum solution to $F_\phi = 0$. Following Coleman, we shall call $\phi_{c\ell}(\vec{x}, t)$ a soliton solution if the following properties are satisfied:

(i) $\mathcal{E}_{\phi = \phi_{c\ell}}$ is finite.

(ii) $E_{\phi = \phi_{c\ell}}(\vec{x}, t)$ is nonsingular for all $\vec{x}$ and t and "localized" for all times t. A solution is localized at any time t if there is a bounded region of space defined by $E_{\phi = \phi_{c\ell}}(\vec{x}, t) \geq \delta$, δ being any arbitrary number fulfilling

$$0 < \delta < \max_{\vec{x}} E_{\phi = \phi_{c\ell}}(\vec{x}, t) \quad .$$

We say that the solution is localized for all times t if the bounded region can be chosen independent of t.

(iii) $\phi_{c\ell}(\vec{x}, t)$ is nonsingular.

(iv) $\phi_{c\ell}(\vec{x}, t)$ is nondissipative.

Concerning (iv), a solution $\phi_{c\ell}(\vec{x}, t)$ is considered dissipative if

$$\lim_{t \to \infty} \max_{\vec{x}} \mathrm{E}_{\phi=\phi_{c\ell}}(\vec{x}, t) = 0 \quad . \tag{9.2}$$

In our discussions, we will be concerned primarily with "static" solutions for which we denote $\phi_{c\ell}(\vec{x}, t)$ by $\Phi_{c\ell}(\vec{x})$. If we exclude the vacuum, then static, nonsingular, localized solutions are automatically nondissipative. Thus for static solitons, we only require (i), (ii) and (iii). We will, however, be interested in static solutions which satisfy yet another requirement; namely that they be

(v) "classically stable".

We call a static solution "classically stable" if under all possible infinitesimal fluctuations $\delta\phi$ of the fields ϕ about the static solution,

$$\mathcal{E}_{\phi_{c\ell}+\delta\phi} \geq \mathcal{E}_{\phi_{c\ell}} \quad . \tag{9.3}$$

Variational modes $\delta_0\phi$ which leave the energy unchanged are said to be "zero frequency modes" or "zero modes". Thus, for zero modes, $\mathcal{E}_{\phi_{c\ell}+\delta_0\phi} = \mathcal{E}_{\phi_{c\ell}}$. Zero modes are generally associated with various symmetries of the soliton. Excluding the zero modes, a classically stable solution is a local minimum of the energy.

It is generally believed that configurations $\phi_{c\ell}$ which are "classically unstable" (that is, local maxima or stationary points of the energy) do not survive as stationary states in the corresponding quantum theory. This is because quantum fluctuations are likely to cause transitions away from any state localized around $\phi_{c\ell}$ and restore the system to its ground state or some other appropriate state of lower energy.

Topological Considerations

Although the question of stability is ultimately a dynamical one, topology often plays an important role in its discussion. We shall now elaborate on this point. Let Q be the set of all finite energy and nonsingular field configurations ϕ at some fixed time t. It is thus the configuration space of the system. A subset Q_1 of Q is said to be (path-)connected if any field ϕ_1 of Q_1 can be continuously deformed to any other field ϕ_1' in Q_1. Two subsets Q_1 and Q_2 of Q are said to be disconnected if any field ϕ_1 of Q_1 cannot be continuously deformed to any field ϕ_2 in Q_2. We shall consider the case where Q has $N > 1$ disconnected components Q_n, each Q_n being connected. Q is the union of all these disconnected components,

$$Q = \bigcup_{n=1}^{N} Q_n.$$

Let $\phi(\vec{x})$ and $\phi'(\vec{x})$ be two fields in Q. As in Chapter 8, we shall say that $\phi(\vec{x})$ and $\phi'(\vec{x})$ are homotopic to each other, and write $\phi \sim \phi'$, if there exists a sequence of fields $\phi^{(\tau)}(\vec{x})$, $0 \leq \tau \leq 1$, which is continuous in τ and $\vec{x}$, with $\phi^{(0)}(\vec{x}) = \phi(\vec{x})$ and $\phi^{(1)}(\vec{x}) = \phi'(\vec{x})$. With this definition, all fields within one component Q_n are "homotopic" to each other, while a field $\phi^{(n)}$ belonging to Q_n is not homotopic to a field $\phi^{(n')}$ belonging to $Q_{n'}$ when $n' \neq n$.

We can treat all our examples assuming that n is countable. We make this assumption below.

What is the physical significance of this classification? Consider the initial conditions $(\phi^{(n)}, d\phi^{(n)}/dt)$ at time $t = 0$ for the equations of motion. (These equations for simplicity are assumed to be second order in time). Assume that $\phi^{(n)} \in Q_n$. After a lapse of time T, suppose that $\phi^{(n)}$ becomes $\phi'^{(n)}$ and $d\phi^{(n)}/dt$ becomes $d\phi'^{(n)}/dt$. Since time evolution is assumed to be a continuous operation, it follows that $\phi'^{(n)}$ is homotopic to $\phi^{(n)}$. Hence, $\phi'^{(n)} \in Q_n$. In other words, the value of n associated with the field $\phi^{(n)}$ is a constant of the motion. Hence, if we define Q_0 to be the sector which contains the vacuum solution ϕ_{vac}, then the configurations $\phi^{(n)} \in Q_n$, $n \neq 0$, cannot be time evolved to ϕ_{vac} or in fact, to any other $\phi^{(m)} \in Q_m$ with $m \neq n$. For such reasons, the sectors $Q_{n \neq 0}$ of field configurations are said to be "topologically stable".

Soliton configurations which fulfill the five properties listed previously and which are in topologically stable sectors are called topological solitons.

Such solitons will be a main topic of interest for us.

Although a theory may admit sectors of topologically stable field configurations, it does not necessarily follow that there exist classically stable soliton solutions, or for that matter, any non-vacuum static solutions to the equations of motion. There are however interesting cases where such soliton solutions do exist as we shall now indicate.

For some theories, there exists a special lower bound on the energy of fields in the various topologically stable sectors. The bound usually involves the topological index n and is generically called the "Bogomol'nyi bound" [73]. [Also see Refs. [74,75].] An example of the form of such a bound is $\mathcal{E}_{\phi \in Q_n} \geq C|n|$ ($C = $ constant). If the Bogomol'nyi bound is the best possible bound and it can be saturated for a nonzero n by a localized field configuration, then this field configuration is classically stable. The latter follows since all smooth fluctuations must either raise the energy or, in the case of zero modes, leave it invariant. It follows in particular that a static field configuration which saturates the bound is a local minimum of the energy and must satisfy the equations of motion. The Bogomol'nyi bound can be saturated in one of the examples we shall study in this Chapter.

As stated earlier, there are cases where there exist topologically stable sectors although no soliton solutions exist. It is sometimes possible to modify the dynamics of the model to cure this situation and obtain a static solution. Modifications may be made by either adding new terms to the Lagrangian or introducing new dynamical variables. In this and subsequent Chapters, we shall see examples of both these approaches.

We next discuss examples of topological solitons in one and two spatial dimensions. The example in one dimension involves a real scalar field. Two examples in two spatial dimensions will be given. The first involves a linear field theory, while the second involves a nonlinear theory. For solitons in the former, the topology of the space $Q^{(\infty)}$ of fields at spatial infinity is relevant. For solitons in the latter, we must study the topology of the space Q of fields defined over all space. An appropiate generalization of this nonlinear theory to three spatial dimensions leads to Skyrme's topological soliton.

9.2 A SOLITON IN ONE DIMENSION

The example we review in one space-one time is that of a real scalar field $\phi(x, t)$. The Lagrangian density is

$$\mathcal{L} = -\frac{1}{2}\partial_\mu\phi\partial^\mu\phi - U(\phi^2) \quad . \tag{9.4}$$

It is invariant under the reflection $\phi \to -\phi$. For $U(\phi^2)$, we take the potential

$$U(\phi^2) = \frac{\lambda}{2}(\phi^2 - a^2)^2 \quad , \quad a^2, \lambda > 0 \quad . \tag{9.5}$$

The energy density associated with this Lagrangian density is

$$E(x, t) = \frac{1}{2}(\partial_0\phi)^2 + \frac{1}{2}(\partial_1\phi)^2 + U(\phi^2) \quad . \tag{9.6}$$

(The subscript ϕ on E has now been omitted.) There are two possible vacuum solutions: $\phi_{vac} = \pm a$. A particular choice for the vacuum breaks the reflection symmetry. Consequently, we say that this symmetry is spontaneously broken.

Let $Q = \{\phi\}$ consist of all finite energy field configurations at a given time. Finiteness of the energy requires that ϕ approaches one of the two vacuum solutions as $x \to \pm\infty$. ϕ need not go to the same value at both $x \to \pm\infty$. There are then four possibilities:

$$Q_{+0}: \quad \phi \to a \quad \text{as} \quad x \to \pm\infty \quad ;$$

$$Q_{-0}: \quad \phi \to -a \quad \text{as} \quad x \to \pm\infty \quad ;$$

$$Q_{+1}: \quad \phi \to a \quad \text{as} \quad x \to \infty, \quad \text{and} \quad \phi \to -a \quad \text{as} \quad x \to -\infty \quad ;$$

$$Q_{-1}: \quad \phi \to -a \quad \text{as} \quad x \to \infty, \quad \text{and} \quad \phi \to a \quad \text{as} \quad x \to -\infty \quad .$$

Thus, $Q = Q_{+0} \cup Q_{-0} \cup Q_{+1} \cup Q_{-1}$. Configurations $\phi^{(a)}$ belonging to Q_a are not homotopic to $\phi^{(b)}$ belonging to Q_b for $a \neq b$. This is because a family of fields $\phi^{(\tau)}$ connecting $\phi^{(a)}$ with $\phi^{(b)}$ as τ is varied will necessarily violate the boundary conditions for some τ and hence are not contained in Q.

The two spaces Q_{+0} and Q_{-0} contain the vacuum solutions $\phi = a$ and $-a$, respectively, while $Q_{+1} = \{\phi^{(1)}\}$ and $Q_{-1} = \{\phi^{(-1)}\}$ do not. The sectors $Q_{\pm 1}$ are topologically stable. Next we show that Q_{+1} and Q_{-1} contain static solutions to the field equation which are classically stable as well as topologically stable.

The field equation following from the Lagrangian (9.4) is

$$[\partial_0^2 - \partial_1^2]\phi + 2\lambda(\phi^2 - a^2)\phi = 0 \quad . \tag{9.7}$$

It has the static solutions

$$\phi = \phi_{c\ell} = \pm a \tanh \mu x \quad , \quad \mu = a\sqrt{\lambda} \quad . \tag{9.8}$$

The solution with the plus (minus) sign is called the kink (antikink) solution and is consistent with the boundary conditions for the fields in Q_{+1} (Q_{-1}). The energy density is localized,

$$E(x,t) = a^2\mu^2 \operatorname{sech}^4 \mu x \quad , \tag{9.9}$$

with total energy $\mathcal{E} = \frac{4}{3}a^2\mu$.

The kink and antikink are classically stable. To see this perturb around the classical solution:

$$\phi(x,t) = \phi_{c\ell}(x) + \eta(x,t) \quad . \tag{9.10}$$

To lowest order, the equation of motion for the fluctuations is

$$[\partial_0^2 - \partial_1^2]\eta = 2\lambda(3\phi_{c\ell}^2 - a^2)\eta \quad . \tag{9.11}$$

Upon expanding η in terms of normal modes,

$$\eta(x,t) = \sum_n \varepsilon_n \eta_n(x) e^{i\omega_n t} \tag{9.12}$$

[where ε_n are chosen so that $\eta_n(x)$ is real], we obtain an eigenvalue equation for $\eta_n(x)$:

$$[-\partial_1^2 + 2\lambda(3\phi_{c\ell}^2 - a^2)]\eta_n(x) = \omega_n^2 \eta_n(x) \quad . \tag{9.13}$$

Classical stability means that there are no negative eigenvalues ω_n^2. To see that $\omega_n^2 \geq 0$, first note that there does exist a zero mode

$$\eta_0(x) = \partial_1 \phi_{c\ell}(x) = \pm a\mu \ \text{sech}^4 \mu x, \qquad (9.14)$$

with $\omega_0 = 0$. Now there is a well known theorem stating that for a one dimensional Schrödinger equation with an arbitrary potential, the bound state eigenfunction associated with the lowest eigenvalue has no nodes. Then since η_0 has no nodes, ω_0^2 corresponds to the lowest eigenvalue and there are no negative eigenvalues.

As stated in the Introduction, zero modes are normally associated with various symmetries of the solution. The mode η_0 is associated with the translational symmetry of the origin of the kink (antikink), since under an infinitesimal translational $x \to x - \varepsilon_0$, the change of $\phi_{c\ell}$ is

$$\delta_0 \phi = \varepsilon_0 \partial_1 \phi_{c\ell} \quad . \qquad (9.15)$$

It is now readily checked that the solutions (9.8) fulfill all the five properties discussed previously. They are thus topological solitons.

9.3 SOLITONS IN TWO DIMENSIONS

In this Section, we look at two different models containing topological solitons in two spatial dimensions. In both models, there is a global symmetry group G which plays an important role, but the way this group acts on the fields is different for the two cases. In the first example the group is linearly realized, while in the second example it is nonlinearly realized.

Model 1

Here we take ϕ to be a scalar doublet of fields (ϕ_1 , ϕ_2) and consider the Lagrangian

$$\mathcal{L} = -\frac{1}{2}\partial_\mu\phi_\alpha\partial^\mu\phi_\alpha - U(|\phi|^2) \quad , \quad |\phi|^2 = \phi_\alpha\phi_\alpha \quad . \tag{9.16}$$

It is invariant under the global $SO(2)$ transformation

$$\phi_\alpha \to R_{\alpha\beta}\phi_\beta \quad , \quad R^T R = 1 \quad . \tag{9.17}$$

If we take the potential U to be

$$U(|\phi|^2) = (|\phi|^2 - a^2)^2 \quad , \quad a^2 , \lambda > 0 \quad , \tag{9.18}$$

the $SO(2)$ symmetry is spontaneously broken by the vacuum. Here the vacuum is given by $\phi_\alpha = (\phi_{vac})_\alpha = $ constant, where $(\phi_{vac})_\alpha$ (and hence the constants) are subject to the constraint

$$((\phi_{vac})_1)^2 + ((\phi_{vac})_2)^2 = a^2 \quad . \tag{9.19}$$

Now the number of different vacua is infinite and can be parametrized by an angle χ:

$$((\phi_{vac})_1 , (\phi_{vac})_2) = (a\cos\chi, a\sin\chi) \quad .$$

In order to pass from one vacuum to another, by for example a semiclassical tunneling process, fields must rotate everywhere in space. This involves an infinite rotational kinetic energy. As a result, the system selects one fixed direction for the ground state, and the $SO(2)$ [or $U(1)$] symmetry is spontaneously broken.

A nonzero (but finite) energy configuration for ϕ is required to fulfill

$$\phi_1^2 + \phi_2^2 \to a^2 \quad , \quad \text{as } |\vec{x}| \to \infty \quad . \tag{9.20}$$

(The rate of approach must be fast enough to guarantee that the total energy $\mathcal{E}$ is finite). Thus, the set $Q^{(\infty)}$ of all fields $\phi_\infty = (\phi_{\infty 1}, \phi_{\infty 2})$ at spatial infinity is made up of mappings to a circle S^1. "Spatial infinity" in a plane also corresponds to a circle $\tilde{S}^1$ (with infinitely large radius):

$$\tilde{S}^1 : x_1^2 + x_2^2 = R^2 \quad , \quad R \to \infty \quad . \tag{9.21}$$

Thus a field ϕ_∞ at spatial infinity (and at a fixed time) maps a circle to a circle:

$$\phi_\infty : \tilde{S}^1 \to S^1 \quad .$$

As we shall see below, the space $Q^{(\infty)}$ falls into an infinite number of disconnected componenents $Q_n^{(\infty)} = \{\phi_\infty^{(n)}\}$, $Q^{(\infty)}$ being the union

$$\underset{n}{\cup} \; Q_n^{(\infty)}$$

of these $Q_n^{(\infty)}$. Let us parametrize the circle $\tilde{S}^1$ by an angle θ , $0 \le \theta < 2\pi$. Then the index n is called the "winding number" and is the number of times that the circle S^1 is covered by a field $\phi_\infty^{(n)}$ as θ ranges from 0 to 2π. We explain these remarks more fully now.

Consider first a "trivial" (or vacuum) solution $\phi_{vac} = (a, \, 0)$. At spatial infinity, the field $(\phi_\infty)_{vac} = \lim_{R \to \infty} \phi_{vac}$ maps all of $\tilde{S}^1$ to the same point $(a, \, 0)$ on S^1. Thus, $(\phi_\infty)_{vac}$ is characterized by zero winding number $(\phi_\infty)_{vac} \in Q_0^{(\infty)}$. To this $(\phi_\infty)_{vac}$, we can associate all maps $\phi_\infty^{(0)}$ which are homotopic to $(\phi_\infty)_{vac}$. The set of all maps $\phi_\infty^{(0)}$ makes up the "trivial sector" $Q_0^{(\infty)}$ of $Q^{(\infty)}$.

A typical field $\phi_\infty^{(1)}$ of the sector $Q_1^{(\infty)}$ is defined by

$$\phi_\infty^{(1)}(\theta) = (a\cos\theta, \; a\sin\theta) \quad , \tag{9.22}$$

where we have suppressed the time argument in the field. In this example, as θ runs from 0 to 2π, all points on S^1 are covered once and only once. $\phi_\infty^{(1)}$ is a typical winding number one map. The equivalence class of maps homotopic to such a winding number one map constitutes the winding number one sector $Q_1^{(\infty)}$.

A typical winding number n map is

$$\phi_\infty^{(n)}(\theta) = (a\cos n\theta, \; a\sin n\theta) \quad . \tag{9.23}$$

n takes the value 0, ± 1, ± 2, ... n cannot take nonintegral values since then $\phi_\infty^{(n)}$ would not be a single valued function on $\tilde{S}^1$ [for which we require that $\phi_\infty^{(n)}(0) = \phi_\infty^{(n)}(2\pi)$]. $Q_n^{(\infty)}$ consists of all maps which are homotopic to $\phi_\infty^{(n)}$.

It is easy to show that it is not possible to continuously deform a field $\phi_\infty^{(n)}$ to a field $\phi_\infty^{(m)}$, if $n \neq m$. For under continuous deformations, n (or m) has to change continuously. But then being an integer it cannot change at all. Thus the fields in $Q_n^{(\infty)}$ are not homotopic to fields in $Q_m^{(\infty)}$ if $n \neq m$.

So far, we have learned that the space $Q^{(\infty)}$ of all fields defined at spatial infinity falls into an infinite number of homotopy classes. The same is then true for the space Q of fields defined over all space, that is $Q = \cup_n Q_n$. Q_n is defined as the space of all configurations $\phi^{(n)}$ whose limit as $|\vec{x}| \to \infty$ is an element $\phi_\infty^{(n)}$ of $Q_n^{(\infty)}$. [Note that for any field $\phi_\infty^{(n)}$ in $Q_n^{(\infty)}$, there is a field $\phi^{(n)}$ defined for all $\vec{x}$ which approaches $\phi_\infty^{(n)}$ as $|\vec{x}| \to \infty$. For example at a given time t, we can set $\phi^{(n)}(\vec{x}, t) = f(r)\phi_\infty^{(n)}(\hat{x})$, where $f(r)$ is any smooth function such that $f(\infty) = 1$, $f(0) = 0$ and $\hat{x} = \vec{x}/r$.]

As stated in the Section 9.1, the physical significance of the integer n associated with the field $\phi(\vec{x}, t) \equiv \phi(x)$ is that it is a constant of the motion. This integer is just the label of the homotopy classes of the fields at a fixed time.

The above discussion shows the existence of topologically stable sectors in the model under consideration. The vacuum is an element of Q_0. The sector Q_n with $n \neq 0$ is topologically stable in the limited sense that a field in Q_n will not evolve in time to the vacuum or to a field in any other sector Q_m with $m \neq n$.

Even though there are topologically stable sectors, none of them contain static solutions to the equations of motion. This is shown by a simple scaling argument of Derrick [76]. Suppose that ϕ_{cl} is a static solution. Its energy $\mathcal{E}$ is the sum of two terms:

$$\mathcal{E} = \mathcal{E}_1 + \mathcal{E}_2 \quad ,$$

$$\mathcal{E}_1 = \frac{1}{2} \int d^D x \; (\partial_i(\phi_{cl})_a)^2 \quad \text{and} \quad \mathcal{E}_2 = \int d^D x \; U[((\phi_{cl})_a)^2] \; ,$$
$$i = 1, 2, \dots D \; . \tag{9.24}$$

Here we have generalized the system to D spatial dimensions. Under a scaling transformation $\phi(\vec{x}, t) \rightarrow \phi_{cl}(\lambda \vec{x}, t)$, we have

$$\mathcal{E} \equiv \mathcal{E}(1) \rightarrow \mathcal{E}(\lambda) = \lambda^{2-D}\mathcal{E}_1 + \lambda^{-D}\mathcal{E}_2 \; . \tag{9.25}$$

Requiring that $\lambda = 1$ corresponds to a minimum of $\mathcal{E}$ yields the condition

$$\frac{d\mathcal{E}(\lambda)}{d\lambda}|_{\lambda=1} = 0 \quad \text{or} \quad (2 - D)\mathcal{E}_1 = D\mathcal{E}_2 \; . \tag{9.26}$$

Since $\mathcal{E}_1, \mathcal{E}_2 \geq 0$, it follows that $\mathcal{E}_1 = \mathcal{E}_2 = 0$ when $D > 2$. This implies that ϕ_{cl} must be the vacuum solution for $D > 2$. In the case $D = 2$, we have $\mathcal{E}_2 = 0$, so that $((\phi_{cl})_1)^2 + ((\phi_{cl})_2)^2 = a^2$ for all $\vec{x}$. This requires that ϕ_{cl} (as $|\vec{x}| \rightarrow \infty$) has zero winding number and hence is in $Q_0^{(\infty)}$. We can prove this result as follows. Let r, θ denote polar coordinates in the plane. For any nonzero r, ϕ_{cl} defines a map of the circle (with coordinate θ) to a circle [because of the condition $((\phi_{cl})_1)^2 + ((\phi_{cl})_2)^2 = a^2$]. The winding number n of this map cannot depend on r, as changing r is a continuous change. As $r \rightarrow 0$, all values of θ represent the same spatial point, so that $n \rightarrow 0$. Hence n is identically zero which shows the result. It follows that a stable static solution with nonzero topological index will not exist for Lagrangians of the form (9.16) for $D \neq 1$.

As indicated in the Section 9.1, such a negative result can often be evaded by suitably changing the dynamics. In our two dimensional example, this can be accomplished by introducing a gauge boson field A_μ, and making the $SO(2)$ [or $U(1)$] global symmetry a local or gauge symmetry. This is done as follows: Let g be an element of the group of maps from space-time to the group $U(1)$. Then we can write $g = e^{i\Lambda}$, where Λ is a real function on space-time. ϕ_α and A_μ transform under g according to

$$\phi_1(x) + i\phi_2(x) \equiv \Phi(x) \rightarrow e^{i\Lambda(x)}\Phi(x) \; ,$$

$$A_\mu(x) \rightarrow A_\mu(x) + \frac{1}{e}\partial_\mu\Lambda(x) \; , \tag{9.27}$$

where e is the "electric" charge. We then change the Lagrangian (9.16) to the gauge invariant expression

$$\mathcal{L}' = -\frac{1}{4}F_{\mu\nu}F^{\mu\nu} - \frac{1}{2}(D_\mu\Phi)^\star D^\mu\Phi - U(|\phi|^2) \qquad (9.28)$$

where $F_{\mu\nu} = \partial_\mu A_\nu - \partial_\nu A_\mu$ and $D_\mu\Phi = \partial_\mu\Phi - ieA_\mu\Phi$. After the scale transformations $\Phi(\vec{x}, t) \to \Phi(\lambda\vec{x}, t)$ and $A_\mu(\vec{x}, t) \to \frac{1}{\lambda}A_\mu(\lambda\vec{x}, t)$, the contribution to the action from the first term in (9.28) goes as λ^{4-D} which thus makes it possible to evade the above no-go theorem. Now along with the condition $|\Phi(\vec{x}, t)| \to a$ as $|\vec{x}| \to \infty$, finiteness of energy demands that

$$D_\mu\Phi \to 0 \quad \text{as} \quad |\vec{x}| \to \infty$$

or

$$A_\mu \to \frac{1}{e}\partial_\mu\chi \quad \text{as} \quad |\vec{x}| \to \infty \qquad (9.29)$$

where $\Phi = |\Phi|e^{i\chi}$. This means that the phase angle χ survives as the only degree of freedom at spatial infinity. The previous topological considerations are clearly still valid. The boundary condition on A_μ in (9.29) in addition leads to the quantization of "magnetic" flux. Indeed Stokes' theorem gives,

$$\int_{\mathbf{R}^2} d^2x\, F_{12} = \int_{\partial\mathbf{R}^2} dx^i A_i = \frac{1}{e}\int_{\partial\mathbf{R}^2} d\chi = \frac{1}{e}\left[\chi|_{\theta=2\pi} - \chi|_{\theta=0}\right] \quad . \qquad (9.30)$$

The right hand side of (9.30) is just $2\pi n/e$ where n is the winding number.

Here we will not attempt to find the explicit nontrivial static solutions of this model. It is known that there exist such solutions with localized magnetic flux [77]. Upon embedding the two dimensional solitons trivially in three dimensions, one obtains vortex lines. These vortices are the relativistic generalization of the vortices that occur in the Ginsburg-Landau theory of superconductivity.

In the following Sections and Chapters, we will have use for the second and third homotopy group, $\pi_2(M)$ and $\pi_3(M)$. These are groups whose elements are the equivalence classes of mappings of two spheres and three spheres, respectively, to M. They are obtained as straightforward generalizations of the discussion in Section 8.2 on $\pi_1(M)$. For details of these generalizations, any one of the many excellent reviews may be consulted.

Model 2

Here we shall consider a typical example of a so-called nonlinear model, namely, the nonlinear σ-model.

By a nonlinear model, we mean loosely a field theory with the following properties:

1. The fields are subject to nonlinear constraints.

2. The Lagrangian and the constraints are invariant under the action of a global symmetry group G.

With this definition, Model 1 is nonlinear since the boundary condition on the fields at spatial infinity is not linear. More precisely, the description "nonlinear" is reserved to those models where the physical fields for all points $\vec{x}$ take values in a manifold M which is not a vector space. (With this more precise definition Model 1 is actually a "linear" model.) The global invariance group G acts transitively on M in most of these models. Here, we shall assume that such is indeed the case. Then by a well known result, M is a homogeneous space for G. If H is the stability group of a point $p \in M$,

$$H = \{h \in G | hp = p\} \quad , \tag{9.31}$$

then M can be identified with the space of left cosets G/H:

$$M = \{gH\} \quad . \tag{9.32}$$

General methods exist for constructing Lagrangians for these theories. We shall illustrate them in Chapter 10. In Chapter 11 we show how Chern-Simons terms can be added to the Lagrangians for nonlinear theories in the case of an odd number of space-time dimensions.

In the previous examples of topological solitons, it was sufficient to look at $Q^{(\infty)}$ (the space of physical fields at spatial infinity) for the topological considerations. For solitons in nonlinear models, it is often necessary to consider the topology of physical fields defined over all space. We demonstrate this for the nonlinear σ-model after first describing the model.

The nonlinear σ-model has $G = SU(2)$ and $H = U(1)$. To be specific, we can regard G as the group of 2×2 unitary matrices of determinant 1 and take H to be $\{e^{i\alpha\tau_3}; \ 0 \le \alpha < 2\pi\}$, where τ_3 is the third Pauli matrix. A map π can be defined from G to the space of left cosets $\{gH\}$. π projects $ge^{i\alpha\tau_3} \in G$, for all α, to the same point in $\{gH\}$. Since α is a continuous parameter, $\{gH\}$ defines a two dimensional manifold which we now show is the surface of a two sphere S^2. For this purpose, define

$$N \equiv g\tau_3 g^\dagger = n_a \tau_a \quad , \tag{9.33}$$

τ_a, $a = 1, 2, 3$, being the three Pauli matrices and $g \in SU(2)$. N is invariant under the transformations $g \to ge^{i\alpha\tau_3}$ (and only under these transformations). Furthermore, N defines a two dimensional manifold since

$$\frac{1}{2} Tr \ N^2 = n_a n_a = 1 \quad . \tag{9.34}$$

Consequently, the map π is defined by $N[\pi(g) = N]$ and the constraint (9.34) implies that $SU(2)/U(1) = S^2$.

The fields in the nonlinear σ-model will be denoted by ϕ_a. They are subject to the constraint $\phi_a(\vec{x}, t)\phi_a(\vec{x}, t) = 1$. Thus, $\phi_a(\vec{x}, t)$ corresponds to n_a in (9.34). The action of $G = SU(2)$ on these fields is

$$\phi_a \to R_{ab}(g)\phi_b \quad , \tag{9.35}$$

where R is an element of the adjoint representation of $SU(2)$ [$R \in SO(3)$]. Note that the constraint is invariant under this action of G. In terms of the two independent degrees of freedom, say ϕ_1 and ϕ_2, the above action of G is nonlinear.

The Lagrangian density should be chosen so that it is invariant under G. The simplest choice is

$$\mathcal{L} = -\frac{\beta}{2}\partial_\mu\phi_a\partial^\mu\phi_a \quad , \qquad (9.36)$$

where β is a constant, and we have suppressed the arguments of the fields. Physically this system describes an infinite ferromagnet with ϕ_a corresponding to the three spin components. Because of the constraint on ϕ_a, the Lagrangian (9.36) does not describe a free system. Interactions of the field with itself are implicit. To see this, we can write $\mathcal{L}$ in terms of two independent degrees of freedom, say ϕ_1 and ϕ_2. Then

$$\mathcal{L} = -\frac{\beta}{2}\frac{2 - \phi_i\phi_i}{1 - \phi_i\phi_i}(\partial_\mu\phi_i)^2 \quad , \qquad (9.37)$$

where the index i is summed over 1 and 2. Eq. (9.37) is illdefined for ϕ_3=0.

The energy density associated with $\mathcal{L}$ is

$$\mathrm{E}(\vec{x},t) = \frac{\beta}{2}\{(\partial_0\phi_a)^2 + (\partial_i\phi_a)^2\} \quad . \qquad (9.38)$$

The vacuum solution is $\phi = \phi_{vac} =$ constant [which is subject to the constraint on ϕ_a]. Now E is invariant under global $SU(2)$ transformations. Thus ϕ_{vac} can be reduced to $(0, 0, 1)$ by the action of the $SU(2)$ group without affecting the energy. This corresponds to alligning all the spins of the ferromagnet towards the direction of the north pole. Only rotations about the third axis leave the spins invariant. Consequently, the global $SU(2)$ symmetry is spontaneously broken to a global $U(1)$.

Next, consider general configurations ϕ with a nonzero, but finite energy. For large $|\vec{x}| \equiv R$, the fields ϕ define a mapping from the circle S^1 with the radius R to S^2. Since $\pi_1(S^2) = 0$, this mapping is homotopic to the mapping of ϕ_{vac} at $|\vec{x}| = R$. [Even if the fundamental group $\pi_1(M)$ of the manifold M is nontrivial, the fields at R would have to be homotopic to the vacuum solution. This is so since ϕ defines a trivial mapping at $r = 0$ and the topological index cannot change as r is continuously varied from $r = 0$ to R.] Finiteness of energy then requires that

$$\phi(\vec{x},t) \to \phi_{vac} \quad , \quad \text{as} \quad r \to \infty \qquad (9.39)$$

and that the rate of approach $\phi \to \phi_{vac}$ is fast enough to guarantee that the energy $\mathcal{E}$ is finite. Again, we choose $\phi_{vac} = (0,0,1)$ without any loss of information. Now ϕ approaches $(0,0,1)$ and not some angle dependent limit at $r = \infty$. Thus, we may think of all points at spatial infinity as being a single point. Such an identification essentially converts the plane $\mathbf{R}^2 = \{(x_1, x_2)\}$ at a constant time to the surface of a two sphere $\tilde{S}^2$. The fields $\phi = (\phi_1, \phi_2, \phi_3)$ are well defined on this $\tilde{S}^2$ in view of the boundary condition.

Let us now discuss these observations in further detail. Let ξ_μ, $\mu = 1, 2, 3$ be the stereographic coordinates associated with $\vec{x}$:

$$\xi_i(\vec{x}) = \frac{2x_i}{r^2 + 1} \quad , \quad i = 1, 2 \quad ; \quad \xi_3(\vec{x}) = \frac{r^2 - 1}{r^2 + 1} \quad ,$$

$$\xi_1(\vec{x})^2 + \xi_2(\vec{x})^2 + \xi_3(\vec{x})^2 = 1 \quad . \tag{9.40}$$

The coordinates ξ_μ span a two sphere $\tilde{S}^2$. They are not globally valid coordinates for $\mathbf{R}^2$ which unlike S^2 is not a compact manifold. The change in topology is coming about because all the "points at spatial infinity" of $\mathbf{R}^2$ which correspond to $r \to \infty$ are mapped to one point of $\tilde{S}^2$, namely the north pole $N = (0,0,1)$ of $\tilde{S}^2$. In reality, the "points at infinity" are not points of $\mathbf{R}^2$ at all. Thus, to get a topologically accurate representation of $\mathbf{R}^2$, we should remove the north pole N from $\tilde{S}^2 : \mathbf{R}^2 = \tilde{S}^2 \setminus \{N\}$.

The topological difference between $\mathbf{R}^2$ and $\tilde{S}^2$ can make a difference for some functions. For example, the function $f(\vec{x}) = |\vec{x}|$ is a continuous function on $\mathbf{R}^2$, but the function obtained by the substitution $x_i = \xi_i/(1 - \xi_3)$ is not a continuous function on $\tilde{S}^2$, becoming infinite at the north pole. Another example is the function $\hat{x} = \vec{x}/|\vec{x}|$ which is continuous on $\mathbf{R}^2$, while its image function on $\tilde{S}^2$ has no well defined limit as the north pole is approached.

However, for functions which approach a constant limit as $r \to \infty$, the change of variable $\vec{x} \to \xi$ does produce a well defined function on $\tilde{S}^2$. In this sense then, because of the boundary condition on ϕ, we can imagine that the space on which the field ϕ is defined is $\tilde{S}^2$.

Thus, the configuration space Q of the nonlinear σ-model is made up of fields ϕ which map $\tilde{S}^2$ to $M = S^2$:

$$\phi : \tilde{S}^2 \rightarrow S^2 \quad . \tag{9.41}$$

The situation is thus analogous to the maps ϕ_∞ in Model 1. It is then plausible to expect that for the nonlinear σ-model, as well, the configuration space Q falls into an infinite number of disconnected components Q_n, with $Q = \cup_n Q_n$. This result is true. Here n is the generalization of the previous winding number associated with ϕ_∞ and is still called the winding number. We can describe the elements of Q_n as follows: There is a map $\phi^{(n)}$ in Q_n (constructed explicitly below) under which exactly n points of $\tilde{S}^2$ are mapped to the same point of S^2 so that when ξ covers $\tilde{S}^2$ once, $\phi^{(n)}(\xi, t)$ covers S^2 n times. Further for this map, all these n coverings of S^2 have the same orientation. The remaining elements of Q_n are all homotopic to $\phi^{(n)}$. The equivalence class of all maps homotopic to $\phi^{(n)}$ is thus Q_n.

Once again, these equivalence classes Q_n can be made into a group under a suitable product. This group is called the second homotopy group and it is denoted by $\pi_2(M)$. Here $M = S^2$. Like $\pi_1(S^1)$, $\pi_2(S^2)$ is isomorphic to the group Z of all integers ["winding numbers"] under addition.

The equivalence class Q_0 contains the vacuum solution $\phi_{vac}(\xi, t) = (0, 0, 1)$. Thus, we can take $\phi^{(0)} = \phi_{vac}$. Q_0 consists of all maps which are homotopic to ϕ_{vac}. An element $\phi^{(1)}$ of Q_1 is obtained by simply setting

$$\phi_\mu^{(1)}(\xi, t) = \xi_\mu \quad , \tag{9.42}$$

t here being fixed. Obviously, for (9.42), the sphere S^2 is covered once as ξ_μ ranges over $\tilde{S}^2$

We now construct an element $\phi^{(n)}$ of $Q^{(n)}$ for arbitrary n. For this we introduce spherical coordinates (Θ, Ψ) for the spatial two sphere $\tilde{S}^2$. Thus,

$$
\begin{aligned}
\xi_1(\vec{x}) &= \sin\Theta\cos\Psi \quad , \\
\xi_2(\vec{x}) &= \sin\Theta\sin\Psi \quad , \\
\xi_3(\vec{x}) &= \cos\Theta \quad , \quad 0 \le \Theta \le \pi \quad , \quad 0 \le \Psi < 2\pi \quad .
\end{aligned} \tag{9.43}
$$

Then a typical element $\phi^{(n)}$ is

$$
\begin{aligned}
\phi_1^{(n)}(\vec{x}, t) &= \sin\Theta \cos n\Psi \quad, \\
\phi_2^{(n)}(\vec{x}, t) &= \sin\Theta \sin n\Psi \quad, \\
\phi_3^{(n)}(\vec{x}, t) &= \cos\Theta \quad,
\end{aligned}
\tag{9.44}
$$

Q_n consisting of all maps homotopic to the above $\phi^{(n)}$.

Again, the significance of the above classification is that since time evolution is a continuous operation, the integer n is a constant of motion. It is useful to have an explicit formula for this conserved quantum number. For this purpose, consider the current

$$
j^\mu = -\frac{1}{8\pi}\varepsilon_{abc}\varepsilon^{\mu\nu\lambda}\phi_a\partial_\nu\phi_b\partial_\lambda\phi_c \quad,
\tag{9.45}
$$

where $\varepsilon^{\mu\nu\lambda}$ is the usual totally antisymmetric tensor, and we have reverted back to the $\vec{x}, t$ coordinates. This current is conserved regardless of what the equations of motion are. For upon taking its divergence, we obtain

$$
\partial_\mu j^\mu = -\frac{1}{8\pi}\varepsilon_{abc}\varepsilon^{\mu\nu\lambda}\partial_\mu\phi_a\partial_\nu\phi_b\partial_\lambda\phi_c \quad.
\tag{9.46}
$$

The right hand side of (9.46) contains the triple scalar product of the three tangent vectors $\partial_0\vec{\phi}$, $\partial_1\vec{\phi}$ and $\partial_2\vec{\phi}$ defined at $(\vec{x}, t)$. When multiplied by $d^2x\,dt$, it represents an infinitesimal volume element at $(\vec{x}, t)$. But because of the constraint on ϕ_a, the tangent vectors at $(\vec{x}, t)$ are forced to lie in a plane. Consequently, the volume element and the right hand side of (9.46) vanishes. So

$$
\partial_\mu j^\mu = 0 \quad.
\tag{9.47}
$$

It follows that the associated charge

$$
B(\phi) = -\frac{1}{8\pi}\int d^2x \,\varepsilon_{abc}\varepsilon_{ij}\phi_a\partial_i\phi_b\partial_j\phi_c
\tag{9.48}
$$

is a constant of the motion [ε_{ij} being the two dimensional antisymmetric symbol with $\varepsilon_{12} = +1$]. Its value is the conserved quantum number; it has the value n when $\phi = \phi^{(n)} \in Q_n$. The factor $-1/8\pi$ is chosen so that $B(\phi)$ is, in fact, an integer.

To see that $B(\phi) = n$, write ϕ_a in terms of spherical coordinates:

$$
\begin{aligned}
\phi_1(\vec{x}, t) &= \sin\theta(\vec{x})\cos\chi(\vec{x}) \quad , \\
\phi_2(\vec{x}, t) &= \sin\theta(\vec{x})\sin\chi(\vec{x}) \quad , \\
\phi_3(\vec{x}, t) &= \cos\theta(\vec{x}) \quad .
\end{aligned}
\tag{9.49}
$$

Then,

$$
B(\phi) = \frac{1}{4\pi} \int \sin\theta(\vec{x}) d\chi(\vec{x}) \wedge d\theta(\vec{x}) \quad .
\tag{9.50}
$$

Since $\frac{1}{4\pi}\sin\theta\, d\chi \wedge d\theta$ is the normalized volume element on the two sphere, $B(\phi)$ indicates the number of times the sphere S^2 is covered as $\vec{x}$ runs over all values and is therefore an integer.

Previously we saw that Derrick's scaling argument rules out the possibility of having nontrivial static solutions to a linear scalar field theory in two (or greater) space dimensions. However, for the nonlinear σ-model with the Lagrangian (9.36), Derrick's argument can only be used to rule out the existence of static solutions in all but two space dimensions. This is because the static energy contains only one term which we denote by $\mathcal{E}_S$. Under $\vec{x} \to \lambda\vec{x}$, it scales like $\mathcal{E}_S \to \lambda^{2-D}E_S$. The minimum value of the energy for this variation is zero in all but $D = 2$ dimensions.

A lower bound on the energy (the "Bogomol'nyi bound" [70,71,73]) for classical solutions can be obtained from the identity

$$
(\partial_i\phi_a \pm \varepsilon_{abc}\varepsilon_{ij}\phi_b\partial_j\phi_c)^2 \geq 0 \quad .
\tag{9.51}
$$

After completing the square, we can write

$$
\frac{2}{\beta}\mathcal{E}_S = \int d^2x (\partial_i\phi_a)^2 \geq 8\pi|n| \quad .
\tag{9.52}
$$

The bound is saturated if

$$
\partial_i\phi_a = \mp\varepsilon_{abc}\varepsilon_{ij}\phi_b\partial_j\phi_c \quad .
\tag{9.53}
$$

A general solution to this equation was obtained by Belavin and Polyakov [75]. Here we shall only look for a spherically symmetric $n = 1$ solution.

Spherical symmetry in two spatial dimensions normally means the following:

$$\varepsilon_{ij}x_i\partial_j\phi_a = 0 \quad . \tag{9.54}$$

This condition is consistent with the constraint on ϕ_a. However, it has the undesired result that all fields satisfying it have $B(\phi) = 0$. This is because the general solution to (9.54) is $\phi_a(\vec{x},t) = \tilde{\phi}_a(r,t)$, so that

$$\partial_i\phi_a = \hat{x}_i\frac{\partial\tilde{\phi}_a}{\partial r} \quad , \quad \hat{x}_i = \frac{x_i}{r} \quad .$$

Upon substituting into the expression for $B(\phi)$, we then obtain the result $B(\phi) = 0$.

We shall now modify our symmetry requirement in order to obtain configurations with $B(\phi) \neq 0$. The stability group H for the vacuum $\phi_{vac} = (0,0,1)$ consists of $U(1)$ rotations of the fields about the third axis. We now require that the vector (ϕ_1, ϕ_2) is invariant under a combined spatial rotation and internal $U(1)$ transformation, that is,

$$\varepsilon_{ij}x_i\partial_j\phi_k + \varepsilon_{ki}\phi_i = 0 \quad ,$$
$$\varepsilon_{ij}x_i\partial_j\phi_3 = 0 \quad , \quad i,j,k = 1,2 \quad . \tag{9.55}$$

The second equation in (9.55) results from the first equation and the constraint on ϕ_a.

A general solution to the modified symmetry requirement is

$$\phi_i = \sin\theta(\cos\psi\ \hat{x}_i + \sin\psi\ \varepsilon_{ij}\hat{x}_j) \quad ,$$
$$\phi_3 = \cos\theta \quad , \tag{9.56}$$

where θ and ψ are functions of the radial variable r. The boundary condition at spatial infinity requires that θ at infinity is an integer multiple of 2π. Further, since $r = 0$ represents a single spatial point, ϕ_a at $r = 0$ must have a unique value and cannot depend on $\hat{x}$. This forces θ at $r = 0$ to be an integral multiple of π. After substituting the above form for ϕ_a into the expression for the winding number, we get

$$B(\phi) = \frac{1}{2}(\cos\theta(\infty) - \cos\theta(0)) \quad . \tag{9.57}$$

The $B(\phi) = 1$ solution is obtained with $\theta(\infty) = 0$ and $\theta(0) = \pi$. Note that $|B(\phi)| > 1$ configurations are incompatible with our symmetry requirement.

As we saw before, the Bogomol'nyi bound is saturated when ϕ_a fulfills a field equation linear in derivatives of ϕ_a. It implies that

$$\frac{d\theta}{dr} + \frac{1}{r}\sin\theta = 0 \quad , \quad \frac{d\psi}{dr} = 0 \quad . \tag{9.58}$$

The solution with the desired boundary conditions is

$$\theta(r) = 2\tan^{-1}\frac{a}{r} \quad , \quad \psi = \psi_0 \quad , \tag{9.59}$$

where a and ψ_0 are constants. If we set $a = 1$ and $\psi_0 = 0$, this solution corresponds exactly to the stereographic projection $\phi_a(\vec{x}, t) = \xi_a(\vec{x})$. For a general a and ψ_0, we have [omitting t in the argument of ϕ_a],

$$\begin{aligned} \phi_i(\vec{x}) &= [R_{\psi_0}]_{ij}\xi_j(\vec{x}/a) \quad , \\ \phi_3(\vec{x}) &= \xi_3(\vec{x}/a) \quad , \end{aligned} \tag{9.60}$$

where we regard ξ_a as a function of $\vec{x}/a$ using (9.40) and R_{ψ_0} is the rotation matrix

$$R_{\psi_0} = \begin{pmatrix} \cos\psi_0 & \sin\psi_0 \\ -\sin\psi_0 & \cos\psi_0 \end{pmatrix} \quad . \tag{9.61}$$

The solution (9.60) can also be written

$$\phi_a(\vec{x}) = \xi_a\left(\frac{\overrightarrow{R_{\psi_0}x}}{a}\right) \quad . \tag{9.62}$$

Since this solution saturates the Bogomol'nyi bound, its classical energy is $4\pi\beta$. Further, since all fluctuations other than those associated with zero modes raise the energy, the solution is classically stable.

A still more general solution than that discussed above is obtained by replacing $\vec{x}$ in (9.60) by $\vec{x} - \vec{x}_0$, where $\vec{x}_0$ is a constant vector. This generalization gives us the freedom to choose the location of the origin $\vec{x}_0$ of the soliton. Now altogether, there are several zero modes obtained by differentiating the solution with respect to $\vec{x}_0, \psi_0$ and a. They correspond, respectively, to: (1)

translations of the soliton center, (2) $U(1)$ rotations and (3) dilations. All of these transformations leave the energy unchanged. (1), (2) and (3) are associated with $\vec{x}_0$, ψ_0 and a respectively. As regards to (2), ψ_0 parametrizes an internal $U(1)$ (or equivalently an external space) rotation. A shift in the value of ψ_0 corresponds to such a transformation.

We will not discuss the model further here, but refer the reader to the literature [see e.g. the reviews by Novikov et al. [81] and Zakrewksi [93]]. In Section 11.2 we briefly examine the collective rotational motion of the soliton obtained by changing ψ_0. There we also add a term, known as the Chern-Simons term, to the Lagrangian given by Eq. (9.36) and briefly discuss the effects of such a term on the properties of the soliton.

Chapter 10

NONLINEAR MODELS AS GAUGE THEORIES

10.1 INTRODUCTION

The following are a few of the reasons for the interest in the nonlinear models defined in Section 9.3:

a) Empirically successful effective Lagrangian models [78,79] are models of this sort.

b) In $1 + 1$ spacetime, many such models admit an infinite number of conservation laws and are examples of completely integrable field theories [80].

c) They have points of resemblance with QCD [59]. For instance, in $1 + 1$ spacetime, some of these models have asymptotic freedom [81] and instantons [75,82]. Thus in $1 + 1$ spacetime, they provide a testing ground for hypotheses regarding QCD [59]. More recently, an analogy has been made between gauge fields and such nonlinear models defined on the space of all contours [83].

d) The axisymmetric Einstein equations are related to the equations for nonlinear models in $1 + 1$ spacetime [84].

e) On including a suitable Wess-Zumino term in the 1+1 dimensional action, they can define conformal field theories [for a review see e.g. Refs. 85] and strings [for a review see e.g. Ref. 86] moving on a group manifold [87]. Such models are often refered to as Wess-Zumino-Novikov-Witten models [88,89]. We shall discuss them very briefly in Section 16.6.

In this Chapter, we will give a systematic method for the construction of Lagrangians for such models. The inclusion of topological terms, such as the Chern-Simons term and the Wess-Zumino term, in the total Lagrangian will be discussed later. Our approach [90] is closely related to the approach of other authors [91,92].

The fields in our Lagrangian will have values in G. Thus these fields are $\{g\}$ with $g(x) \in G$. Further, the Lagrangian will be invariant under the gauge transformations

$$g(x) \rightarrow g(x)h(x) \quad , \qquad h(x) \in H \subset G \quad . \tag{10.1}$$

Therefore the gauge invariant (physical) fields have values in G/H. It is the latter which are the fields of the usual nonlinear models. Note that the principal fibre bundle structure $H \rightarrow G \rightarrow G/H$ occurs naturally in our approach.

The Lagrangian is made of some simple differential forms (Maurer - Cartan forms) defined on G. For our purposes, these forms can be described as follows. We identify (the compact semisimple Lie group) G with any one of its faithful unitary representations and denote the latter as well by G. Let $\{L(\rho)\}$ be a basis for the Lie algebra $\underline{G}$ of G with the properties

$$L(\rho)^\dagger = L(\rho) \quad ,$$
$$\text{Tr } L(\rho)L(\sigma) = \delta_{\rho\sigma} \quad , \quad \rho, \sigma \in \{1, 2, \ldots, [G]\} \quad . \tag{10.2}$$

Here $[G]$ is the dimension of G. For $\alpha \leq [H]$, the generators $L(\alpha)$ are taken to span the Lie algebra $\underline{H}$ of H and are called $T(\alpha)$:

$$L(\alpha) = T(\alpha) \quad , \quad \alpha \leq [H] \quad . \tag{10.3}$$

The remaining generators are called $S(i)$:

$$L(i) = S(i) \quad , \quad [H] + 1 \leq i \leq [S] \quad . \tag{10.4}$$

Note the commutation relations

$$[L(\rho), L(\rho)] = i \, \eta_{\rho\sigma\lambda} L(\lambda) \quad , \tag{10.5}$$

where

$$[T(\alpha), T(\beta)] = i \, C_{\alpha\beta\gamma} \, T(\gamma) \quad , \tag{10.6a}$$

$$[T(\alpha), S(i)] = i \, \overline{C}_{\alpha ij} \, S(j) \quad , \tag{10.6b}$$

$$[S(i), S(j)] = i[D_{ij\alpha} \, T(\alpha) + \overline{D}_{ijk} \, S(k)] \quad . \tag{10.6c}$$

The absence of T's on the right side of (10.6b) is due to

$$\begin{aligned} \text{Tr } T(\gamma)[T(\alpha), S(i)] &= \text{Tr } S(i)[T(\gamma), T(\alpha)] \\ &= 0 \end{aligned} \tag{10.7}$$

by (10.2) and (10.6a). Further, since

$$\text{Tr } S(j)[T(\alpha), S(i)] = \text{Tr } T(\alpha)[S(i), S(j)] \quad , \tag{10.8}$$

we have the equality

$$\overline{C}_{\alpha ij} = D_{ij\alpha} \quad . \tag{10.9}$$

Let

$$\omega_\mu(g) = g^{-1}\partial_\mu g \quad , \quad g = g(x) \quad . \tag{10.10}$$

Under the gauge transformation (10.1), it transforms as follows:

$$\omega_\mu(gh) = h^{-1}\omega_\mu(g)h + h^{-1}\partial_\mu h \quad , \quad h = h(x) \in H \quad . \tag{10.11}$$

We can write ω_μ as a sum of two parts:

$$\begin{aligned} \omega_\mu &= A_\mu + B_\mu \quad , \\ A_\mu(g) &= T(\alpha) \text{ Tr } T(\alpha)g^{-1}\partial_\mu g \quad , \\ B_\mu(g) &= S(i) \text{ Tr } S(i)g^{-1}\partial_\mu g \quad . \end{aligned} \tag{10.12}$$

Then

$$A_\mu(gh) = h^{-1}A_\mu(g)h + h^{-1}\partial_\mu h \quad , \quad B_\mu(gh) = h^{-1}B_\mu(g)h \quad . \tag{10.13}$$

To find this result, we have used the facts that

$$h\,T(\alpha)h^{-1} = (Ad\ h)_{\beta\alpha}\,T(\beta) \quad ,$$
$$h\,S(i)h^{-1} = D(h)_{ij}\,S(j) \quad , \tag{10.14}$$

[cf. (10.6a) and (10.6b)] where $Ad\ h$ and $D(h)$ are orthogonal [by (10.2)]:

$$\begin{aligned}
[Ad\ h]^{-1} &= [Ad\ h]^T \\
&= Ad\ h^{-1} \quad , \tag{10.15} \\
[D(h)]^{-1} &= [D(h)]^T \\
&= D(h^{-1}) \quad . \tag{10.16}
\end{aligned}$$

From (10.12), A_μ is seen to be $\underline{H}$-valued. By (10.13), it transforms like a gauge potential for the gauge group H.

All known Lagrangians for nonlinear models are constructed from A_μ and B_μ. Examples are (up to constants)

$$\mathcal{L}_1 = -\text{Tr}\ B_\mu B^\mu \quad , \tag{10.17}$$

$$\mathcal{L}_2 = -\frac{1}{4}\ \text{Tr}\ F_{\mu\nu}(A)F^{\mu\nu}(A) \quad , \tag{10.18}$$

$$F_{\mu\nu}(A) \equiv \partial_\mu A_\nu - \partial_\nu A_\mu + [A_\mu, A_\nu] \quad . \tag{10.19}$$

These are invariant under the gauge transformation (10.1).

10.2 EXAMPLES OF NONLINEAR MODELS

For familiar choices of G and H, we now show how to reduce such Lagrangians to more conventional forms.

Example 1

Let $G = SU(2)$ and $H = U(1)$ with

$$T(1) = \sigma_3/\sqrt{2} \quad ,$$
$$S(i+1) = \sigma_i/\sqrt{2} \quad , \quad i \in \{1,2\} \quad .$$

(10.20)

Here σ_ρ, $\rho = 1,2,3$, are Pauli matrices. Note that $G/H = S^2$. Let us introduce a triplet of nonlinear fields ϕ_α, $\alpha = 1,2,3$ via

$$\frac{1}{\sqrt{2}}\sigma_\alpha\phi_\alpha \equiv \Phi = \frac{1}{\sqrt{2}}g\sigma_3 g^\dagger \quad .$$

(10.21)

Clearly,

$$\phi_\alpha\phi_\alpha = \text{Tr } \Phi^2 = 1 \quad .$$

(10.22)

We now reduce the Lagrangian (10.17) to the standard form $-\frac{1}{2}(\partial_\mu\phi_\alpha)(\partial^\mu\phi_\alpha)$. We have

$$\text{Tr } B_\mu B^\mu = \text{Tr } I_\mu I^\mu \quad , \quad I_\mu = -g B_\mu g^\dagger$$

(10.23)

while a little algebra shows that

$$\frac{i}{2}\varepsilon_{\alpha\beta\gamma}\phi_\alpha\partial_\mu\phi_\beta\sigma_\gamma = I_\mu \quad .$$

(10.24)

Using

$$\phi_\alpha\partial_\mu\phi_\alpha = 0$$

(10.25)

which follows from (10.22), we find

$$\mathcal{L}_1 = -\frac{1}{2}(\partial_\mu\phi_\alpha)(\partial^\mu\phi_\alpha) \quad .$$

(10.26)

Thus $\mathcal{L}_1$ is identical (up to a multiplicative factor) to the Lagrangian (9.36) describing the nonlinear model of Section 9.3.

Example 2

Let $G = SU(2) \times SU(2)$, with H being the diagonal $SU(2)$ subgroup. We can write

$$G = \left\{ g \equiv \begin{pmatrix} g_1 & 0 \\ 0 & g_2 \end{pmatrix} \right\} \quad , \quad g_i \in SU(2) \quad ,$$
$$H = \left\{ \begin{pmatrix} h & 0 \\ 0 & h \end{pmatrix} \right\} \quad , \quad h \in SU(2) \quad ,$$

(10.27)

and

$$T(\alpha) = \frac{1}{2}\begin{pmatrix} \sigma_\alpha & 0 \\ 0 & \sigma_\alpha \end{pmatrix} \quad , \quad \alpha \in \{1,2,3\} \quad ,$$

$$S(i) = \frac{1}{2}\begin{pmatrix} \sigma_i & 0 \\ 0 & -\sigma_i \end{pmatrix} \quad , \quad i \in \{1,2,3\} \quad . \tag{10.28}$$

The usual nonlinear $SU(2) \times SU(2)$ chiral model is associated with this G and H as we now show. We may remark that $G/H = S^3$.

Define τ_μ $(\mu = 0,1,2,3)$ by $\tau_0 = 1_{2\times2}$, $\tau_i = i\sigma_i$ and set

$$g\gamma_0 g^\dagger = \gamma_\alpha \psi_\alpha \quad , \quad \gamma_\mu = \begin{pmatrix} 0 & \tau_\mu \\ \tau_\mu^\dagger & 0 \end{pmatrix} \quad . \tag{10.29}$$

The fields ψ_α fulfill

$$\psi_\alpha \psi_\alpha = 1 \tag{10.30}$$

and thus span S^3.

If as before we set $I_\mu = -gB_\mu g^{-1}$, some algebra shows that

$$I_\mu = iS(i)[\psi_i \partial_\mu \psi_0 - \psi_0 \partial_\mu \psi_i] + i\varepsilon_{ijk}T(i)\psi_j \partial_\mu \psi_k \quad . \tag{10.31}$$

Thus

$$\mathrm{Tr}\, B_\mu B^\mu = \mathrm{Tr}\, I_\mu I^\mu = -\partial_\mu \psi_\alpha \partial^\mu \psi_\alpha \tag{10.32}$$

Example 3

There is some interest in models where the fields take values in the space $\mathbf{CP}^{n-1}$ [93; the large n-limit is discussed in Refs. [82 and 94]]. The latter can be defined as follows: Let

$$z = (z_1, z_2, \ldots, z_n) \quad , \quad \sum_\alpha |z_\alpha|^2 = 1 \quad , \tag{10.33}$$

where z_α is complex. Thus z is a point on S^{2n-1}. Now if we identify z and $e^{i\theta}z$ for all real θ, the result is $\mathbf{CP}^{n-1}$. The latter is a manifold of real dimension $2(n-1)$.

The description of $\mathbf{CP}^{n-1}$ in terms of G and H is as follows: Let $G = U(n)$ and $H = U(n-1) \times U(1)$. Then $G/H = \mathbf{CP}^{n-1}$. The proof is elementary. If $G = \{g\}$, with g written in the defining representation of $U(n)$, we can identify $z_\alpha = g_{\alpha 1}$. Eq. (10.33) follows from the unitarity of G. Under the right action of the group $g \to gh$, $h \in U(n)$, z transforms according to

$$z_\alpha \to z_\alpha h_{11} + g_{\alpha i} h_{i1} \ , \quad i = 2, 3, \ldots, n \quad . \tag{10.34}$$

z is left invariant under this transformation if we set $h_{\alpha 1}$, and hence $h_{1\alpha}$, equal $\delta_{\alpha 1}$. With these conditions, h defines a representation of $U(n-1)$. Thus H is the little group of the manifold obtained by identifying z and $e^{i\theta} z$ for all θ. Consequently, $\mathbf{CP}^{n-1} = G/H$.

For the Hermitian generators of $U(n)$, we may take

$$\begin{aligned}
L_-(\alpha, \beta)_{\rho\sigma} &= \frac{i}{2}(\delta_{\alpha\rho}\delta_{\beta\sigma} - \delta_{\beta\rho}\delta_{\alpha\sigma}) \ , \\
L_+(\alpha, \beta)_{\rho\sigma} &= \frac{1}{2}\left(\delta_{\alpha\rho}\delta_{\beta\sigma} + \delta_{\beta\sigma}\delta_{\alpha\sigma} - \frac{4}{n}\delta_{\alpha\beta}\delta_{\rho\sigma}\right) \quad .
\end{aligned} \tag{10.35}$$

Here $L_\pm(\alpha, \beta) = \pm L_\pm(\beta, \alpha)$, and we must replace the trace condition in (10.2) by

$$\begin{aligned}
\mathrm{Tr}\ L_\pm(\alpha, \beta) L_\pm(\rho, \sigma) &= \frac{1}{2}(\delta_{\alpha\rho}\delta_{\beta\sigma} \pm \delta_{\alpha\sigma}\delta_{\beta\rho}) \ , \\
\mathrm{Tr}\ L_+(\alpha, \beta) L_-(\rho, \sigma) &= 0 \quad .
\end{aligned} \tag{10.36}$$

$\underline{H}$ is spanned by

$$\begin{aligned}
T_\pm(i,j) &= L_\pm(i,j) \ , \quad i, j = 2, 3, \ldots, n \ , \\
T_0 &= \sum_\alpha L_+(\alpha, \alpha) \quad .
\end{aligned} \tag{10.37}$$

Substituting into $I_\mu = -gB_\mu g^\dagger$, we find

$$[I_\mu]_{\rho\lambda} = \partial_\mu(z_\rho z_\lambda^\star) - 2z_\sigma z_\lambda^\star \partial_\mu(z_\rho z_\sigma^\star) \quad . \tag{10.38}$$

Thus

$$\mathrm{Tr}\ B_\mu B^\mu = \mathrm{Tr}\ I_\mu I^\mu = -2|\partial_\mu z_\alpha|^2 + 2|z_\alpha^\star \partial_\mu z_\alpha|^2 \tag{10.39}$$

which agrees with the conventional form for the Lagrangian of the $\mathbf{CP}^{n-1}$ model:

$$\mathcal{L} = -\frac{1}{2}|D_\mu z_\alpha|^2 \quad , \quad D_\mu \equiv \partial_\mu - z_\alpha^\star \partial_\mu z_\alpha \quad . \tag{10.40}$$

Chapter 11

THE CHERN-SIMONS TERM

11.1 INTRODUCTION

In an odd number of space-time dimensions D, there is a gauge invariant action which can be written down in a Yang-Mills theory or a nonlinear model which is not of the form (10.17) or (10.18). This term, known as the Chern-Simons term, has the following properties:

i) The corresponding action S_{cs} is independent of the space-time metric. This is because it can be written as the integral of a D form Ω_D on a D dimensional space-time manifold M.

$$S_{cs} = \int_M \Omega_D \quad , \tag{11.1}$$

Ω_D is a functional of Yang-Mills potentials A_μ (and their derivatives) which take values in some Lie algebra $\underline{H}$. For nonlinear models, A_μ are defined in (10.12). From (11.1), S_{sc} is invariant under diffeomorphisms of the space-time manifold M.

ii) S_{cs} (although not Ω_D) is invariant under gauge transformations (10.13) which are connected to the identity mapping.

iii) S_{cs} is linear in time derivatives. Consequently, static solutions to the equations of motion following from an action S_0 are also solutions to the equations of motion following from the action $S = S_0 + S_{cs}$.

iv) If d denotes exterior derivative, then

$$d\Omega_D = \text{Tr}[\underbrace{F \wedge F \wedge \cdots \wedge F}_{(D+1)/2 \text{ times}}] \equiv \text{Tr} \left[F^{(D+1)/2} \right] \quad , \tag{11.2}$$

where F is the curvature two-form $F = dA + A^2$ and A is the connection one form $A = A_\mu(x) \, dx^\mu$. Note that the existence of Ω_D fullfilling (11.2) follows at least locally from Poincaré's lemma because $\text{Tr}[F^N]$ is closed. The latter follows from the Bianchi identity $DF \equiv dF + A \wedge F - F \wedge A = 0$ and

$$d\text{Tr}[F^N] = N \ \text{Tr}[DF \wedge F^{N-1}] \tag{11.3}$$

In recent years the Chern-Simons term has found utility in quite a few areas of theoretical physics:

a) It was found to give a description of anomalies [28-32]. In fact, eq. (11.2) can be reexpressed in a manner which demonstrates the breakdown in the conservation of the axial U(1) current of QCD. [Cf. Chapter 12.1.] For D=3, we can write $\Omega_3 = \epsilon_{\mu\nu\lambda\sigma} K^\mu dx^\nu \wedge dx^\lambda \wedge dx^\sigma$, where K^μ represents the axial current. Then Eq. (11.2) implies that $\partial_\mu K^\mu \propto \epsilon^{\mu\nu\lambda\sigma} \text{Tr}[F_{\mu\nu} F_{\lambda\sigma}]$, which corresponds to the quantum anomaly of the axial current. [Cf. Eq. (12.7).]

b) When the Chern-Simons action S_{cs} is included in a Yang-Mills theory in D=3, and the total action is $S = S_0 + S_{cs}$ where $S_0 = 1/4 \ \text{Tr}[F_{\mu\nu} F^{\mu\nu}]$, then S describes a system with massive vector bosons. S_{cs} is hence said to induce a "topological mass" in the system [95].

c) When the term θS_{cs} is added to the 2+1 dimensional nonlinear σ−model discussed in Sec. 9.3, the solitons of the model acquire nonstandard spin and statistics [96-98,55]. θ is an arbitrary coefficient, and under a 2π rotation of soliton, or an exchange of two solitons, the phase of the soliton wavefunction changes by an amount which is proportional to θ. Such novel "fractional statistics" have been found useful in the study of the quantum Hall effect [99].

There is also some speculation that such a mechanism may play a role in the theory of high temperature superconductivity [see e.g. Refs. [100]].

d) When the Chern-Simons action is considered alone, i.e. the total action of the system is just $S = S_{cs}$, a novel quantum theory results, and it was shown to be exactly solvable. Theories of this type, which involve no metric, are known as "topological field theories" [101]. They are invariant under a large group of symmetries, corresponding to the diffeomorphism group, which can reduce the number of physical (gauge invariant) observables to a finite number. Solutions to topological field theories have led to some recent advances in the mathematical theory of knots [102]. Also, gravity in 2+1 dimensions was shown to be a theory of this type, as the Einstein-Hilbert action in 2+1 dimensions is equivalent to a Chern-Simons action written for the Poincaré group ISO(2,1) [103,104].

Related discussions on the Chern-Simons theory can be found in Refs. [105].

After writing the general expression for the Chern-Simons action in what follows, we shall apply it to the nonlinear models in Section 2. There we show that the solitons of the 2+1 dimensional nonlinear σ-model discussed in Sec. 9.3 acquire nonstandard spin when S_{cs} is included in the total action of the system.

Following Zumino [106], we define a one-parameter family of two forms F_τ parametrized by τ, where

$$F_\tau = \tau \, dA + \tau^2 A^2 \quad . \tag{11.4}$$

Thus $F_{\tau=1} = F$. We can solve Eq. (11.2) for Ω_D using F_τ:

$$\Omega_D = \frac{D+1}{2} \int_0^1 d\tau \, \mathrm{Tr} \, [A \wedge F_\tau^{(D-1)/2}] \tag{11.5}$$

To show that (11.5) is a solution to (11.2) use the Bianchi identity $D_\tau F_\tau \equiv dF_\tau + \tau(A \wedge F_\tau - F_\tau \wedge A) = 0$ and

$$d\Omega_D = \frac{D+1}{2} \int_0^1 d\tau \; \text{Tr} \; [dA \wedge F_\tau^{(D-1)/2} - A \wedge d(F_\tau^{(D-1)/2})] \tag{11.6}$$

$$= \frac{D+1}{2} \int_0^1 d\tau \; \text{Tr} \; [(dA + 2\tau A^2) \wedge F_\tau^{(D-1)/2} - A \wedge D_\tau(F_\tau^{(D-1)/2})] \quad . \tag{11.7}$$

The last term in (11.7) vanishes since

$$D_\tau(F_\tau^N) = D_\tau F_\tau \wedge F_\tau^{N-1} + F_\tau \wedge D_\tau F_\tau \wedge F_\tau^{N-2} + \cdots + F_\tau^{N-1} \wedge D_\tau F_\tau = 0 \quad ,$$

while the remaining terms can be written as

$$\int_0^1 d\tau \frac{\partial}{\partial \tau} \; \text{Tr} \; [F_\tau^{(D+1)/2}] \quad , \tag{11.8}$$

from which eq. (11.2) follows.

Next we evaluate (11.5) for a few cases:

$$D = 3 : \quad \Omega_3 = \text{Tr}[A \wedge dA + \frac{2}{3} A^3] \tag{11.9}$$

$$D = 5 : \quad \Omega_5 = \text{Tr}[A \wedge (dA)^2 + \frac{3}{2} A^3 \wedge dA + \frac{3}{5} A^5] \tag{11.10}$$

$$D = 7 : \quad \Omega_7 = \text{Tr}[A \wedge (dA)^3 + \frac{8}{5} A^3 \wedge (dA)^2 + \frac{4}{5} A \wedge dA \wedge A^2 \wedge dA +$$
$$+ 2A^5 \wedge dA + \frac{4}{7} A^7] \tag{11.11}$$

In terms of space-time components, the D=3 Chern-Simons action is

$$S_{cs} = -\frac{1}{4\pi^2} \int_M d^3x \; \epsilon^{\mu\nu\lambda} \; \text{Tr}[A_\mu \partial_\nu A_\lambda + \frac{1}{3} A_\mu [A_\nu, A_\lambda]] \quad , \tag{11.12}$$

where we have redefined the action by an overall normalization constant. Under a gauge tranformation of the form,

$$A_\mu \rightarrow g^{-1} A_\mu g + g^{-1} \partial_\mu g \quad ,$$

(11.12) changes according to

$$S_{cs} \rightarrow S_{cs} + \frac{1}{12\pi^2} \int_M d^3x \; \text{Tr}[\; \epsilon^{\mu\nu\lambda} g^{-1} \partial_\mu g \; g^{-1} \partial_\nu g \; g^{-1} \partial_\lambda g]$$
$$- \frac{1}{4\pi^2} \int_M d^3x \; \text{Tr} \; [\epsilon^{\mu\nu\lambda} \partial_\mu (A_\nu \; \partial_\lambda g g^{-1})] \quad , \tag{11.13}$$

where g is a field which takes values in a Lie group H with Lie algebra $\underline{H}$.

By Stokes theorem, the last term in (11.13) can be expressed as an integral over the boundary ∂M (if it has one) of the three-manifold M. It vanishes after we require the boundary condition

$$g \to 1 \quad \text{as } \partial M \text{ is approached (at a suitable rate)} \quad , \qquad (11.14)$$

1 being the identity element of H. Such a condition is commonly imposed in gauge theories.

Now, the second term in (11.13) does not in general vanish. As is discussed in Section 13.1, if we take the manifold M to be a three-sphere S^3, then the second term in (11.13) is an integer (actually, it is normalized to be two times an integer). [Alternatively, if M is the manifold $\mathbf{R}^3$ and we demand that $g \to 1$ as time goes to $\pm\infty$, as well as at spatial boundaries, then g defines a mapping from S^3 to H. So again, the second term in (11.13) is an integer.] This (even) integer denotes an element of the third homotopy group of H, $\pi_3(H)$. It is zero only for maps $g : M \to H$ which are homotopic to the trivial map $g(x) \equiv 1$. Thus, the Chern-Simons action is invariant only under gauge transformations $g(x)$ which are connected to the identity. However, semiclassical arguements do not require that the total action S is invariant, but rather that $\exp(\mathrm{i}S)$ is invariant. Then if $\pi_3(H)$ is not trivial and we assume that θS_{cs} is the only term in S which is not invariant under large gauge transformations, θ must take on discrete values (π times an integer).

Similar considerations apply for arbitrary D. [Cf. Ref. 106.] Although $d\Omega_D$ is invariant under all possible gauge transformations $g : M \to H$, Ω_D can change by a closed D form whose integral over M need not be zero. Provided we once again impose boundary condition (11.14), this integral takes the form $\int_M \mathrm{Tr}[(g^{-1}dg)^D]$. For $M = S^D$, the latter is associated with the Dth homotopy group of H, $\pi_D(H)$.

11.2 THE CHERN-SIMONS TERM IN THE 2+1 DIMENSIONAL NONLINEAR σ-MODEL

Here we introduce a term θS_{cs} in the action for the nonlinear σ model described in Section 9.3. This term is shown to induce nonstandard spin to the topological solitons of the model. Here θ is an arbitrary constant and since $D = 3$, S_{cs} is given by (11.12).

Before restricting ourselves to the nonlinear σ model, let us begin with a general G/H model as discussed in Chapter 10. For such models A_μ is written in terms of fields g with values in G according to (11.12). Upon substituting (10.12) into the Chern-Simons action (11.12), we find

$$S_{cs} = \frac{1}{12\pi^2} \int_M d^3x \; \epsilon^{\mu\nu\lambda} \, \text{Tr} \, [g^{-1}\partial_\mu g \; g^{-1}\partial_\nu g \; g^1 \partial_\lambda g + I_\mu I_\nu I_\lambda] \quad , \qquad (11.15)$$

where I_μ is defined in Eqs. (10.12) and (10.23). Again if the $M = S^3$, the first term in the trace has the topological significance of yielding elements of $\pi_3(G)$. On the other hand, the meaning of the second term in the trace is obscure to us. It does vanish for the nonlinear σ model. This is obvious since the $I_\mu(x)$'s are tangent vectors on G/H (which is S^2 for the nonlinear σ-model). Since $\text{Tr}[\epsilon^{\mu\nu\lambda} I_\mu I_\nu I_\lambda] \, d^3x$ corresponds to a 3-volume element, it must vanish on the two-dimensional manifold S^2. [It also vanishes for the other two examples of nonlinear models discussed in Chapter 10. This is due to the fact that the structure constants $\overline{D}_{ijk}$ are zero for these cases.]

To obtain (11.15) from (11.9), write $\omega \equiv g^{-1}dg = A + B$, where $A \equiv A_\mu dx^\mu$, $B \equiv B_\mu dx^\mu$ and A_μ and B_μ are defined in (10.12). Then $\text{Tr}[A \wedge dA] = \text{Tr}[A \wedge d\omega] = \text{Tr}[B \wedge \omega^2 - \omega^3]$, since $d\omega = -\omega^2$. Therefore

$$\Omega_3 = \text{Tr}[-\frac{1}{3} \, \omega^3 - \frac{2}{3} \, B^3 - B \wedge \omega^2 + 2B^2 \wedge \omega] \quad .$$

Using $\text{Tr}[B \wedge A^2] = 0$, the last two terms in paranthesis reduce to B^3. Finally, upon applying the definition $I = -gBg^{-1}$, we obtain the desired result (11.15).

S_{cs} vanishes for all static topological solutions since it is linear in time derivatives. On the other hand, it can give a nonzero contribution for solitons undergoing certain collective motions. We show this below for the topological solitons of the nonlinear σ-model exhibited in Section 9.3.

The static solutions to the equations of motion for the nonlinear σ-model in two spatial dimensions were given in (9.60). They involve an arbitrary constant ψ_0 denoting the orientation of the soliton. In what follows we shall consider rotating solitons, so we shall elevate ψ_0 to a dynamical degree of freedom and consider it to be time dependent. We show that S_{cs} is nonzero for topological solitons which undergo rotation.

The solution (9.60) is formulated in terms of the gauge invariant variables ϕ_α which span $S^2 = SU(2)/U(1)$, where $\phi_\alpha\sigma_\alpha = g\sigma_3 g^{-1}$. In order to substitute into (11.15), we need to express the soliton solution in terms of the variables g which take values in $SU(2)$. As we discussed previously, this is not possible globally because the Hopf bundle is nontrivial. That is, in general, there exists no global section $g = g(\phi_\alpha)$. This fact, however, need not concern us, as the Chern-Simons term is gauge invariant (at least for gauge transformations which are connected to the identity), so it should be possible to reexpress S_{cs} in terms of gauge invariant quantitites. We show this below for rotating solitons.

Let g_0 (locally) represent the static soliton solution with corresponding orientation angle ψ_0 equal to zero. To obtain a solution with nonzero ψ_0, write

$$g = \exp\{i\psi_0\,\sigma_3/2\}g_0 \quad . \tag{11.16}$$

In order to consider rotating solitons, let ψ_0 be time (x^0) dependent. Upon substituting (11.16) into (11.15) and using Stoke's theorem, we find

$$S_{cs} = -\frac{i\Delta\psi_0}{8\pi^2}\int_{S^1}\mathrm{Tr}[\sigma_3\,dg_0g_0^{-1}] \quad , \tag{11.17}$$

where $\Delta\psi_0 = \psi_0(x^0 = +\infty) - \psi_0(x^0 = -\infty)$ and the integral is over the spatial boundary, which we take to be a circle of infinite radius. Now introduce the notation $\phi^{(g)}$, where

$$\sigma_\alpha\phi_\alpha^{(g)} \equiv g\sigma_3 g^{-1} \quad . \tag{11.18}$$

Then we can write

$$S_{cs} = -\Delta\psi_0 B(\phi^{(g_0^{-1})})/2\pi \quad , \tag{11.19}$$

where the integer B is defined in eq. (9.48). $B(\phi^{(g_0)})$ is the soliton winding number, it fulfills

$$B(\phi^{(g_0^{-1})}) = -B(\phi^{(g_0)}) \quad . \tag{11.20}$$

More generally,

$$B(\phi^{(gh)}) = B(\phi^{(g)}) + B(\phi^{(h)}) \quad . \tag{11.21}$$

This relation is easily proven upon applying the identity

$$B(\phi^{(g)}) = \frac{i}{4\pi} \int_{S^1} \text{Tr}[g^{-1}dg \; \sigma_3] \quad , \tag{11.22}$$

where S^1 again denotes the circle at spatial infinity. It gives,

$$B(\phi^{(gh)}) - B(\phi^{(h)}) = \frac{i}{4\pi} \int_{S^1} \text{Tr}[g^{-1}dg \; h\sigma_3 h^{-1}] \quad . \tag{11.23}$$

Since the left hand side of (11.23) is an integer, then so is the right hand side. The latter is a functional of h (as well as g) defined on the circle at infinity. Because $\pi_1(SU(2)) = 0$, h on S^1 is homotopic to the identity map $h = 1$. Since the right hand side of (11.23) is an integer it must have the same value for all maps h. Finally, upon substituting the trivial map h=1 into the right hand side of (11.23), the latter becomes $B(\phi^{(g)})$, hence providing (11.21). Eq. (11.20) is obtained by setting $h = g_0^{-1}$.

Thus to conclude, for a soliton of winding number one which undergoes a 2π rotation as x^0 evolves from $-\infty$ to $+\infty$, $S_{cs} = 1$. If the total action for the system is

$$S = S_0 + \theta S_{cs} \quad , \tag{11.24}$$

and we assume that S_0 is quadratic (or of higher order) in time derivatives of the fields, then upon evaluating S for an adiabatic rotation of the soliton by 2π we find $S = \theta$. Semiclassically, we could thus argue that the value of Feynman path integral for such a process is $e^{i\theta}$. This result indicates that the angular momentum associated with the winding number one soliton is $\theta/2\pi$.

Alternatively, we can obtain the same result by using the collective coordinate approach of Bowick, Karabali and Wijewardhana [98]. In this approach we substitute the ansatz (11.16) into the total action $S = \int dx^0 \, L$, to obtain the effective Lagrangian

$$L = \frac{1}{2} I(\phi^{(g_0)}) \dot{\psi}_0^2 + \frac{\theta}{2\pi} B(\phi^{(g_0)}) \dot{\psi}_0 - M(\phi^{(g_0)}) \qquad (11.25)$$

$M(\phi)$ and $I(\phi)$ can be interpreted as the mass and the moment of inertia, respectively, associated with the field configuration ϕ. For the fields $\phi^{(g_0)}$ corresponding to the soliton solution (9.60), [and the action S_0 obtained from (9.36)], $M = 4\pi\beta$. (On the other hand, it is not too hard to show that I for the fields $\phi^{(g_0)}$ corresponding to the soliton solution (9.60), is infinite. We shall however ignore this issue here, and instead assume that some long range cutoff is present in the integral for the moment of inertia. A discussion of this issue can be found in Ref. 107.) The momentum conjugate to the angular variable ψ_0 is

$$p = I(\phi^{(g_0)}) \dot{\psi}_0 + \frac{\theta}{2\pi} B(\phi^{(g_0)}) \qquad (11.26)$$

The corresponding Hamiltonian is

$$H = M(\phi^{(g_0)}) + \frac{[p - \theta B(\phi^{(g_0)})/2\pi]^2}{2I(\phi^{(g_0)})} \qquad . \qquad (11.27)$$

The angular momentum J is canonically conjugate to the angular variable ψ_0, but this defines it only up to an additive constant. Thus J = p + constant. To fix this constant we can require that J=0 when the Hamiltonian equations imply that ψ_0 is a constant. Then from

$$\{\psi_0, H\} = [p - \theta B(\phi^{(g_0)})/2\pi]/I(\phi^{(g_0)}) \qquad , \qquad (11.28)$$

we conclude that

$$J = p - \theta B(\phi^{(g_0)})/2\pi \qquad . \qquad (11.29)$$

The spectrum of the operator $J = -i\partial/\partial\psi_0 - \theta B(\phi^{(g_0)})/2\pi$ is ambiguous. This is so because the configuration space S^1 of the one-dimensional rotator is infinitely connected, its univeral covering space being R^1. As we have seen in Chapter 8, quantization of such a system is ambiguous. Using the discussion of that Chapter, it is readily verified that the eigenvalues of $-i\partial/\partial\psi_0$ are of the form K + integer, where the constant K depends on the specific choice of quantization. If we choose the quantization where wave functions are single valued functions of ψ_0, then we can set K=0. In such a case, the spectrum of J is of the form $-\theta B(\phi^{(g_0)})/2\pi$+integer.

PART III

SKYRMIONS

Chapter 12

THE EFFECTIVE LAGRANGIAN FOR QCD

12.1 INTRODUCTION

Quantum Chromodynamics (QCD) [for a review see e.g. Refs. 59] is currently accepted to be the theory which describes the strong interactions of elementary particles. In this Chapter, we shall briefly review the global symmetries of QCD and indicate the arguments which lead one from QCD to the effective Lagrangian for the description of strong interactions at low energies. The effective Lagrangian which emerges is a generalization of the SU(2)×SU(2) chiral Lagrangian introduced in Chapter 10. Skyrmions occur as solitons in this effective Lagrangian and constitute the major topic of interest in this Part.

QCD is a gauge theory based on the "colour" group $SU(3)$, this group is commonly denoted as $SU(3)_C$. The fields in QCD are the quark fields $q_a^\alpha(\alpha = 1, 2, 3;\ a = 1, 2, \ldots, N_f)$ and the Yang-Mills potentials $A_\mu^m(m = 1, 2 \ldots, 8)$. The quark field is a colour triplet, and its index α is transformed by $SU(3)_C$.

The index a is a flavour index, and we assume that there are N_f flavours altogether.

In the standard $SU(3)_C \times SU(2) \times U(1)$ model of strong and weak interactions, the masses of the quarks are induced by the Higgs fields which are responsible for the spontaneous breakdown of the weak group $SU(2) \times U(1)$ to the $U(1)$ group of electromagnetism. The quarks are therefore massless when these Higgs fields are ignored. Since the "current-algebra" masses of the up quark, $u = q_1$, and the down quark, $d = q_2$, are small compared to the pion mass [$m_\pi \approx 137\ MeV$] or the QCD scale parameter Λ_{QCD} [$\Lambda_{QCD} \approx 200\ MeV$], the approximation which ignores the Higgs fields and quark masses is a reasonable first approximation to the dynamics involving only the first two quark flavours. Although the "current-algebra" mass of $s = q_3$ is of the order of 200 MeV, we can hope that the generalization to three quarks also yields a reasonable first approximation. It is doubtful, however, that we can similarly ignore the masses of the remaining quarks $(c, b, t, \ldots)$ in view of their very large masses in comparison with the pion mass and Λ_{QCD}. [The mass of e.g. c is $\approx$ 1.5 GeV.]

This Chapter treats the chiral symmetry properties of QCD. These symmetries exist in QCD only in the zero quark mass limit. For reasons of generality, we shall often phrase our discussion for any N_f as though quarks of all flavours have zero mass and chiral symmetry is a good symmetry. However, in view of the preceding remarks, the resultant phenomenological inferences are not realistic if they involve the heavy quarks.

We may note here that it is not quite accurate to state that quarks have zero mass in the absence of weak interactions if their bare masses are zero. Spontaneous breakdown of chiral symmetry will in general contribute to the masses of the quarks. However, in the absence of weak interactions, the QCD Lagrangian will have chiral symmetry, and for us this is the significant point.

In the absence of the weak interactions, the QCD Lagrangian density reads

$$\mathcal{L}_{QCD} = -\bar{q}_a \gamma_\mu D^\mu q_a - \frac{1}{4} F^m_{\mu\nu} F^{\mu\nu m} \quad , \tag{12.1}$$

where $D_\mu = \partial_\mu - i\lambda_m A^m_\mu$, λ_m are the $SU(3)$ Gell-Mann matrices, D_μ is the covariant derivative and $F^m_{\mu\nu}$ are the components of the field tensors for the colour group.

In the terms of the left and right handed quark fields

$$q_L = \frac{1+\gamma_5}{2} q \quad , \quad q_R = \frac{1-\gamma_5}{2} q \quad , \tag{12.2}$$

the quark part of the Lagrangian density $\mathcal{L}_{QCD}$ becomes

$$-\bar{q}_{La} \gamma_\mu D^\mu q_{La} - \bar{q}_{Ra} \gamma_\mu D^\mu q_{Ra} \quad . \tag{12.3}$$

This Lagrangian density and hence $\mathcal{L}_{QCD}$ are invariant under separate unitary transformations on the flavour indices of the left handed and right handed quarks: it has the symmetry of the group $U(N_f) \times U(N_f) \equiv U(N_f)_L \times U(N_f)_R = \{(u_L, u_R)\}$ under which the quarks transform according to the rule

$$q_L \rightarrow u_L q_L \quad , \quad q_R \rightarrow u_R q_R \quad . \tag{12.4}$$

We may note that not all the transformations in (12.4) need be symmetries of the full quantum theory however. Some of them may not be quantum symmetries because of anomalies or of spontaneous symmetry breakdown.

Let us now briefly consider the various subgroups of the group $U(N_f) \times U(N_f)$.

i) $\underline{U(1)_V}$: This is the subgroup which transforms q_L and q_R by the same phase:

$$q_L \rightarrow e^{i\theta} q_L \quad , \quad q_R \rightarrow e^{i\theta} q_R \quad . \tag{12.5}$$

The conserved charge associated with this symmetry is the baryon number. This symmetry is exact in QCD in the sense that it is not spoiled either by anomalies or by spontaneous symmetry breakdown in the full quantum theory.

ii) $\underline{U(1)_A}$: This is the so-called "chiral" $U(1)$ group. Under this group, q_L and q_R transform by opposite phases: .

$$q_L \to e^{i\theta} q_L \quad , \quad q_R \to e^{-i\theta} q_R \quad \text{or} \quad q \to e^{i\gamma_5\theta} q \quad . \tag{12.6}$$

This symmetry, however, is not preserved when $\mathcal{L}_{QCD}$ is quantized, but is broken by an anomaly. More precisely, the current $J_\mu^5 = i\bar{q}\gamma_\mu\gamma_5 q$ which is associated with this transformation, and which is conserved according to näive manipulations with $\mathcal{L}_{QCD}$, is in fact not conserved in the full quantum theory. Rather, it fulfills the equation

$$\partial^\mu J_\mu^5 = -\frac{iN_f}{16\pi^2} \varepsilon^{\mu\nu\alpha\beta} F_{\mu\nu}^m F_{\alpha\beta}^m \quad . \tag{12.7}$$

iii) $\underline{G \equiv SU(N_f)_L \times SU(N_f)_R}$: This is the subgroup $\{(L, R)\}$ of $U(N_f)_L \times U(N_f)_R$ where L and R are restricted to be matrices of determinant unity. There are many indications that in the full quantum theory, this symmetry group G is spontaneously broken to the subgroup $H = \{(V, V)\}$. The group H is $SU(N_f)$. It is the vector subgroup of G which transforms q_L and q_R by the same amount. For two flavours, H is the $SU(2)$ of isospin and for three flavours, it is the $SU(3)$ of eightfold way.

Because G is spontaneously broken to H, by Goldstone's theorem, there are $N_f^2 - 1$ massless mesons in QCD. For two flavours, these Goldstone modes are identified with the three pions, while for three flavours, these modes are identified with the pseudoscalar octet.

12.2 THE QCD EFFECTIVE CHIRAL LAGRANGIAN

The effective Lagrangian density $\mathcal{L}$ emerges when we attempt to construct a model which describes the dynamics of these Goldstone modes. Let us list the properties we require of this Lagrangian density:

a) Firstly, $\mathcal{L}$ must be invariant under $G = SU(N_f)_L \times SU(N_f)_R$, this property being the analogue of the G-invariance of the QCD Lagrangian. Thus, $\mathcal{L}$ is to be constructed from a (multicomponent) field ϕ which is transformed by G, $\mathcal{L}$ being invariant under these transformations.

b) Secondly, ϕ should have exactly $N_f^2 - 1$ degrees of freedom per space-time point. This is a requirement of minimality: we want to describe the dynamics of the Goldstone modes and only of these modes. Thus, the effective Lagrangian we discuss involves the approximation where the couplings of the Goldstone modes to other excitations are ignored. It is possible to improve this model by introducing for instance fields for vector and axial vector mesons, but we shall not discuss these modifications.

c) Finally, we require that the subgroup of G which leaves any value of the fields invariant is (isomorphic to) exactly H and no more. If this can be arranged, then we would have nicely built in the spontaneous symmetry breakdown $G \to H$ in the geometry of the field space itself. For, in the ground state, the field will have some constant value, ϕ_0 say, for all x. The symmetry group of the ground state is thus H, signifying the breakdown $G \to H$.

Requirements b) and c) determine the space of values of the field to be the coset space G/H. Instead of explaining this theorem in generality, we exhibit G/H in the case of our interest, and verify that b) and c) are fulfilled. The model is the generalization of Example 2 in Chapter 10 from $N_f = 2$ to an arbitrary number of flavours.

For $G = SU(N_f)_L \times SU(N_f)_R = \{(L, R)\}$ and $H = \{(V, V)\}$, the manifold G/H is the same as the manifold of the group $SU(N_f)$. This follows easily by observing that a typical coset is $(L, R)H = (LR^\dagger, \mathbf{1})(R, R)H = (LR^\dagger, \mathbf{1})H$ since $(R, R) \in H$. Here $\mathbf{1}$ is the unit matrix. Hence the cosets are in one to one correspondence with the elements $LR^\dagger$ of $SU(N_f)$ and the field ϕ can be identified with the field U where $U(\vec{x}, t)$ is an element of $SU(N_f)$. The group G acts on U according to the rule

$$U \to LUR^\dagger \ . \tag{12.8}$$

148

The dimension of $SU(N_f)$ is N_f^2-1. Therefore, requirement b) is fulfilled.

As regards c), we can observe that a typical value of $U(\vec{x},t)$ is the unit matrix $\mathbf{1}$. The little group H leaving $\mathbf{1}$ invariant is $H = \{(V,V)\}$ since $V\mathbf{1}V^\dagger = \mathbf{1}$. There is nothing special about this value $\mathbf{1}$ for U, since the little group of $U(\vec{x},t)$ is $\{(U(\vec{x},t)VU(\vec{x},t)^\dagger, V)\}$ which is also isomorphic to H. Thus, condition c) is also verified.

We must now construct a chirally invariant action density for U in order to meet requirement a). We can construct several such action densities. The general construction uses the "left invariant" Maurer-Cartan "form" (or covariant vector)

$$V_\mu^L = U^\dagger \partial_\mu U \qquad . \tag{12.9}$$

It transforms under (L,R) according to

$$V_\mu^L \to RV_\mu^L R^\dagger \tag{12.10}$$

so that it is invariant under $SU(N_f)_L$.

Equally well, we can use the "right invariant" Maurer-Cartan covector

$$V_\mu^R = \partial_\mu U U^\dagger \tag{12.11}$$

which under (L,R) is transformed to

$$V_\mu^R \to LV_\mu^R L^\dagger \qquad . \tag{12.12}$$

Thus, V_μ^R is invariant under $SU(N_f)_R$.

Note that any chirally invariant polynomial $f(V_\mu^L)$ in V_μ^L is equal to $f(V_\mu^R)$ and conversely. This is because

$$V_\mu^R = UV_\mu^L U^\dagger \tag{12.13}$$

so that

$$f(V_\mu^R) = f(UV_\mu^L U^\dagger) = f(V_\mu^L) \quad , \tag{12.14}$$

where the last step follows because f is an invariant polynomial in V_μ^L. [Here, we assume that $f(V_\mu^L)$ contains no derivatives of V_μ^L other than those which can be written in terms of V_μ^L using say the identity $\partial_\mu V_\nu^L - \partial_\nu V_\mu^L + [V_\mu^L, V_\nu^L] = 0$.]

We can now write down invariant Lagrangian densities using V_μ^L. A typical example, and one of the greatest relevance to us, is

$$\mathcal{L} = \mathcal{L}_0 + \mathcal{L}_1 \quad,$$

$$\mathcal{L}_0 = \frac{F_\pi^2}{16} \operatorname{Tr} \{V_\mu^L V^{L\mu}\} \quad, \quad \mathcal{L}_1 = \frac{1}{32e^2} \operatorname{Tr} \{[V_\mu^L, V_\nu^L]^2\} \quad . \quad (12.15)$$

$\mathcal{L}_0$ is the N_f flavour generalization of the Lagrangian (10.32). Because $\mathcal{L}$ describe an "effective" theory, we shall not worry about problems of renormalizability. It is known that at the level of a quantum field theory, there are counterterms to be added to (12.15) which are not of the form of $\mathcal{L}_0$ and $\mathcal{L}_1$. In principal, an infinite number of such terms compatible with the symmetries of QCD can be present in the effective Lagrangian.

The standard approximation made in effective theories is to keep only a finite number of such terms. As will be shown in the following Chapter, the term $\mathcal{L}_0$ alone is insufficient for demonstrating the existence of classically stable topological solitons. It was shown by Skyrme [108] that topological solitons do appear in the model containing both $\mathcal{L}_0$ and $\mathcal{L}_1$. Although the term $\mathcal{L}_1$ is not unique for accomplishing this purpose, it is the standard choice in the literature and it shall be utilized extensively in the following Chapters. $\mathcal{L}_0$ and $\mathcal{L}_1$ are however unique in that they are the only terms compatible with the symmetries of QCD which are at most quadratic in terms involving time derivatives. This is a convenient (but not necessary) feature for performing a semiclassical analysis of the effective Lagrangian. (Actually, as has been recently shown [109], the term $\mathcal{L}_0$ alone can lead to solitons in the quantum theory. In such a model quantum fluctuations play a central role in the stabilization of the soliton. We will not be discussing the model in the text.)

The constant F_π (known as the pion decay constant) in $\mathcal{L}_0$ can be determined using weak interaction theory from the $\pi^+ \to \mu^+ + \nu_\mu$ reaction rate. The result is

$$F_\pi = 186.4 \text{ MeV} \quad . \quad (12.16)$$

The second term $\mathcal{L}_1$ in this action density is called the Skyrme term. It will be discussed in the next Chapter.

We note that the model described above is a nonlinear model in the sense described in Chapters 9 and 10. The nonlinear constraints on the fields U are the unitarity and unimodularity conditions. These constraints, as well as the Lagrangian, are invariant under the action of G.

We conclude this Chapter with a brief description of how (12.15) is used to describe low energy pion dynamics. When we are interested only in phenomena to leading order in the pion momenta, then it is sufficient to approximate $\mathcal{L}$ by $\mathcal{L}_0$, since $\mathcal{L}_0$ contains the least number of derivatives of U. Let us make this approximation and further take $N_f = 2$. Then to describe low energy dynamics, we write

$$U(\vec{x}, t) = \exp\{\frac{2i}{F_\pi}\pi_a(\vec{x}, t)\tau_a\} \quad , \quad a = 1, 2, 3 \quad , \tag{12.17}$$

where τ_a are the Pauli matrices. Expanding U in powers of π_a and inserting in $\mathcal{L}_0$, we find,

$$\mathcal{L}_0 = \frac{1}{2}(\partial_\mu \pi_a)^2 + \frac{2}{F_\pi^2}(\pi_a \partial_\mu \pi_b - \pi_b \partial_\mu \pi_a)^2 + \ldots \tag{12.18}$$

The fields π_a are to be identified with the three pion fields. The lowest order term is the usual free meson Lagrangian, while the next order term describes the four point vertex for pions from which $\pi - \pi$ scattering lengths can be computed in the tree approximation. Higher order terms can similarly be used to compute low-energy multipion scattering amplitudes. The model is in reasonable agreement with low energy $\pi - \pi$ scattering data to leading order in the center of mass momentum.

The Lagrangian density $\mathcal{L}_0$ describes massless pions. In a more realistic theory of π-mesons, the pion mass m_π should be included. This can be done here by adding the term

$$\frac{F_\pi^2 m_\pi^2}{16} \, \mathrm{Tr} \, (U + U^\dagger - 2 \times \mathbf{1}) \tag{12.19}$$

to $\mathcal{L}_0$. [$\mathbf{1}$ is the 2×2 unit matrix.] To lowest order in the π's, this term gives $m_\pi^2 \pi^2/2$ while additional interactions are introduced in $\mathcal{L}_0$ by this term at higher orders. This term "explicitly" breaks $SU(2)_L \times SU(2)_R$ to the isospin subgroup $H = SU(2)$, and hence does not satisfy a). It does not have a significant qualitative effect on the physics of Skyrmions although it does have some quantitative consequences for this physics. We shall, therefore, ignore it in our qualitative discussions.

Chapter 13

SKYRME SOLITONS FOR TWO FLAVOURS

13.1 INTRODUCTION

We are now ready to show how the effective Lagrangian density $\mathcal{L}$ of Chapter 12 admits solitonic solutions and discuss some of their properties when the number of flavours is two. After giving arguments supporting the existence of topological solitons in this and the following Section, we specialize to the spherically symmetric soliton in in Section 3. The semiclassical quantization for the winding number one soliton or "Skyrmion" is explained in Section 4. The quantization procedure is seen to have ambiguities upon which we shall elaborate more fully in subsequent Chapters.

The solitons are consequences of the topology of the space of fields U. This topology in turn is sensitive to boundary conditions at spatial infinity which are consequences of requiring that the energy integral is finite. The energy functional takes the form

$$\int d^3x \left\{ -\frac{F_\pi^2}{16} \text{Tr } V_0^2 - \frac{F_\pi^2}{16} \text{Tr } V_i^2 - \frac{1}{16e^2} \text{Tr } \{[V_0, V_i]^2\} + \right.$$
$$\left. -\frac{1}{32e^2} \text{Tr } \{[V_i, V_j]^2\} \right\} \quad . \tag{13.1}$$

Here $V_\mu = V_\mu^L$ denotes the left invariant Maurer-Cartan form. It is a sum of positive terms. [Note that L_0 and L_i are antihermitean.] A sufficient condition for its existence is

$$U(\vec{x},t) \to U_0 \quad \text{as} \quad r \to \infty \quad , r^2 = \vec{x} \cdot \vec{x} \quad , \tag{13.2}$$

where U_0 is a constant matrix and the rate of approach $U \to U_0$ is fast enough to guarantee that $E < \infty$ [say $U_0^{-1}U = [\mathbf{1} + 0(1/r^2)]$ as $r \to \infty$].

The action density $\mathcal{L}$ is invariant under global chiral rotations. The constant U_0 can be reduced to $\mathbf{1}$ by the global chiral rotation $U \to U_0^{-1}U$ without affecting the form of $\mathcal{L}$. The orientation of U_0 to the value $\mathbf{1}$ by such a chiral rotation involves no loss of information. It corresponds in the case of the infinite ferromagnet to the familiar operation where we align the spins in the ground state along some axis. Therefore, hereafter, we choose the boundary condition

$$U(\vec{x},t) \to \mathbf{1} \quad \text{as} \quad r \to \infty \quad . \tag{13.3}$$

We can now show, following the arguments of Section 9.3, that the configuration space of this model consists of an infinite number of disconnected pieces. We repeat those arguments once more here for the convenience of the reader.

The field U approaches $\mathbf{1}$ and not some angle dependent limit as $r \to \infty$. Thus, we may think of all points at spatial infinity as being identified to a single point. Such an identification converts the Euclidean space $\mathbf{R}^3$ of the coordinates $\vec{x}$ at a constant time to the three sphere $\tilde{S}^3$. The field U is well defined on this $\tilde{S}^3$ in view of the boundary condition.

More formally, as in (9.40) we can introduce stereographic coordinates $\xi_\mu(\vec{x})$ to define this $\tilde{S}^3$:

$$\vec{\xi}(\vec{x}) = \frac{2\vec{x}}{r^2+1} \quad , \quad \xi_4(\vec{x}) = \frac{r^2-1}{r^2+1} \quad ,$$

$$\vec{\xi}(\vec{x})^2 + \xi_4(\vec{x})^2 = 1 \quad , \quad \vec{x} = \vec{\xi}(\vec{x})/(1 - \xi_4(\vec{x})) \quad . \tag{13.4}$$

The coordinates ξ_μ span a three sphere $\tilde{S}^3$. As we explained in Chapter 9 while discussing Model 2, they are not globally valid coordinates for $\mathbf{R}^3$ which unlike $\tilde{S}^3$ is not a compact manifold. If we want a topologically accurate representation of $\mathbf{R}^3$, we should remove the north pole N from $\tilde{S}^3$: $\mathbf{R}^3 \approx \tilde{S}^3 \backslash N$. Now while the topological difference between $\mathbf{R}^3$ and S^3 can make a difference for some functions, for functions of the type U which approach a constant limit as $r \to \infty$, the change of variables $x \to \xi$ does produce a well defined function on $\tilde{S}^3$. We may thus regard the field U as defined on $\tilde{S}^3$.

Thus, the configuration space Q of the two flavour chiral model is made up of fields U which map $\tilde{S}^3$ to the group $SU(2)$:

$$U : \quad \vec{x} \to U(\vec{x}) \in SU(2) \quad ,$$

$$U(\vec{x}) \to \mathbf{1} \quad , \quad \text{as} \quad r \to \infty \quad . \tag{13.5}$$

(The time variable has been temporarily suppressed.) Our task is to explain the topology of Q and make it plausible that they fall into an infinite number of disconnected components Q_n, Q being the union $\bigcup_n Q_n$ of these Q_n.

The group $SU(2)$, which is the space of values of the maps U, consists of 2×2 unitary matrices of determinant 1. Any such matrix can be written as

$$n_0 \mathbf{1} + i\vec{\tau} \cdot \vec{n} \quad , \tag{13.6}$$

where

$$n_0^2 + \vec{n}^2 = 1 \quad , \tag{13.7}$$

and τ_i are the Pauli matrices. From this equation, we see that the group $SU(2)$ is identical topologically to the three sphere S^3.

Thus, the fields U are functions on $\tilde{S}^3$ (coordinated by $\vec{x}$ or ξ) and take values in S^3. If U_0 and U_1 are two such fields, we shall say that they are homotopic to each other, and write $U_0 \sim U_1$ if we can continuously deform U_0 to U_1 respecting the boundary condition (13.5). That is to say, $U_0 \sim U_1$ if there exists a sequence of maps $U_{(\tau)}$ from $\tilde{S}^3$ to S^3 compatible with the boundary condition (13.5) which is continuous in $\vec{x}$ and τ and fulfills $U_{(0)} = U_0$, and $U_{(1)} = U_1$.

Consider first the "trivial" field $U^{(0)}$ which maps all of $\vec{x}$ to the same point $\mathbf{1}$ of $SU(2)$. To this $U^{(0)}$, we can associate all maps $\tilde{U}^{(0)}$ which are homotopic to $U^{(0)}$. The set of all such maps makes up the trivial sector Q_0 of Q. The usual analysis of pion physics using the chiral model utilizes the sector Q_0.

A typical field $U^{(1)}$ of the sector Q_1 is the function

$$\cos \theta(r)\mathbf{1} + i\vec{\tau} \cdot \hat{x} \sin \theta(r) \quad ,$$

$$\theta(0) = \pi \quad , \quad \theta(\infty) = 0 \quad , \quad \frac{d\theta(r)}{dr} \equiv \theta'(r) < 0 \quad . \tag{13.8}$$

Because of the condition $\theta'(r) < 0$, θ monotonically decreases from π to 0 as r increases from 0 to ∞ and takes each value between π and 0 exactly once. Thus, if we regard $(\cos \theta(r), \hat{x} \sin \theta(r))$ as being the polar coordinates of a point on S^3, then as $\vec{x}$ ranges over all its values, this point assumes every value on S^3 once and exactly once. In other words, $SU(2)$ is covered exactly once under this map $U^{(1)}$. It is a typical "winding number one" map. The equivalence class of maps homotopic to such a winding number one map constitute the winding number one sector Q_1 of Q.

Under charge conjugation, the field U in general transforms as follows:

$$U \to U^T \quad . \tag{13.9}$$

The transform of the map $U^{(1)}$ by charge conjugation is a winding number -1 map, and the charge conjugates of all the members of Q_1 constitute the winding number -1 sector Q_{-1}. The members of Q_{-1} are not homotopic to members of Q_1 as will be revealed in the course of further discussion.

A typical map of winding number $n(= 0, \pm 1, \pm 2, \ldots)$ map is simply the nth power $U^{(1)n}$ of the winding number 1 map. The class Q_n consists of all maps homotopic to $U^{(1)n}$. We may remark in this connection that U^T and U^{-1} are homotopic, so that the definition of Q_{-1} given in this paragraph and in the last paragraph are equivalent.

As was the case in Chapter 9, the integer n (here associated with the field U) is a constant of the motion. A field U is associated with the integer n if $U \in Q_n$. It is practically and conceptually useful to have an explicit formula for this conserved quantum number. For this purpose, we first prove that the current

$$J_\mu = \frac{1}{24\pi^2} \varepsilon_{\mu\nu\lambda\rho} \text{ Tr } V^\nu V^\lambda V^\rho \tag{13.10}$$

is conserved (regardless of the details of equations of motion !):

$$\partial^\mu J_\mu \equiv 0 \quad . \tag{13.11}$$

The proof is as follows. Since

$$\partial_\mu V_\nu - \partial_\nu V_\mu + [V_\mu, V_\nu] = 0 \quad , \tag{13.12}$$

$\partial^\mu J_\mu$ is composed of four terms of which a typical one is

$$\frac{1}{24\pi^2} \varepsilon_{\mu\nu\lambda\rho} \text{ Tr } V^\mu V^\nu V^\lambda V^\rho \quad . \tag{13.13}$$

Let us relabel the indices of the ε tensor according to $\mu\nu\lambda\rho \to \nu\lambda\rho\mu$. Then on using $\varepsilon_{\nu\lambda\rho\mu} = -\varepsilon_{\mu\nu\lambda\rho}$ and the invariance of the trace under cyclic permutation of matrices, this becomes

$$-\frac{1}{24\pi^2} \varepsilon_{\mu\nu\lambda\rho} \text{ Tr } V^\nu V^\lambda V^\rho V^\mu = -\frac{1}{24\pi^2} \varepsilon_{\mu\nu\lambda\rho} \text{ Tr } V^\mu V^\nu V^\lambda V^\rho . \tag{13.14}$$

Thus this term, and likewise the remaining three terms, are zero. Hence $\partial^\mu J_\mu \equiv 0$.

It follows that the associated charge

$$B(U) = \frac{1}{24\pi^2} \int d^3x \ \varepsilon_{ijk} \text{ Tr } V_i V_j V_k \tag{13.15}$$

is a constant of motion. Its value is in fact the conserved quantum number n where $U \in Q_n$.

The factor $1/24\pi^2$ is put in so that B takes on integer values. In due course, we will see how to get this normalization.

From the formula for the winding number (13.15) it is easy to check that the operation of charge conjugation (13.9) does change a winding number n map into a winding number $-n$ map. This is so since $B(U^T) = -B(U)$.

13.2 THE SIZE OF THE SOLITON AND THE SKYRME TERM

As we mentioned in the last Chapter, effective Lagrangians are notoriously nonunique. They are not restricted by requirements of renormalizability and hence are allowed to contain terms in U with any number of derivatives. When we examine low energy phenomena, we can imagine expanding the Lagrangian in powers of these derivatives. The leading terms of the Lagrangian govern the leading terms in momentum in low energy scattering amplitudes, like pion scattering lengths.

In the preceding discussion, we have exhibited the Skyrme term which involves four derivatives. It is not the leading term, but *one* of the next to leading terms. For example, the interaction proportional to

$$\mathcal{L}_I = \text{Tr } V_\mu V^\mu V_\nu V^\nu \qquad (13.16)$$

involves only four derivatives, but it is not part of the Skyrme term. It is thus natural to wonder what is so special about the Skyrme interaction as compared to all the other interactions. Here we shall explain its role in leading to massive classically stable soliton solutions.

Let us first examine the situation wherein we idealize the Lagrangian density by the leading term

$$\mathcal{L}_0 = \frac{F_\pi^2}{16} \text{Tr } V_\mu V^\mu \quad . \qquad (13.17)$$

The pion decay constant F_π has the dimension of energy. If a solitonic ansatz is characterized by a size R, its energy will thus be a positive constant times $F_\pi R$ and the ground state corresponds to the limit $R \to 0$. In other words, the leading term $\mathcal{L}_0$ alone cannot support a stable soliton of nonzero size and energy, rather any such configuration of U will dissipate energy by radiation of pion field and will shrink towards zero size and zero energy in course of time. [Quantum mechanically, however, this can be prevented due to quantum fluctuations [109].]

158

Another way to show that $\mathcal{L}_0$ does not admit a nontrivial static classical solution is to use Derrick's scaling argument. Note that $\mathcal{L}_0$ scales in the same way as the Lagrangian density describing the nonlinear σ-model of Section 9.3. There we saw that this scaling property was sufficient to show that nontrivial static solutions exist only in two spatial dimensions. Thus, the model described by $\mathcal{L}_0$ has no static solution except $U = 1$ in three dimensions.

When the Skyrme term is also included in the Lagrangian, there is an elegant lower bound to the energy of the soliton. The bound is the analogue of the Bogomol'nyi bound we considered earlier although it predates the latter by many years. It is based on the observation that

$$\int d^3x \; \text{Tr} \; \left\{ \frac{F_\pi}{4} V_i \pm \frac{1}{8e} \varepsilon_{ijk}[V_j, V_k] \right\}^2 \leq 0 \quad . \tag{13.18}$$

[To show this result, notice that V_i and $\varepsilon_{ijk}[V_j, V_k]$ are antihermitean matrices and that for any antihermitean matrix A, $Tr\, A^2 \leq 0$.] From this, we arrive at the bound

$$- \int d^3x \; \text{Tr} \; \left\{ \frac{F_\pi^2}{16} V_i^2 + \frac{1}{32e^2}[V_i, V_j]^2 \right\} \geq \frac{3\pi^2 F_\pi}{e} |B(U)| \tag{13.19}$$

due to Skyrme. The left side is the potential energy of the field U. The bound thus shows that in the presence of the Skyrme term, the soliton energy and mass are bounded from below.

With Skyrme's term, the size R of the soliton cannot also be zero in the ground state. The contribution to energy, ΔE, from the Skyrme term is of the form

$$\Delta E = \frac{c_2}{e^2 R} \tag{13.20}$$

due to dimensional reasons, where c_2 is a constant. Therefore, the total energy E, i.e.

$$E = c_1 F_\pi^2 R + \frac{c_2}{e^2 R} \, , \tag{13.21}$$

has a minimum for a nonzero R. [The two constants c_1 and c_2 here are positive.]

It is the nice feature of stabilizing the soliton at a nonzero energy and radius that accords the Skyrme term its special status.

There are, however, other terms involving four derivatives as we have emphasized before. There is no good argument which suggests that these are less important than the Skyrme term. We can also add terms involving more than four derivatives. There is no good argument to suggest that such terms are ignorable either. For example, the so called large N_C limit of QCD [110] fails to show that higher derivative terms are down by powers of N_C as compared to the leading terms.

Despite these criticisms, we will approximate the action density by $\mathcal{L}_0 + \mathcal{L}_1$ when discussing the solitons in view of the nice features of this model. On the other hand, we should be aware that the approximation $\mathcal{L}_0 + \mathcal{L}_1$ of the action density is hard to justify, and therefore that the predictions based on this approximation should be viewed with caution.

We conclude this Section by noting that there is no nontrivial solution which saturates the bound (13.18). A configuration saturating the bound would have to satisfy

$$V_i = \mp \frac{1}{eF_\pi} \varepsilon_{ijk} V_j V_k \quad . \tag{13.22}$$

Upon taking the divergence of this equation, the right hand side vanishes. To see this, use the identity $\partial_i V_j - \partial_j V_i + [V_i, V_j] = 0$. Thus

$$\partial_i V_i = 0 \quad . \tag{13.23}$$

However, this is precisely the equation of motion obtained by minimizing the static energy for $\mathcal{L}_0$ without the additional contribution coming from $\mathcal{L}_1$. It is obtained from $\mathcal{L}_0$ by making the variation

$$\delta U = U(i\varepsilon_a \tau_a) \tag{13.24}$$

which implies the variation

$$\delta U^\dagger = -i\varepsilon_a \tau_a U^\dagger \quad , \tag{13.25}$$

where $\varepsilon_a = \varepsilon_a(\vec{x}, t)$ are infinitesimal parameters. These are the most general variations possible for a field valued in $SU(2)$. For these variations, the change in the static energy E of $\mathcal{L}_0$ is

$$\delta E = -\frac{iF_\pi^2}{8} \int d^3x \; \partial_i \varepsilon_a \; \mathrm{Tr} L_i \tau_a \tag{13.26}$$

which yields the result $\partial_i L_i = 0$ for $\delta E = 0$. Since as we saw earlier $\mathcal{L}_0$ does not admit static solitonic solutions, we see that the bound (13.19) cannot be saturated.

Although the bound cannot be saturated for $|B| \geq 1$, static solutions to the field equations are known to exist for $|B| = 1$. In Section 3 we shall discuss the "spherically symmetric" static $B = 1$ solution.

13.3 THE "SPHERICALLY SYMMETRIC" ANSATZ

In the solitonic number zero sector, the field which takes on the constant value **1** is the field of highest symmetry consistent with our boundary condition. It is fully Poincaré invariant and provides a classical description of the vacuum state.

When the winding number B is not zero, the field U cannot be translationally invariant. A translationally invariant field is a constant and corresponds to $B = 0$.

When $B \neq 0$, U cannot be rotationally invariant either. This is because for the latter such a U depends only on the radial distance r. Therefore

$$U^\dagger \partial_i U = \hat{x}_i U^\dagger \frac{\partial U}{\partial r}$$

and

$$\varepsilon_{ijk}(U^\dagger \partial_i U)(U^\dagger \partial_j U)(U^\dagger \partial_k U) = 0 \quad . \tag{13.27}$$

Like the solitons of the nonlinear σ-model of Section 9.3, there are, however, fields which are invariant under a modified rotational symmetry. A field with such a symmetry is invariant under the combined spatial *and* isospin rotations:

$$-i(\vec{x} \times \vec{\nabla})_i \, U(\vec{x}) + [\frac{\tau_i}{2} \, , \, U(\vec{x})] = 0 \quad .\tag{13.28}$$

This is analogous to the conditions (9.55).

The general solution of (13.28) is

$$U(\vec{x}) = U_c(\vec{x}) \equiv \cos\theta(r)\mathbf{1} + i\vec{\tau} \cdot \hat{x} \sin\theta(r) \quad .\tag{13.29}$$

The winding number for this field need not to be zero when the boundary conditions are for instance as in Eq.(13.8) as we have already seen. We shall discuss more general boundary conditions below. We must, however, first settle the boundary conditions on θ.

Our one requirement on U and hence on U_c is that it reduces to $\mathbf{1}$ at $r = \infty$. Therefore, $\cos\theta(\infty) = 1$ or $\theta(\infty) = 2\pi \times$ integer. With the redefinition $\theta(r) \to \theta(r) - \theta(\infty)$ [which does not affect the form of the ansatz], we can thus assume that

$$\theta(\infty) = 0 \quad .\tag{13.30}$$

We also require the field U to be well defined as $r \to 0$. If $\sin\theta(0) \neq 0$, U will approach an $\hat{x}$ dependent and hence ill defined value at $\vec{x} = 0$ as is seen by taking the limit $r \to 0$. So we also impose the requirement $\sin\theta(0) = 0$. Hence

$$\theta(0) = n\pi \quad , \quad n = 0 \quad , \quad \pm1, \pm2, \ldots \quad .\tag{13.31}$$

The winding number $B(U_c)$ for the ansatz U can be calculated by brute force. The answer is

$$\begin{aligned} B(U_c) &= \frac{1}{\pi}[\theta(0) - \theta(\infty)] \\ &= n \quad . \end{aligned}\tag{13.32}$$

Thus, any value of $B(U_c)$ can be realized using fields with the generalized spherical symmetry (13.28).

We can now estimate the ground state energy of the $B = 1$ solitonic sector using the Skyrme Lagrangian density $\mathcal{L}_0 + \mathcal{L}_1$. The energy of a static configuration for $\mathcal{L}_0 + \mathcal{L}_1$ is

$$E(U) = -\int d^3x \operatorname{Tr}\left\{\frac{F_\pi^2}{16}V_i^2 + \frac{1}{32e^2}[V_i, V_j]^2\right\} \quad . \tag{13.33}$$

We expect the ground state to be described by a configuration U with the maximum possible symmetry [111]. As mentioned previously, this expectation seems to be borne out by more formal arguments for $B = 1$. We can thus estimate the ground state energy by minimizing the static energy functional $E(U)$ using the "spherically symmetric" ansatz. For this ansatz, the integral after some calculation reduces to

$$E(U_c) = \frac{\pi}{2}F_\pi^2 \int_0^\infty dr\, r^2 \left\{\theta'^2 + \frac{8}{r^2}\sin^2\theta\left[\frac{1}{4} + \frac{1}{e^2 F_\pi^2}\left(\theta'^2 + \frac{\sin^2\theta}{2r^2}\right)\right]\right\} \quad . \tag{13.34}$$

On introducing the dimensionless radial coordinate

$$\tilde{r} = F_\pi e r \quad , \tag{13.35}$$

F_π and e become overall factors in the energy expression:

$$E(U_c) = \frac{\pi}{2}\frac{F_\pi}{e} \int_0^\infty d\tilde{r}\, \tilde{r}^2 \left\{\left(\frac{d\theta}{d\tilde{r}}\right)^2 + \frac{8\sin^2\theta}{\tilde{r}^2}\left[\frac{1}{4} + \left(\frac{d\theta}{d\tilde{r}}\right)^2 + \frac{\sin^2\theta}{2\tilde{r}^2}\right]\right\} \quad . \tag{13.36}$$

One can thus numerically estimate the minimum M of $E(U_c)$ by variational, relaxation or other methods independently of the values of F_π and e. The result is

$$M \cong 36.4\,\frac{F_\pi}{e} \quad . \tag{13.37}$$

This result is consistent with the Bogomol'nyi bound $M \geq 3\pi^2 F_\pi/e$.

We can now turn to the quantization of the soliton and an explanation of the conjectures of Skyrme. We note here that, according to Skyrme, the quantum soliton has spin so that the soliton mass contains a centrifugal term in addition to the static mass M. Thus, it is this corrected mass which is to be compared with a suitable experimental mass.

13.4 SEMICLASSICAL QUANTIZATION OF THE SKYRMION

We start this Section with a discussion of the zero modes associated with the classical ansatz U_c.

The ansatz U_c was constructed such that it is invariant under combined spatial and isospin rotations. It is not invariant separately under spatial rotations or under isospin transformations.

Under a spatial rotation $R \in SO(3)$, it becomes

$$\cos \theta(r)\mathbf{1} + i\tau_j R_{ji} \hat{x}_i \sin \theta(r) \quad . \tag{13.38}$$

Since the Lagrangian is invariant under spatial rotations, the energy of the configuration for every such R is degenerate with the energy for $R_{ij} = \delta_{ij}$.

Under a flavour rotation by $A \in SU(2)$, U_c transforms to

$$A \left[\cos \theta(r)\mathbf{1} + i\vec{\tau} \cdot \hat{x} \sin \theta(r)\right] A^\dagger \quad . \tag{13.39}$$

(13.39) is distinct from (13.29) unless $A = \pm \mathbf{1}$. Further, since the Lagrangian is invariant under flavour rotations, these configurations characterized by different A are degenerate in energy.

If A is an element of $SU(2)$, there is a well known two to one map $A \to R(A)$ from $SU(2)$ onto $SO(3)$. It is given by the formula

$$A\tau_i A^\dagger = \tau_j R_{ji}(A) \quad , \tag{13.40}$$

where τ_i are the Pauli matrices. Comparison of (13.38) and (13.39) now shows that these two degenerate configurations are in fact the same.

These configurations fulfill the boundary condition $U \to \mathbf{1}$ at spatial infinity and are thus physically allowed configurations.

Thus, the ansatz U_c has three rotational zero modes associated with A. It also has the usual translational zero modes since the energy is unchanged under the replacement $\vec{x} \to \vec{x} - \vec{x}_0$. We will have little to say about the translational zero modes hereafter.

Let ψ_A be the quantum state peaked around the classical ground state $AU_cA^\dagger$. Then the quantum ground state is not any one of these ψ_A's. It is rather a superposition of all such states. We want to determine this quantum ground state by semiclassical reasoning.

Since the zero modes responsible for the differentiation of classical and quantum ground states are the modes generated by flavour rotations, the simplest approximation to the quantum ground state is to retain just these modes and quantize them. We expect such an approximation to be good for low lying excitations. It amounts to the substitution

$$U(\vec{x}, t) = A(t)U_c(\vec{x})A(t)^\dagger \tag{13.41}$$

in the Lagrangian and quantization of A. That is, we retain just the "zero frequency modes", "zero modes" or "collective coordinates" and quantize them.

An approximation of this sort which retains just a few modes out of a possible infinite number of modes requires justification. It has been the subject of some criticism. We shall, however, proceed with our calculations using this approximation.

On substituting (13.41) in (12.15), we find that

$$L = \int d^3x \mathcal{L}$$

reduces to

$$L = -\frac{1}{2}a(U_c) \, \mathrm{Tr} \, (A^\dagger \dot{A})^2 - E(U_c) \quad,$$

$$\begin{aligned} a(U_c) &= \frac{4}{3}\pi F_\pi^2 \int_0^\infty dr \, r^2 \sin^2\theta \, + \, \frac{16\pi}{3e^2} \int_0^\infty dr \, r^2 \sin^2\theta \left(\theta'^2 + \frac{\sin^2\theta}{r^2}\right) \\ &\cong \frac{447}{F_\pi e^3} \quad. \end{aligned} \tag{13.42}$$

We show below that the quantization of L is ambiguous. It can be quantized in three different ways, which we refer to as methods i), ii) and iii). In method i), the states of the system are all bosonic. In method ii), they are all fermionic. In method iii), both bosonic and fermionic states occur in the spectrum. [Here "bosonic" and "fermionic" refer to the angular momenta associated with the states. We will not consider issues involving statistics of these states. Thus it is perhaps more appropriate to call them "tensorial" and "spinorial", respectively.]

Quantization Methods i) and ii)

For methods i) and ii), we rewrite (13.42) in terms of $R(A)$ using (13.40). It is a small exercise in group theory to show from the identity

$$A^{-1}\dot{A} = \frac{\tau_i}{2}\operatorname{Tr}\tau_i A^{-1}\dot{A} \qquad (13.43)$$

that

$$R(A)^{-1}\dot{R}(A) = \theta_i \operatorname{Tr}\tau_i A^{-1}\dot{A} \quad, \qquad (13.44)$$

where θ_i are the spin one angular momentum matrices:

$$[\theta_i]_{jk} = -i\varepsilon_{ijk} \quad . \qquad (13.45)$$

Let us substitute (13.44) in (13.42) to find a Lagrangian for $R(A)$. As A runs over $SU(2)$, $R(A)$ runs over all of $SO(3)$, so that we can drop the dependence on A of $R(A)$ and call it R and regard this R (a generic element of $SO(3)$) as the point of the configuration space. The Lagrangian now becomes

$$L = -\frac{1}{8}a(U_c)\operatorname{Tr}(R^{-1}\dot{R})^2 - E(U_c) \quad, \quad R(t) \in SO(3) \; . \qquad (13.46)$$

This Lagrangian describes a rigid rotator with its orientation specified by the $SO(3)$ matrix R. The phase space for this system is well known. The role of the configuration space variables q_i is played by the elements R_{ij} of R. Their Poisson brackets (PB's) vanish:

$$\{R_{ij}, R_{k\ell}\} = 0 \quad . \qquad (13.47)$$

The role of the momentum variables is played by the "left" $SU(2)$ generators L_α. (The Greek indices, like the Latin indices run from 1 to 3.) Their PB's are those of angular momenta:

$$\{L_\alpha, L_\beta\} = \varepsilon_{\alpha\beta\gamma} L_\gamma \quad . \tag{13.48}$$

Further, they act from the left on R in the sense that their PB's with R correspond to left multiplication by $i\theta_\alpha$:

$$\{L_\alpha, R_{ij}\} = i[\theta_\alpha R]_{ij} \quad . \tag{13.49}$$

Of course, we can replace L_α by "right" $SU(2)$ generators R_α which act from the right on the matrix R. Thus, if

$$R_\alpha = L_\beta R_{\beta\alpha} \quad , \tag{13.50}$$

then

$$\{R_\alpha, R_\beta\} = -\varepsilon_{\alpha\beta\gamma} R_\gamma \quad , \tag{13.51}$$

$$\{R_\alpha, R_{ij}\} = i[R\theta_\alpha]_{ij} \quad . \tag{13.52}$$

The last equation here follows easily on using the identity $R\theta_\alpha R^T = R\theta_\alpha R^{-1} = \theta_\beta R_{\beta\alpha}$. Using it, we can deduce that $\{R_\alpha, R_\beta\}$ and $-\varepsilon_{\alpha\beta\gamma} R_\gamma$ have the same PB's with R_{ij} and thus deduce (13.51). [The elements R_{ij} of the matrix R and the right rotation generators R_α are not to be confused with each other.] In a similar way, we can deduce that

$$\{L_\alpha, R_\beta\} = 0 \quad . \tag{13.53}$$

The physical meaning of L_α and R_α follows from their PB's with U. Since

$$U(\vec{x}) = \cos\theta(r) + i\tau_j R_{ji}\hat{x}_i \sin\theta(r) \quad , \tag{13.54}$$

we find, using (13.45),

$$\{L_\alpha, U(\vec{x})\} = i\left[\frac{\tau_\alpha}{2}, U(\vec{x})\right] \quad ,$$

$$\{R_\alpha, U(\vec{x})\} = (\vec{x} \times \vec{\nabla})_\alpha U(\vec{x}) \quad . \tag{13.55}$$

[We have suppressed the time variable here.] In other words, L_α is the generator of $SU(2)$ flavour (isospin) rotations and R_α is what is normally called the angular momentum.

With this interpretation in hand, we can easily find their expressions in terms of R and $\dot{R}$. Thus, L_α is the constant of motion associated with the variation

$$\delta R = \varepsilon i \theta_\alpha R \quad , \tag{13.56}$$

where ε is a small parameter. Making ε time dependent, we find,

$$\delta L = -\frac{ia(U_c)}{4} \text{ Tr } (\theta_\alpha \dot{R} R^{-1})\dot{\varepsilon} \quad . \tag{13.57}$$

Thus, by Noether's theorem,

$$L_\alpha = -\frac{ia(U_c)}{4} \text{ Tr } (\theta_\alpha \dot{R} R^{-1}) \quad . \tag{13.58}$$

Similarly, for R_α, the variation is

$$\delta R = \varepsilon i R \theta_\alpha \quad . \tag{13.59}$$

Since

$$\delta L = -\frac{ia(U_c)}{4} \text{ Tr } (\theta_\alpha R^{-1} \dot{R})\dot{\varepsilon} \tag{13.60}$$

for this variation, we find

$$R_\alpha = -\frac{ia(U_c)}{4} \text{ Tr } (\theta_\alpha R^{-1} \dot{R}) \quad . \tag{13.61}$$

We have thus got the "momentum" variables in terms of R and $\dot{R}$.

We have yet to write down the Hamiltonian. Since the Lagrangian consists only of kinetic energy terms [modulo the constant piece $-E(U_c)$], the Hamiltonian H is this term written in terms of R and L_α or R and R_α [modulo the constant piece $E(U_c)$]. Using the identities

$$\frac{1}{2}\theta_\alpha \text{ Tr } \theta_\alpha \dot{R} R^{-1} = \dot{R} R^{-1} \quad ,$$
$$\frac{1}{2}\theta_\alpha \text{ Tr } \theta_\alpha R^{-1} \dot{R} = R^{-1} \dot{R} \quad , \tag{13.62}$$

we thus find

$$H = \frac{2}{a(U_c)} L_\alpha L_\alpha + E(U_c) = \frac{2}{a(U_c)} R_\alpha R_\alpha + E(U_c) \quad . \qquad (13.63)$$

We now turn to the quantization of this system. On passage to quantum theory, we replace R, L_α and R_α by quantum operators $\hat{R}$, $\hat{L}_\alpha$ and $\hat{R}_\alpha$. Their commutators $[\cdot\,,\,\cdot]$ follow from the PB's of the classical variables by the usual prescription of Dirac:

$$[\hat{R}_{ij}\,,\,\hat{R}_{kl}] = 0 \quad ,$$
$$[\hat{L}_\alpha\,,\,\hat{R}_{ij}] = -[\theta_\alpha \hat{R}]_{ij} \quad ,$$
$$[\hat{L}_\alpha\,,\,\hat{L}_\beta] = i\varepsilon_{\alpha\beta\gamma}\hat{L}_\gamma \quad ,$$
$$[\hat{R}_\alpha\,,\,\hat{R}_{ij}] = -[\hat{R}\theta_\alpha]_{ij} \quad ,$$
$$[\hat{R}_\alpha\,,\,\hat{R}_\beta] = -i\varepsilon_{\alpha\beta\gamma}\hat{R}_\gamma \quad ,$$
$$[\hat{L}_\alpha\,,\,\hat{R}_\beta] = 0 \quad . \qquad (13.64)$$

Quantization consists in finding a representation of this algebra of observables on a Hilbert space. It is generally assumed that this representation must be irreducible. There are, however, two inequivalent irreducible representations of this algebra, so that quantization is ambiguous. Let $\mathcal{H}^{(B)}$ and $\mathcal{H}^{(F)}$ be the Hilbert spaces on which this algebra is realized. The space $\mathcal{H}^{(B)}$ is spanned by states with $I = J$ where both I and J are integral. Here $I(I+1)$ and $J(J+1)$ are eigenvalues of $L_\alpha L_\alpha$ and $R_\alpha R_\alpha$, respectively. These states are tensorial. This quantization is the same as the usual quantization of the rigid rotator with I and J being interpreted as the body fixed and space fixed angular momenta. The space $\mathcal{H}^{(F)}$, on the other hand, is spanned by states with $I = J = 1/2,\ 3/2,\ldots$ These states are thus spinorial.

We now give the details of these quantizations.

Method i)

Consider functions ψ from the group $SO(3)$ to complex numbers. Thus, for each $R \in SO(3)$, $\psi(R)$ is a complex number. The scalar product between two such functions is defined to be

$$(\psi_1, \psi_2) = \int_{SO(3)} d\mu(R)\psi_1(R)^*\psi_2(R) \quad , \tag{13.65}$$

where $d\mu(R)$ is the invariant measure on $SO(3)$. The Hilbert space $\mathcal{H}^{(B)}$ is the space of these functions with this scalar product.

Quantization consists in finding operators $\hat{R}_{ij}$, $\hat{L}_\alpha$ and $\hat{R}_\alpha$ such that the commutations relations (CR's) are fulfilled. The first two operators are defined by the rules

$$[\hat{R}_{ij}\psi](R) = R_{ij}\psi(R) \quad ,$$
$$[e^{i\varepsilon_\alpha \hat{L}_\alpha}\psi] = \psi[e^{-i\varepsilon_\alpha \theta_\alpha}R] \quad , \tag{13.66}$$

where ε_α are constants.

The infinitesimal form of (13.66) gives the differential realization of $\hat{L}_\alpha$. The rule (13.66) is suggested by the fact that $\hat{L}_\alpha$ generates *left* rotations on R. Similarly $\hat{R}_\alpha$ are defined by

$$[e^{i\varepsilon_\alpha \hat{R}_\alpha}\psi](R) = \psi[Re^{-i\varepsilon_\alpha \theta_\alpha}] \quad . \tag{13.67}$$

It is possible to check the CR's from these definitions. An even more transparent way to understand these definitions, which makes the CR's obvious, involves introducing a basis for $\mathcal{H}^{(B)}$. We know from the Peter-Weyl theorem [54] that any function $\psi \in \mathcal{H}^{(B)}$ has the expansion

$$\psi(R) = \sum_{L=0}^{\infty} \sum_{M=-L}^{L} a_{MN}^L D_{MN}^L(R) \quad , \tag{13.68}$$

where the matrices $\{D^L(R)\}$ $(L = 0, 1, 2, \ldots)$ furnish the $(2L+1)$ dimensional irreducible representation of the rotation group. Thus, the functions D_{MN}^L form a basis for $\mathcal{H}^{(B)}$. On these basis functions, (13.66) and (13.67) reduce to

$$[\hat{R}_{ij}D_{MN}^L](R) = R_{ij}D_{MN}^L(R) \quad ,$$
$$\left[\hat{L}_\alpha D_{MN}^L\right](R) = \left[J_\alpha^{(L)}\right]_{M'M} D_{M'N}^L(R) \quad , \tag{13.69}$$

$$[\hat{R}_\alpha D^L_{MN}](R) = -[J^{(L)}_\alpha]_{N'N} D^L_{MN'}(R) \quad , \tag{13.70}$$

where the matrices $J^{(L)}_\alpha$ ($\alpha = 1,2,3$) represent the angular momenta in the $(2L+1)$ dimensional representation. We may, in fact, regard these equations as defining the operators on $\mathcal{H}^{(B)}$.

It is easy to verify the CR's from these equations. Further, since $\hat{R}_{ij}$ transforms like a vector (in the index i for $\hat{L}_\alpha$, in the index j for $\hat{R}_\alpha$), the state $\hat{R}_{ij} D_{MN}$ is a superposition of $D^{L'}$'s with $L' = L+1$ and $L-1$. A state of integer L is thus mapped to states of integer L under the action of $\hat{R}_{ij}$ and of $\hat{L}_\alpha$ and $\hat{R}_\alpha$ and the Hilbert space $\mathcal{H}^{(B)}$ is invariant under the action of all observables.

From (13.70), we see that $\hat{L}^2$ and $\hat{R}^2$ are equal on D^L_{MN}:

$$\hat{L}^2 D^L_{MN} = \hat{R}^2 D^L_{MN} = L(L+1) D^L_{MN} \quad . \tag{13.71}$$

Recalling that $\hat{L}_\alpha$ and $\hat{R}_\alpha$ are isospin and spin (or rather total angular momentum) operators, we thus see that L has the meaning of isospin or total angular momentum of the state D^L_{MN}. Since L is integral, all these states are tensorial.

Method ii)

This method is quite similar to Method i), except that we choose a different Hilbert space $\mathcal{H}^{(F)}$. A basis for this space is $\{D^J_{\rho\alpha}\}$ where the matrix $D^J(g)$, $g \in SU(2)$ and $J = 1/2, \; 3/2, \; \ldots$, represents g in the $(2J+1)$ dimensional irreducible representation of $SU(2)$. Note that J runs only over half integral values. The scalar product is defined by

$$(D^J_{\rho\sigma}, D^{J'}_{\rho'\sigma'}) = \int_{SU(2)} d\mu(g) D^J_{\rho\sigma}(g)^\star D^{J'}_{\rho'\sigma'}(g) \quad , \tag{13.72}$$

where $d\mu(g)$ is the invariant $SU(2)$ measure.

The operator $\hat{R}_{ij}$ is defined using the correspondence $g \rightarrow R(g)$ from $SU(2)$ to $SO(3)$:

$$[\hat{R}_{ij}D^J_{\rho\sigma}](g) = R(g)_{ij}D^J_{\rho\sigma}(g) \quad . \tag{13.73}$$

The remaining operators are defined in exact analogy to Method i):

$$[\hat{L}_\alpha D^J_{\rho\sigma}](g) = [J^{(J)}_\alpha]_{\rho'\rho}D^J_{\rho'\sigma}(g) \quad ,$$
$$[\hat{R}_\alpha D^J_{\rho\sigma}](g) = -[J^{(J)}_\alpha]_{\sigma'\sigma}D^J_{\rho\sigma'}(g) \quad . \tag{13.74}$$

The commutation rules are easily verified. Further, since $R(g)_{ij}$ adds integer angular momentum to the states, $\hat{R}_{ij}\mathcal{H}^{(F)} \subseteq \mathcal{H}^{(F)}$. Thus, $\mathcal{H}^{(F)}$ is closed under the action of all the observables.

The spectra of spin and isospin are calculated as in Method i) with the result $J = I = 1/2,\ 3/2,\ldots$ Thus $\mathcal{H}^{(F)}$ contains only spinorial states.

Method iii)

We now turn to the third method of quantization. In this method, we return to the original form (13.42) of the Lagrangian where the configuration space Q is $SU(2)$ rather than $SO(3)$. The phase space T^*Q is thus spanned by A_{ij}, L_α where

$$\{A_{ij}, A_{k\ell}\} = 0 \quad ,$$
$$\{L_\alpha, A_{ij}\} = i\left[\frac{\tau_\alpha}{2}A\right]_{ij} \quad ,$$
$$\{L_\alpha, L_\beta\} = \varepsilon_{\alpha\beta\gamma}L_\gamma \quad . \tag{13.75}$$

(The PB's involving the right $SU(2)$ generators R_α can be deduced from these equations.) The associated quantum operators fulfill the commutation relations

$$[\hat{A}_{ij}, \hat{A}_{k\ell}] = 0 \quad ,$$
$$[\hat{L}_\alpha, \hat{A}_{ij}] = -\left[\frac{\tau_\alpha}{2}\hat{A}\right]_{ij} \quad ,$$
$$[\hat{L}_\alpha, \hat{L}_\beta] = i\varepsilon_{\alpha\beta\gamma}\hat{L}_\gamma \quad , \tag{13.76}$$

which are similar to those in (13.64). In (13.64), we have also listed the commutators involving the right angular momentum operators $\hat{R}_\alpha = \hat{L}_\beta R(\hat{A})_{\beta\alpha}$ from which the corresponding commutators relevant for the present case can be deduced.

The crucial difference of this approach from the first two approaches resides in the structure of the configuration space Q: it is now $SU(2)$ and not $SO(3)$. Thus, if in imitation of our earlier activity, we try to realize the algebra of these observables on functions $f(A)$ on $SU(2)$ according to the equations

$$[\hat{A}_{ij}f](A) = A_{ij}f(A) \quad ,$$

$$[\hat{L}_\alpha f](A) = \lim_{\varepsilon_\alpha \to 0} \frac{1}{i\varepsilon_\alpha}[f(e^{-i\varepsilon_\alpha \tau_\alpha/2}A) - f(A)] \quad ,$$

$$[\hat{R}_\alpha f](A) = \lim_{\varepsilon_\alpha \to 0} \frac{1}{i\varepsilon_\alpha}[f(Ae^{-i\varepsilon_\alpha \tau_\alpha/2}) - f(A)] \quad , \tag{13.77}$$

then we *cannot* restrict these functions to have either integer values of J or half integer values of J. That is, they cannot belong exclusively to either $\mathcal{H}^{(B)}$ or $\mathcal{H}^{(F)}$. The reason is that A has $J = 1/2$, being equal to $D^{1/2}(A)$. Therefore, $\hat{A}_{ij}$ maps $\mathcal{H}^{(B)}$ to $\mathcal{H}^{(F)}$ and $\mathcal{H}^{(F)}$ to $\mathcal{H}^{(B)}$: $\hat{A}_{ij}\mathcal{H}^{(B)} \subseteq \mathcal{H}^{(F)}, \hat{A}_{ij}\mathcal{H}^{(F)} \subseteq \mathcal{H}^{(B)}$. We can see this result in another way. Functions in $\mathcal{H}^{(B)}$ are even functions of A. This is because a 2π rotation does not change functions in $\mathcal{H}^{(B)}$ and therefore $f(-A) = f(Ae^{2\pi i \tau_3/2}) = f(A)$ if $f \in \mathcal{H}^{(B)}$. Similarly, functions in $\mathcal{H}^{(F)}$ are odd functions of A. Since A_{ij} is odd in A, the stated results follow. As a consequence, we need the full space $\mathcal{H}$ of all square integrable functions on $SU(2)$ to realize the algebra of observables in Method iii). This Hilbert space $\mathcal{H}$ contains *both* tensorial and spinorial states.

Actually, Method iii) can be ruled out for the following reason. The configuration space Q of the two flavour Skyrme model consists of maps of S^3 to $SU(2)$, $Q = $ Maps $\{S^3, SU(2)\}$. This space is known to be two fold connected. [This follows from $\pi_4(SU(2)) = \mathbf{Z}_2$ (see e.g. Refs. [112,113]).] The configuration space in Methods i) and ii) is $SO(3)$ which too is two fold connected. This is, however, not the case in Method iii) since $SU(2)$ is simply connected. Thus, in Method iii), the topology of the unapproximated Skyrme model is incorrectly represented so that this method is not appropriate for the description of the Skyrmion.

Chapter 14

PRELIMINARY DISCUSSION OF SKYRME'S PROPOSALS

14.1 INTRODUCTION

Skyrme proposed in his original work that the solitons with the winding number $B = 1$ should be quantized as fermions [Method ii)]. The wave functions can then be characterized by $I = J = 1/2, 3/2, \ldots$ For $B = 1$, Skyrme wished to identify the states with $I = J = 1/2$ with the nucleon doublet and the states with $I = J = 3/2$ with the states of the Δ resonance. He also suggested that the winding number $B(U)$ is the baryon number so that the topological sector Q_n with $B(U) = n$ on quantization describes states with baryon number n. This suggestion is consistent with N and Δ having unit baryon number. Further, their antiparticles will belong to the sector Q_{-1}.

These proposals of Skyrme [108] were largely ignored until the late seventies, perhaps because of their novelty. Exceptions to this were Refs. [112,113] where the possibility of having spinorial quantum states was discussed in detail. The revival of interest in Skyrme's proposal was due to Pak and Tze [114], and by Gipson and Tze [115], who cast many of Skyrme 's ideas in modern language and studied their implications for weak interactions.

There are two important qualitative ingredients in Skyrme's proposals which merit emphasis. They are the following: a) The identification of winding number with baryon number, b) The choice of the quantization scheme wherein the quantum solitons with $B = 1$ are fermions. The justification of these assumptions came rather late and was the prelude to the general acceptance of the ideas of Skyrme. We will not discuss these qualitative issues in this Chapter, however, but postpone them to the next Chapter. Here we shall instead discuss some phenomenological aspects of Skyrme's model for two flavours. We shall also discuss certain problems associated with the definition of electric charge in this model.

14.2 PHENOMENOLOGICAL COMMENTS

In accordance with Skyrme's conjectures, the nucleon is to be identified with the $I = J = 1/2$ Skyrmion and Δ with the $I = J = 3/2$ Skyrmion, both with $B = 1$. Their classical static energy is M (Eq. (13.37)) while their rotational energies follow from the Hamiltonian of Eq. (13.63). Their masses are thus

$$m_N = M + \frac{3}{2a(U_c)} \quad , \tag{14.1}$$

$$m_\Delta = M + \frac{15}{2a(U_c)} \quad . \tag{14.2}$$

These equations relate the masses m_N and m_Δ to the two parameters F_π and e of the theory. F_π is a known experimental number, whereas experiment cannot reliably fix a value for e. Thus, one possible approach to study the implications of these equations is to first compute the value of e with the experimental values of F_π and of one of the masses, say m_N, as inputs. Knowing F_π and e, we can then compute the other mass m_Δ and compare with the experimental value. Following this procedure, (13.1) gives a quartic equation for e of the form

$$e^4 - Ae + B = 0 \quad , \quad A, B > 0 \quad . \tag{14.3}$$

However, this equation has no real roots for the experimental values of m_N and F_π since the minimum of the function $f(e) = e^4 - Ae + B$ can be shown to be positive and rather large $[\approx 2727]$ for $m_N = 939$ Mev and $F_\pi = 186.4$ MeV. The same difficulty appears if we try computing e from the equation for m_Δ using the experimental value of 1232 MeV for the Δ mass.

An alternative approach which was adopted by Adkins, Nappi and Witten [116], is to compute F_π and e from the experimental values of m_N and m_Δ. The result is

$$F_\pi \cong 131 \text{ MeV} \quad , \quad e \cong 5.51 \quad . \tag{14.4}$$

The result for F_π is approximately 30% off from the experimental value. Although this result for F_π is not impressive, it is also not a discouraging result, especially at this primitive stage of the development of the model. In fact, considering the crudeness of the model and the approximations, it may perhaps be regarded as encouraging. In Chapter 18, we will discuss this result along with some of the other phenomenological consequences of the model.

14.3 ELECTRIC CHARGE FOR TWO FLAVOURS

At this stage, a more serious objection can be raised to the theory. This objection received essentially no attention in the literature. It concerns the values of electric charges.

In introducing electromagnetism, we must describe the coupling of the chiral field to a $U(1)$ gauge field. This is done by extending the model so that a $U(1)$ subgroup of the global symmetry group becomes a local symmetry group. The eigenvalues of the generator of this $U(1)$ group are to be interpreted as electric charges. In nature, we know that the electric charge operator $\hat{Q}$ and the third component of isospin $\hat{I}_3$ can be simultaneously diagonalized. For the pions, $\hat{Q} = \hat{I}_3$. In view of this fact, let us assume that the charge operator has a 2×2 matrix representation

$$Q = \begin{pmatrix} q_1 & 0 \\ 0 & q_2 \end{pmatrix} \tag{14.5}$$

and that when acting on the field $U(x)$, it induces the following transformation:

$$U(x) \rightarrow e^{ie_0 \Lambda Q} U(x) e^{-ie_0 \Lambda Q} = e^{ie_0 \Lambda \tau_3 (q_1 - q_2)/2} U(x) e^{-ie_0 \Lambda \tau_3 (q_1 - q_2)/2} \quad , \tag{14.6}$$

where e_0 is the electromagnetic coupling constant. Note that the right hand side of Eq. (14.6) is, in fact, a rotation of $U(x)$ induced by I_3. Eq. (14.6) corresponds to a symmetry of the Lagrangian density $\mathcal{L}$. The Noether current associated with this symmetry is:

$$\frac{1}{e_0} J^{em}_\mu(x) = -\frac{iF_\pi^2}{8} \operatorname{Tr} L_\mu(Q - U^\dagger Q U) - \frac{i}{8e^2} \operatorname{Tr} [L^\nu, Q - U^\dagger Q U][L_\mu, L_\nu] \quad , \tag{14.7}$$

where $L_\mu = U^\dagger \partial_\mu U$. It is this current which couples linearly to the electromagnetic potential in the $U(1)$ gauge theory of electromagnetism. This gauge theory is obtained in the usual way, that is by replacing ordinary derivatives by covariant derivatives:

$$\partial_\mu U \rightarrow D_\mu U = \partial_\mu U - ie_0 A_\mu [Q, U] \quad . \tag{14.8}$$

The experimental values for the charges of π's, N's and Δ's constrain the diagonal elements q_1 and q_2. We now show that this leads to an inconsistency. Upon substituting (13.41) into the current J^{em}_μ and expanding in the pion fields, we get

$$J^{em}_\mu = -ie_0(q_1 - q_2)(\pi_- \partial_\mu \pi_+ - \pi_+ \partial_\mu \pi_-) + \ldots \tag{14.9}$$

The $\pi^\pm$ charges of $\pm e_0$ then require that

$$q_1 - q_2 = 1 \quad . \tag{14.10}$$

Next we compute the charges of the N and Δ states using the semiclassical approximation wherein we write $U(\vec{x}, t) = A(t) U_c(\vec{x}) A(t)^\dagger$. Upon substituting this into the zeroth component of the current and performing a space integration, we then get for the $B = 1$ charge,

$$Q_{B=1} = \int d^3x \, J_0^{em}(\vec{x}, t) = -e_0 a(U_c) \, \text{Tr} \, \dot{A} A^\dagger Q \quad . \tag{14.11}$$

This equation can be rewritten in terms of the left generators L_α:

$$Q_{B=1} = e_0 L_\alpha \, \text{Tr} \, \tau_\alpha Q \quad . \tag{14.12}$$

Now, since Q is a diagonal matrix, we have

$$Q_{B=1} = e_0 (q_1 - q_2) L_3 \quad . \tag{14.13}$$

Upon using (14.10) we get $Q_{B=1} = e_0 L_3$. Since L_3 is the third component of the isospin operator and since protons and neutrons do not have electric charges proportional to $\pm 1/2$, this result is in clear contradiction with facts.

The above discussion presents an obstacle for the model since it involves a criticism independent of the values of F_π and e. It can be resolved by adding the constant $1/2$ to the charge operator, but this procedure at this stage is ad hoc. [This $1/2$ is, of course, the contribution of hypercharge to charge in the Gell-Mann-Nishijima formula [117].]

The above problem, as also the quantization ambiguity discussed previously, can be resolved by going to the three flavoured Skyrme model. The solution to the quantization ambiguity was originally discovered by Witten [27,118] and will be discussed in Chapter 15. Electric charges in the three flavoured Skyrme model are examined in Chapter 16.

Chapter 15

BARYON NUMBER AND SPIN OF THE SKYRMION

15.1 INTRODUCTION

In the previous Chapter, it was pointed out that two important qualitative questions regarding Skyrme's proposals concerned the identification of
baryon number with winding number and the choice of Method ii) for quantization which ensured that the solitons were spinorial. In this Chapter, we
will discuss the theoretical answers to these questions. (A third question concerning the result that the two-flavoured model gave the wrong values for the
electric charges for the Skyrmion will be dealt with in the next Chapter.) In
Section 2 we discuss the identification of the winding number with the baryon
number. The justification for the choice of Method ii) for quantization will
involve generalizing the Skyrme model to accomodate three or more flavours.
We take up this generalization in Section 3. It also involves introducing a
new term, the "Wess-Zumino" term, in the Lagrangian, to be discussed in
Section 4. The semiclassical quantization of the resultant action is carried out
in Section 5.

It may be mentioned here that the question of whether the Skyrmions obey Fermi statistics has been discussed in the literature [see Sorkin in Ref.[24]]. We will not review these papers in this book however.

15.2 WINDING NUMBER IS BARYON NUMBER

The demonstration of this result is due to Balachandran et al. [119]. It is based on some results of Goldstone and Wilczek [120]. The physical idea is to imagine that the system under consideration is the interacting system of quarks and solitons. (Gluons are not important in this context.) Thus, when we talk about the Skyrme soliton, we should take into account the effects of the filled Dirac sea of quarks. The properties of this Dirac sea can be studied in an approximate way by investigating the dynamics of the second quantized quark field in the presence of an externally prescribed *classical* chiral field U. We shall then see that due to the presence of this external field, the Dirac sea is "polarized" and carries a baryonic current exactly equal to the topological current J_μ. This baryonic current is defined in the conventional way in terms of the quark field.

Let us consider the case where there are N_c colours and the colour group is $SU(N_c)$. The expression for the baryon current of the quarks is [59]

$$J_\mu^{(B)} = \frac{i}{N_c}\bar{q}\gamma_\mu q \quad , \tag{15.1}$$

where the factor $1/N_c$ reflects the fact that each quark has baryon number $1/N_c$. We want to calculate the expectation value

$$< J_\mu^{(B)} >= K_\mu(U) \tag{15.2}$$

in the ground state or the "Dirac vacuum" when the quark field q is coupled to the given classical configuration U. This coupling, for instance, can be described by the quark Lagrangian density

$$\mathcal{L}_Q = -\bar{q}\gamma \cdot \partial q - m[\bar{q}_L U q_R + \text{h. c.}] \quad,$$

$$q_L = \frac{1 + \gamma_5}{2}q \quad , \qquad q_R = \frac{1 - \gamma_5}{2}q \quad . \tag{15.3}$$

Notice that U couples only to the flavour indices of the quarks. Here we need not fix the number of flavours N_f.

We can analyse the qualitative properties of K_μ without detailed calculations. Baryon number conservation tells us that

$$\partial^\mu K_\mu = 0 \quad . \tag{15.4}$$

This equation must be *valid independent of the dynamics of U* since $\mathcal{L}_Q$ conserves baryon number regardless of what this dynamics is. The general solution for such a K_μ is

$$K_\mu(U) = \xi J_\mu(U) + \varepsilon_{\mu\nu\lambda\rho}\partial^\nu \psi^{\lambda\rho}(U) \quad , \tag{15.5}$$

where ξ is a constant and $\psi^{\lambda\rho}$ is any function of U. We recall here that we have earlier shown that the topological current J_μ is conserved regardless of the dynamics of U. Thus, $K_\mu(U)$ also has this feature.

Now although J_μ is conserved, it cannot be written as "the dual of a curl" on the compactification $\tilde{S}^3$ of space $\mathbf{R}^3$. That is, we cannot write

$$J_\mu(U) = \varepsilon_{\mu\nu\lambda\rho}\partial^\nu \phi^{\lambda\rho} \quad . \tag{15.6}$$

If we attempt to solve for $\phi^{\lambda\rho}$ in (15.6), then $\phi^{\lambda\rho}$ will not approach a constant as $r \to \infty$ and will not therefore be a function on $\tilde{S}^3$. This is because if it did approach a constant, the integral

$$\int_{\tilde{S}^3} d^3x \; \varepsilon_{ijk}\partial^i \phi^{jk}$$

will vanish by Stokes' theorem since $\tilde{S}^3$ is a compact manifold without a boundary. On the other hand

$$\int J_0 d^3x$$

takes on any integer value for suitable U. Thus, J_μ is not the dual of a curl. Since J_μ is not a due of a curl, it is important to exhibit both the terms in the expression (15.5) for K_μ.

It is a mathematical theorem that this expression for K_μ is the most general expression for any current on $\tilde{S}^3$ which is conserved independently of the dynamics of U.

The baryon number of the Dirac vacuum is

$$\int K_0 \, d^3x = \xi B(U) \quad . \tag{15.7}$$

It is completely determined by the proportionality constant ξ. It is unaffected by $\psi^{\lambda\rho}$ since $\psi^{\lambda\rho}$ is a smooth function on $\tilde{S}^3$ at a given time (by hypothesis) and consequently the integral

$$\int_{\tilde{S}^3} d^3x \; \varepsilon_{ijk} \partial^i \psi^{jk}$$

vanishes. Thus, when we compute the induced baryon charge, we have to appeal to quantum field theory only to fix the value of the constant ξ.

We now describe a heuristic calculation of ξ using the properties of the axial anomaly. The literature may be consulted for more precise arguments [28-32]. Let us change variables in (15.3) as follows:

$$Q_L = q_L \quad , \quad Q_R = U q_R \quad . \tag{15.8}$$

It then becomes

$$\mathcal{L}_Q = -\overline{Q}_L \gamma \cdot \partial Q_L - \overline{Q}_R \gamma_\lambda [\partial^\lambda + U \partial^\lambda U^\dagger] Q_R - m[\overline{Q}_L Q_R + \text{h. c.}] \quad . \tag{15.9}$$

For the purposes of argument, it is convenient to replace (15.9) by the more general Lagrangian density

$$\mathcal{L}_Q = -\overline{Q}_L \gamma \cdot \partial Q_L - \overline{Q}_R \gamma_\lambda [\partial^\lambda + A^\lambda] Q_R - m[\overline{Q}_L Q_R + \text{h. c.}] \quad , \tag{15.10}$$

where A_λ is a generic (externally prescribed) gauge potential for the group $SU(2)$. For this Lagrangian, we have the divergence equations

$$\partial^\mu[i\overline{Q}_L\gamma_\mu Q_L] = im[\overline{Q}_L Q_R + \text{h. c.}] \quad , \tag{15.11}$$

$$\partial^\mu[i\overline{Q}_R\gamma_\mu Q_R] = im[\overline{Q}_L Q_R + \text{h. c.}] - \frac{iN_c}{32\pi^2}\varepsilon^{\mu\nu\alpha\beta} \, \text{Tr} \, F_{\mu\nu}(A)F_{\alpha\beta}(A), \tag{15.12}$$

where the contribution of the axial $U(1)$ anomaly is shown in (15.12). The factor N_c in this contribution reflects the fact that there are N_c colours. On adding (15.11) and (15.12), and using a well known and easily verified identity for $\varepsilon^{\mu\nu\alpha\beta} \, \text{Tr} \, F_{\mu\nu}F_{\alpha\beta}$, we find,

$$\partial_\mu[i\bar{q}\gamma^\mu q + \frac{iN_c}{8\pi^2}\varepsilon^{\mu\nu\alpha\beta} \, \text{Tr} \, (A_\nu\partial_\alpha A_\beta + \frac{2}{3}A_\nu A_\alpha A_\beta)] = 0 \quad . \tag{15.13}$$

We now take the expectation value of (15.13) in the fermion ground state and assume that it is legitimate to set the bracket in (15.13) equal to zero for such an expectation value. The result then is

$$< J_\mu^{(B)} > = -\frac{i}{8\pi^2}\varepsilon_{\mu\nu\alpha\beta} \, \text{Tr} \, [A^\nu\partial^\alpha A^\beta + \frac{2}{3}A^\nu A^\alpha A^\beta] \quad . \tag{15.14}$$

The right hand side of (15.14) yields the components of the dual of the Chern-Simons three-form [Cf. Chapter 11]. For the problem at hand, where A_λ is $U\partial_\lambda U^\dagger$, (15.14) shows that

$$< J_\mu^{(B)} > = J_\mu \quad \text{and} \quad \xi = 1 \quad . \tag{15.15}$$

The result (15.15) supports Skyrme's conjectures.

Actually, Skyrme's conjectures require only the equality of the total baryon charge and the winding number, and does not require that the current densities $< J_\mu^{(B)} >$ and J_μ are equal. Thus, we can weaken the assumption made in the passage from (15.13) to (15.14) as follows: The general solution of the expectation value of (15.13) in the fermion ground state is

$$< J_\mu^{(B)} > = -\frac{i}{8\pi^2}\varepsilon_{\mu\nu\alpha\beta} \, \text{Tr} \, [A^\nu\partial^\alpha A^\beta + \frac{2}{3}A^\nu A^\alpha A^\beta] + \varepsilon_{\mu\nu\alpha\beta}\partial^\nu\tilde{\psi}^{\alpha\beta}(A) \tag{15.16}$$

for some function $\tilde{\psi}^{\alpha\beta}$. Thus, for $A_\lambda = U\partial_\lambda U^\dagger$, we have

$$< J_\mu^{(B)} > = J_\mu + \varepsilon_{\mu\nu\alpha\beta}\partial^\nu\hat{\psi}^{\alpha\beta}(U) \quad , \quad \hat{\psi}^{\alpha\beta}(U) \equiv \tilde{\psi}^{\alpha\beta}(UdU^\dagger) \quad . \tag{15.17}$$

As we discussed earlier, for the validity of Skyrme's conjectures, it is enough that $\hat{\psi}^{\alpha\beta}(U)$ is well defined on $\tilde{S}^3$ at each fixed time. Thus, the assumption we have to make is this property of $\tilde{\psi}^{\alpha\beta}(U)$.

Incidentally, the equality of baryonic and topological charges has been shown to all orders in the expansion in powers of $1/m$ by Balachandran et al. [115] and by Gipson [117] using different methods.

15.3 SKYRMION IS SPINORIAL

The demonstration of this beautiful result is due to Witten [27] and involves the following surprising procedure: we assume that there are at least three flavours. For $N_f = 3$, where the field U takes values in $SU(3)$, we generalize the Skyrme ansatz by writing

$$U_c = \left[\begin{array}{ccc} \cos\theta(r)\mathbf{1}_{2\times2} + i\vec{\tau}\cdot\hat{x}\sin\theta(r) & & 0 \\ & & 0 \\ 0 & 0 & 1 \end{array} \right] . \qquad (15.18)$$

where $\mathbf{1}_{2\times2}$ is the two-by-two identity matrix. When $\theta(r)$ minimizes the energy functional (13.36) with $\theta(0) = \pi$ and $\theta(\infty) = 0$, (15.18) corresponds to a winding number one soliton or the Skyrmion. When we introduce collective coordinates $A(t)$ [where $A(t)$ is now an element of $SU(3)$], the Lagrangian $L(A, \dot{A})$ has a *gauge* symmetry as we shall see below. That is, there is a *time dependent* group of symmetries under which L changes by a total derivative:

$$L \rightarrow L + \frac{d\psi}{dt} . \qquad (15.19)$$

It is well known that whenever there is a time dependent group of symmetries, it does not lead to conservation laws, but rather to constraints on the phase space in classical theory and to constraints on the allowed states in quantum theory. Well known examples are electrodynamics and chromodynamics where the gauge groups which leave the Lagrangians invariant involve time (as well as space) dependent functions. The associated constraints are represented by the Gauss laws of the two theories. There are numerous such examples in particle mechanics as well as has been discussed previously.

Thus, the existence of a gauge symmetry in $L(A, \dot{A})$ means that the physically allowed states are subject to constraints. Here there is actually only one constraint because the gauge group is obtained by gauging a certain $U(1)$ group. It uniquely picks out one single quantization scheme and thereby removes the quantization ambiguities we have discussed.

The Lagrangian $L(A, \dot{A})$ is not the one we obtain by reducing the Skyrme Lagrangian (12.15) by introducing collective coordinates. In addition to this reduced Lagrangian (which we shall now call L_{SK}), $L(A, \dot{A})$ contains an additional piece L_{WZ}. This term is absent for $N_f = 2$. It alters the precise nature of the constraint and influences the choice of the quantization scheme. [But we still have a constraint in its absence and a unique (but different) quantization scheme.] The necessity of L_{WZ} requires considerations involving QCD, which will be explained later. In this Chapter, we shall merely take L_{WZ} for granted and explore its consequences.

In Chapter 17, we shall see another curious fact. For $N_f = 3$, there is another soliton (not the Skyrmion) for which there is a (three-fold) quantization ambiguity in the approximation where we quantize only the collective coordinates A. This ambiguity is analogous to the two flavour quantization ambiguity encountered in Chapter 13. It is now removed by assuming that there are four flavours ($N_f = 4$) and embedding the soliton in $SU(4)$. But there now appears yet another soliton which has a four fold quantization ambiguity (in the same approximation) which is removed by putting it inside $SU(5)$. In fact for any N_f, there is one soliton with an N_f-fold quantization ambiguity (in the preceding "collective coordinate" approximation) which is removed by putting it in $SU(N_f + 1)$.

Let us now discuss how the gauge symmetry appears. The Skyrme Lagrangian is

$$\int d^3x \left\{ \frac{F_\pi^2}{16} \text{ Tr } V_\mu V^\mu + \frac{1}{32e^2} \text{ Tr } [V_\mu, V_\nu]^2 \right\} \quad , \quad V_\mu = U^\dagger \partial_\mu U \ . \tag{15.20}$$

In this we insert the identity

$$U = A(t)U_c(\vec{x})A(t)^{-1} \quad , \tag{15.21}$$

where A is the collective coordinate, integrate on $\vec{x}$ and find L_{SK}. [For $N_f = 2$, L_{SK} was called L.] We give its explicit form shortly.

The dynamical variable in the chiral model is U and not $A(t)$ and U_c separately. Therefore, the equations of motion from $L = L_{SK} + L_{WZ}$ must be castable in a form which involves only U. Thus, if there is a transformation on A which leaves U invariant, it should leave these equations of motion invariant. It follows that L can change only by a total derivative of the form $d\psi/dt$ under such a transformation. [In fact, L_{SK} will not change even by such a term since it is derived by integrating out $\vec{x}$ in (15.20) and (15.20) involves only U. This however is not true for L_{WZ}.] The time dependent (gauge) transformation of A which leaves U invariant is not hard to find. Consider

$$A(t) \to A(t)e^{iY\beta(t)} \quad , \quad Y = \frac{1}{3}\begin{pmatrix} 1 & 0 & 0 \\ 0 & 1 & 0 \\ 0 & 0 & -2 \end{pmatrix} \ . \tag{15.22}$$

Since Y commutes with U_c, U is invariant under (15.22) which, therefore, defines the gauge group we were after. It is based on the $U(1)$ subgroup generated by the hypercharge operator Y. We shall call this $U(1)$ subgroup as $U(1)_Y$. It acts on the *right* of A.

For the two flavour model, the two fold quantization ambiguity was due to the fact that the physical states could belong to either one of the two inequivalent representations of the group $\mathbf{Z}_2 = \{(1, -1)\}$. That is, they could be either even or odd functions of A. Now let us regard the two flavour model as an appropriate restriction of the three flavour model. Since

$$e^{i3\pi Y} = \begin{pmatrix} -1 & & \\ & -1 & \\ & & 1 \end{pmatrix} \quad , \tag{15.23}$$

the restriction of the three flavour gauge group to two flavours contains $\mathbf{Z}_2$. The gauge constraint for three flavours uniquely determines the value of $\exp[i3\pi Y]$ on the physical states. Consequently, the quantization ambiguity of the two flavour model disappears when it is regarded as the restriction of the three flavour model.

We have yet the unfinished task of indicating the derivation of L_{SK}. We recall the definition of L_{SK}. With $U = A(t)U_c(t)A(t)^{-1}$, it is given by

$$L_{SK}(A, \dot{A}) = \int d^3x \left\{ \frac{F_\pi^2}{16} \, \text{Tr} \, V_\mu V^\mu + \frac{1}{32e^2} \, \text{Tr} \, [V_\mu, V_\nu]^2 \right\} \quad ,$$

$$V_\mu = U^\dagger \partial_\mu U \quad .$$

We can write this in the form

$$L_{SK}(A, \dot{A}) = \frac{1}{4} I_{\alpha\beta}(U_c)[\text{Tr} \, \lambda_\alpha A^{-1} \dot{A}][\text{Tr} \, \lambda_\beta A^{-1} \dot{A}] \quad ,$$

$$I_{\alpha\beta} = -\frac{F_\pi^2}{8} I_{\alpha\beta}^{(2)}(U_c) - \frac{1}{16e^2} I_{\alpha\beta}^{(4)}(U_c) \quad ,$$

$$I_{\alpha\beta}^{(2)} = \int d^3x [2\delta_{\alpha\beta} - \frac{1}{2} \, \text{Tr} \, \lambda_\alpha (U_c \lambda_\beta U_c^\dagger + U_c^\dagger \lambda_\beta U_c)] \quad ,$$

$$I_{\alpha\beta}^{(4)} = \int d^3x \, \text{Tr} \, \{[\lambda_\alpha - U_c^\dagger \lambda_\alpha U_c \, , \, U_c^\dagger \partial_i U_c][\lambda_\beta - U_c^\dagger \lambda_\beta U_c \, , \, U_c^\dagger \partial_i U_c]\} \quad .$$

$$(15.24)$$

Further reduction of (15.24) is tedious, but straightforward. The result is the following expression:

$$L_{SK} = -\frac{a(U_c)}{8}[\text{Tr} \, \lambda_i A^{-1} \dot{A}]^2 - \frac{b(U_c)}{8}[\, \text{Tr} \, \lambda_a A^{-1} \dot{A}]^2 - E(U_c) \quad ,$$

$$a(U_c) = \frac{2}{3}\pi F_\pi^2 \int_0^\infty dr \, r^2 \, \sin^2 \theta$$

$$+ \frac{8\pi}{3e^2} \int_0^\infty dr \, r^2 \, \sin^2 \theta(\theta'^2 + \frac{1}{r^2}\sin^2 \theta) \quad ,$$

$$b(U_c) = \frac{\pi}{2} F_\pi^2 \int_0^\infty dr \, r^2 (1 - \cos \theta)$$

$$+ \frac{\pi}{2e^2} \int_0^\infty dr \, r^2 (1 - \cos \theta)(\theta'^2 + \frac{2}{r^2}\sin^2 \theta) \quad .$$

$$(15.25)$$

Here λ_α are the Gell-Mann matrices and α runs from 1 to 8. The index i is summed over 1, 2, 3, a is summed over 4, 5, 6, 7 and $E(U_c)$ is the classical energy [Cf. Eqs. (13.36) and (13.37)]. Numerically,

$$a(U_c) \cong \frac{447}{F_\pi e^3} \quad , \quad b(U_c) \cong \frac{156}{F_\pi e^3} \quad , \tag{15.26}$$

where we have used the specific winding number one numerical solution for $\theta(r)$. Note that under the $U(1)_Y$ gauge transformation,

$$\text{Tr } \lambda_\alpha A^{-1}\dot{A} \rightarrow \text{Tr } e^{i\beta Y}\lambda_\alpha e^{-i\beta Y} A^{-1}\dot{A} + i\dot{\beta} \text{ Tr } \lambda_\alpha Y \quad . \tag{15.27}$$

The last term is zero for the index $\alpha \leq 7$. On using this fact, it is not difficult to show that L_{SK} is unaltered by the gauge group, as expected.

Earlier, for $N_f = 2$, we have argued that the chiral model is characterized by an infinite number of topological sectors labelled by the winding number $B(U)$. We remark here for completeness that these sectors continue to exist for any $N_f \geq 3$, since $\pi_3[SU(N_f)] = \mathbf{Z}$ for $N_f \geq 2$. The winding number $B(U)$ is given by the familiar formula (13.15) with U now being valued in $SU(N_f)$. The identification of the baryon and topological charges is also correct for any $N_f \geq 2$.

As we have remarked earlier, the Lagrangian for the collective coordinate A involves a term L_{WZ} in addition to L_{SK}. We now turn to a preliminary discussion of L_{WZ}.

15.4 THE WESS-ZUMINO TERM IN COLLECTIVE COORDINATES

The analysis of the structure and properties of the Wess-Zumino term when written in terms of U involves fairly sophisticated concepts. (See Chapters 6 and 16). However, it is quite simple to write down and comprehend after we reduce it by the substitution $U = AU_cA^\dagger$ to a Lagrangian for A. This reduced Lagrangian is

$$L_{WZ}(A, \dot{A}) = -\frac{i}{2}N_c B(U_c) \text{ Tr } Y A^{-1}\dot{A} \quad , \tag{15.28}$$

where N_c is the number of colours. Note its similarity to the Lagrangian treated in Chapter 3, [i.e. the second term in (3.19)].

Under the $U(1)$ gauge transformation, we have

$$
\begin{aligned}
L_{WZ}(A, \dot{A}) &\rightarrow -\frac{i}{2}N_c B(U_c)[\text{Tr } (e^{i\beta Y}Y e^{-i\beta Y}A^{-1}\dot{A}) + \text{Tr } (iY^2\dot{\beta})] \\
&= L_{WZ}(A, \dot{A}) + \frac{1}{2}N_c B(U_c) \text{ Tr } Y^2 \dot{\beta} \\
&= L_{WZ}(A, \dot{A}) + \frac{1}{3}N_c B(U_c)\dot{\beta} \quad ,
\end{aligned}
\tag{15.29}
$$

so that L_{WZ} has the property (15.19).

Let $\hat{Y}$ be the operator which generates this hypercharge transformation in quantum theory. Thus, if $\hat{A}$ is the operator which describes A, then

$$[\hat{Y}, \hat{A}_{ij}] = i\hat{A}_{ik}Y_{kj} \quad . \tag{15.30}$$

Since L_{SK} is invariant under the $U(1)_Y$ gauge transformation, it follows from Noether's theorem that $\hat{Y} = N_c B(U_c)/3$. That is, on any allowed quantum state ψ,

$$\hat{Y}\psi = \frac{1}{3}N_c B(U_c)\psi \quad . \tag{15.31}$$

[If L_{WZ} were absent, (15.31) would be replaced by $\hat{Y}\psi = 0$. As one can see from the discussion below, this constraint would also enforce a unique quantization scheme.]

Consider the Hilbert space $\mathcal{H}$ of functions on $SU(3)$ with the scalar product

$$(\phi_1, \phi_2) = \int_{SU(3)} d\mu(A)\phi_1(A)^\star \phi_2(A) \quad , \quad A \in SU(3) \quad , \tag{15.32}$$

190

where $d\mu(A)$ is the Haar measure on $SU(3)$. On these functions, $\hat{A}$ acts according to the rule

$$[\hat{A}_{ij}\phi](A) = A_{ij}\phi(A) \quad , \qquad (15.33)$$

while the action of $\hat{Y}$ is the infinitesimal form of

$$[e^{i\beta\hat{Y}}\phi](A) = \phi(Ae^{i\beta Y}) \quad . \qquad (15.34)$$

The subspace $\mathcal{H}_0$ of $\mathcal{H}$ which fulfills the constraint (15.31) is evidently given by

$$\mathcal{H}_0 = \left\{ \psi \in \mathcal{H} \mid \psi(Ae^{i\beta Y}) = \psi(A)e^{\frac{i\beta}{3}N_cB(U_c)} \right\} \quad . \qquad (15.35)$$

Since $e^{i6\pi Y} = \mathbf{1}$ and $B(U_c)$ can take any integer value, this equation shows that N_c must be an integer.

The center $\mathcal{C}$ of a group G is the set of all elements in G which commute with every element of G. The center of $SU(3)$ is $\mathbf{Z}_3$. In the defining representation, it is the following set:

$$\mathcal{C} = \{\mathbf{1}, z\mathbf{1}, z^2\mathbf{1} \mid z = e^{2\pi i/3}\} \quad . \qquad (15.36)$$

In any irreducible representation (IRR) of $SU(3)$, $z\mathbf{1}$ is represented by $z^k\mathbf{1}$, where k is an integer. Since $z^3 = 1$, k is defined only modulo 3. Thus, every IRR is characterized by an integer k (mod 3) which is called its triality type. The space $\mathcal{H}_0$ is the direct sum of all IRR's in $\mathcal{H}$ of triality type $N_cB(U_c)$ (mod 3). We see this from the fact that $e^{i2\pi Y}$ is $z\mathbf{1}$ and from the constraint equation on vectors in $\mathcal{H}_0$. For $N_c = 3$, $\mathcal{H}_0$ is thus of triality type zero for any value of B. This accords nicely with the situation in three colour QCD with colour confinement. For in this model, the transformation $q \to zq$ on the quark fields is generated also by the center of the *colour* $SU(3)$ and hence is a part of the *colour* group. Thus, on colour singlets, it necessarily acts trivially so that colour singlets are characterized by $k = 0$ (mod 3).

We can also construct an elegant proof of the fact that for $N_c = 3$, the $B = 1$ states are spinorial. As we saw earlier, spatial rotations are described by the transformations

$$A \to Ah \quad , \qquad (15.37)$$

where h denotes the following embedding of $SU(2)$ in the defining representation in $SU(3)$:

$$h = \begin{pmatrix} h_{2\times 2} & \\ & 1 \end{pmatrix} \quad , \quad h_{2\times 2} \in SU(2) \quad .$$

The rotation of A by angle 2π is thus

$$A \rightarrow A \begin{pmatrix} -1 & & \\ & -1 & \\ & & 1 \end{pmatrix} = A e^{i3\pi Y} \quad . \tag{15.38}$$

Under this transformation, by the constraint equation, a wave function behaves as follows:

$$\psi \rightarrow \psi e^{i\pi N_c B(U_c)} \quad . \tag{15.39}$$

Thus, ψ changes sign for $N_c = 3$ and $B = 1$ and the corresponding states have half integer angular momenta.

It is useful to know the explicit wave functions which describe the low lying states in the $B = 1$ sector. We shall now describe the construction of these wave functions. They will also provide information on the precise $SU(3)$ representation content and angular momenta of these states.

The IRR's of $SU(3)$ are labelled by two integers p and q. An IRR is denoted by (p,q). The singlet for example is $(0,0)$, the octet is $(1,1)$ and the decuplet is $(3,0)$. The element A of $SU(3)$ is represented in the IRR (p,q) by the operator $D^{(p,q)}(A)$. A basis for the vector space on which this IRR acts is $\{|I, I_3, Y>\}$. The meaning of I, I_3 and Y are as follows: Let $\hat{I}_\alpha$ ($\alpha = 1, 2, 3$) denote the generators of the $SU(2)$ subgroup of $SU(3)$ with the standard commutation realtions $[\hat{I}_\alpha, \hat{I}_\beta] = i\varepsilon_{\alpha\beta\gamma}\hat{I}_\gamma$. [In the fundamental representation, this $SU(2)$ acts on the first two components of the vectors.] Let $\hat{Y}$ denote the operator representing hypercharge. Then

$$\hat{I}^2 |I, I_3, Y> = I(I+1)|I, I_3, Y> \quad ,$$
$$\hat{I}_3 |I, I_3, Y> = I_3 |I, I_3, Y> \quad ,$$
$$\hat{Y} |I, I_3, Y> = Y |I, I_3, Y> \quad . \tag{15.40}$$

With these definitions, it is known that a basis for the Hilbert space $\mathcal{H}$ is the collection of all matrix elements

$$< I, I_3, Y | D^{(p,q)}(A) | I', I_3', Y' > \equiv D^{(p,q)}_{(I,I_3,Y)(I',I_3',Y')}(A) \qquad (15.41)$$

when (p,q) and all the remaining indices take on all possible values.

The operators $D^{(p,q)}$ fulfill the group property

$$D^{(p,q)}(FG) = D^{(p,q)}(F)D^{(p,q)}(G) \quad , \quad F,G \in SU(3) \quad . \qquad (15.42)$$

It follows that $\mathcal{H}_0$ is spanned by the functions

$$< I, I_3, Y | D^{(p,q)}(A) | I', I_3', \frac{1}{3}N_c B(U_c) >$$
$$= D^{(p,q)}_{(I,I_3,Y)(I',I_3',\frac{1}{3}N_c B(U_c))}(A) \quad . \qquad (15.43)$$

Let us first consider how flavour $SU(3)$ acts on these wave functions. Since $U \to A'UA'^\dagger$ when we transform A to $A'A$, $A \to A'A$ represents $SU(3)$ transformation. Under this transformation,

$$D^{(p,q)}_{(I,I_3,Y)(I',I_3',Y')}(A) \to D^{(p,q)}_{(I,I_3,Y)(I'',I_3'',Y'')}(A')$$
$$\times D^{(p,q)}_{(I'',I_3'',Y'')(I',I_3',Y')}(A) \quad . \qquad (15.44)$$

Thus, these wave functions are transformed by the IRR (p,q) of $SU(3)$, and the transformation is carried by the first group of indices (I, I_3, Y).

The action of spatial rotation on A is $A \to Ah$. If $D^{(I')}(h)$ is the matrix which represents h in the $2I'+1$ dimensional IRR of $SU(2)$ in the representation in which the third component of isospin is diagonal, we find that under spatial rotation,

$$D^{(p,q)}_{(I,I_3,Y)(I',I_3',Y')}(A) \to D^{(p,q)}_{(I,I_3,Y)(I',I_3'',Y')}(A) \; D^{(I')}_{I_3''I_3'}(h) \quad . \qquad (15.45)$$

Thus, I' represents (physical) angular momentum and I_3' its third component.

Let us now specialise to the case $N_c = 3$ and $B(U_c) = 1$ so that $Y' = N_c B(U_c)/3 = 1$. With this restriction on Y', the lowest allowed IRR's are the octet $(1,1)$ and decuplet $(3,0)$. The preceding discussion shows that the wave functions from these IRR's transform as octets and decuplets under flavour $SU(3)$. What about their spins? Clearly they are given by the allowed values of I' when $Y' = 1$. When Y' is 1, I' is 1/2 for the octet and 3/2 for the decuplet. Thus, we have an octet of spin 1/2 states and a decuplet of spin 3/2 states. The former is identified with the spin 1/2 baryon octet and the latter with the spin 3/2 baryon decuplet.

The model predicts an infinite number of excitations labelled by (p, q). However, since the model is expected to be valid only for low energy phenomenology, this prediction must be viewed with caution. The existence of higher excitations is one of the many aspects that the Skyrme model shares with the well known strong coupling model.

15.5 CANONICAL QUANTIZATION

We have already discussed all the important qualitative features which arise when $L(A, \dot{A})$ is quantized. However, it may still be of interest to derive these results in a systematic manner using canonical methods. We now give the details of such a calculation using methods already described in Chapter 5. Explicit expressions for the energy eigenvalues of the states in the $B = 1$ sector will also be given.

Since $A \in SU(3)$, it can be locally parametrized by eight variables $\xi = (\xi_1, \xi_2, \ldots, \xi_8)$. Thus, A in the Lagrangian is $A[\xi(t)]$ and $\dot{A}$ is $(\partial A/\partial \xi_\alpha)\dot{\xi}_\alpha$, while the momentum conjugate to ξ_α is

$$\pi_\alpha = \frac{\partial L}{\partial \dot{\xi}_\alpha} = -a(U_c)[A^\dagger \dot{A}]_i \left[A^\dagger \frac{\partial A}{\partial \xi_\alpha}\right]_i - b(U_c)[A^\dagger \dot{A}]_a \left[A^\dagger \frac{\partial A}{\partial \xi_\alpha}\right]_a$$

$$- \frac{i}{\sqrt{3}} N_c B(U_c) \left[A^\dagger \frac{\partial A}{\partial \xi_\alpha}\right]_8 \quad . \tag{15.46}$$

Again i takes values $1, 2, 3$, a takes values $4, 5, 6, 7$ and α takes values from 1 to 8. Here

$$[A^\dagger \dot{A}]_\alpha = \frac{1}{2} \operatorname{Tr} \lambda_\alpha A^\dagger \dot{A} \quad . \tag{15.47}$$

Now from general theory [see Chapter 5], one knows that there exists a nonsingular matrix $N(\xi) = [N_{\alpha\beta}(\xi)]$ such that

$$i A \lambda_\alpha = \frac{\partial A}{\partial \xi_\beta} N_{\beta\alpha} \quad . \tag{15.48}$$

Upon writing

$$R_\alpha = -\pi_\beta N_{\beta\alpha} \tag{15.49}$$

and using the standard canonical Poisson brackets (PB's)

$$\begin{aligned} \{\xi_\alpha, \xi_\beta\} &= \{\pi_\alpha, \pi_\beta\} = 0 \quad , \\ \{\xi_\alpha, \pi_\beta\} &= \delta_{\alpha\beta} \quad , \end{aligned} \tag{15.50}$$

we find

$$\begin{aligned} \{A_{\alpha\beta}, A_{\gamma\delta}\} &= 0 \quad , \\ \{R_\alpha, A_{\beta\gamma}\} &= i[A\lambda_\alpha]_{\beta\gamma} \quad , \\ \{R_\alpha, R_\beta\} &= -2 f_{\alpha\beta\gamma} R_\gamma \quad . \end{aligned} \tag{15.51}$$

The expression for R_α in terms of coordinates and velocities follows from (15.46) and (15.49):

$$R_\alpha = i a(U_c)[A^\dagger \dot{A}]_i \delta_{i\alpha} + i b(U_c)[A^\dagger \dot{A}]_a \delta_{a\alpha} + \frac{N_c}{\sqrt{3}} B(U_c) \delta_{\alpha 8} \quad . \tag{15.52}$$

It implies the existence of the primary constraint

$$\chi = R_8 + \frac{1}{\sqrt{3}} N_c B(U_c) \approx 0 \quad . \tag{15.53}$$

This is the constraint which generates the gauge symmetry (15.22).

The Hamiltonian is

$$H = \pi_\alpha \dot{\xi}_\alpha - L + \kappa\chi \quad , \tag{15.54}$$

where κ is a Lagrange multiplier. [See Chapter 2 for a review of the theory of constrained Hamiltonian systems.] Substituting for π_α and L, this reduces to

$$H = \frac{R_i R_i}{2a(U_c)} + \frac{R_a R_a}{2b(U_c)} + E(U_c) + \kappa\chi \quad . \tag{15.55}$$

Since $\{\chi, H\} = 0$, there are no secondary constraints.

The PB's are replaced by commutators following standard rules when we pass on to quantum mechanics. If c is a classical variable, let us denote the corresponding operator by $\hat{c}$. We find

$$[\hat{A}_{\alpha\beta}, \hat{A}_{\gamma\delta}] = 0 \quad ,$$
$$[\hat{R}_\alpha, \hat{A}_{\beta\gamma}] = -[\hat{A}\lambda_\alpha]_{\beta\gamma} \quad ,$$
$$[\hat{R}_\alpha, \hat{R}_\beta] = -2i\, f_{\alpha\beta\gamma}\hat{R}_\gamma \quad . \tag{15.56}$$

Quantization consists in finding an irreducible representation (IRR) of the appropriate algebra of observables, bearing in mind the constraint (15.34), $\hat{A}^\dagger\hat{A} = 1$ and the requirement $\hat{R}_\alpha = \hat{R}_\alpha^\dagger$. Such quantization has essentially been carried out in the previous pages. We shall briefly repeat that discussion here for completeness. The calculation is done in two steps. In the first step, we ignore the constraint $\hat{\chi}$ and realize the algebra (15.56) on a Hilbert space $\mathcal{H}$. In the second step, we impose the constraints on states, thereby reducing $\mathcal{H}$ to a subspace $\mathcal{H}_0$. The observables $\hat{\theta}$ [that is gauge invariant operators] are required to commute with the first class constraint $\hat{\chi}$. They have then the property of leaving $\mathcal{H}_0$ invariant, $\hat{\theta}\mathcal{H}_0 \subseteq \mathcal{H}$.

The Hilbert space $\mathcal{H}$ we start with is the space of functions on $SU(3)$ with the scalar product

$$(\psi, \phi) = \int_{SU(3)} d\mu(A)\, \psi(A)^\star\phi(A) \quad , \tag{15.57}$$

where $d\mu(A)$ is the invariant measure on $SU(3)$ and $\psi, \phi \in \mathcal{H}$. On this space, the operator $\hat{A}$ is realized by the rule

$$[\hat{A}_{\alpha\beta}\psi](A) = A_{\alpha\beta}\psi(A) \quad , \tag{15.58}$$

while $\hat{R}_\alpha$ is defined by the relation

$$[e^{i\theta_\alpha \hat{R}_\alpha}\psi](A) = \psi(Ae^{-i\theta_\alpha\lambda_\alpha}) \quad, \tag{15.59}$$

for all constants θ_α.

We next turn to the constraint (15.53). It reduces the space $\mathcal{H}$ to the subspace $\mathcal{H}_0$ of functions which fulfill

$$\psi(Ae^{-i\theta_8\lambda_8}) = e^{-\frac{i}{\sqrt{3}}N_c B(U_c)\theta_8}\psi(A) \quad. \tag{15.60}$$

The construction of a basis for $\mathcal{H}_0$ and the $SU(3)$ and spin spectrum of states have been discussed earlier.

The observables in the system must commute with the first class constraint $\hat{\chi}$. Examples of observables are

$$D^{(p,q)}_{(I,I_3,Y)(I',I_3',0)}(\hat{A}) \quad \text{and} \quad \hat{L}_\alpha = \hat{R}_\beta D^{(1,1)}_{\beta\alpha}(\hat{A}^{-1}) \quad. \tag{15.61}$$

Note that these $\hat{L}_\alpha$ are the generators of the flavour $SU(3)$.

The energy eigenvalues can be obtained by writing

$$\sum_{a=4}^{7} \hat{R}_a \hat{R}_a = \sum_{\alpha=1}^{8} \hat{R}_\alpha \hat{R}_\alpha - \sum_{i=1}^{3} \hat{R}_i \hat{R}_i - \hat{R}_8^2 \quad, \tag{15.62}$$

and knowing the value of the quadratic Casimir invariant in the IRR (p,q). The answer for the wave function

$$D^{(p,q)}_{(I,I_3,Y)(I',I_3',N_c B(U_c)/3)}(A)$$

is

$$
\begin{aligned}
M_{(p,q)I',Y'} = {} & E(U_c) + \left[\frac{1}{a(U_c)} - \frac{1}{b(U_c)}\right] 2I'(I'+1) \\
& + \frac{1}{2b(U_c)}\left[\frac{4}{3}\{p^2 + q^2 + 3(p+q) + pq\} - 3Y'^2\right] \quad.
\end{aligned}
\tag{15.63}
$$

The spin I' of the octet (decuplet) is $1/2$ $(3/2)$. Their masses are thus

$$M_8 = E(U_c) + \frac{3}{2a(U_c)} + \frac{3}{b(U_c)} \quad ,$$
$$M_{10} = E(U_c) + \frac{15}{2a(U_c)} + \frac{3}{b(U_c)} \quad .$$

$$(15.64)$$

For further discussion of the phenomenology of the model, we refer the reader to Chapter 18 and the literature [58,122].

Chapter 16

MORE ON THE WESS-ZUMINO TERM

16.1 INTRODUCTION

In the preceding Chapter, we have encountered the Wess-Zumino term for the chiral model. However, we discussed it only in the context of collective coordinates. In that Chapter we did not discuss its form and properties in the full field theoretic context or the derivation of $L_{WZ}(A, \dot{A})$ from field theory. We also made no attempt to argue for the necessity of such a term except for showing that it leads to the desired quantization for the Skyrmion. We turn to these tasks in this Chapter. The discussion will be relatively brief, especially as regards the differential geometric aspects of this term. Chapters 6 and 7 contain a more exhaustive discussion of this term from the point of view of principal fibre bundles and symmetry aspects.

It was emphasized by Witten [118] that there are processes involving pseudoscalar mesons which are not forbidden by the symmetries of QCD, but which are nonetheless forbidden by any action of the form

$$\int d^4x \mathcal{L} \quad , \tag{16.1}$$

where $\mathcal{L}$ is a smooth function of $U(x)$ and its derivatives. The effective vertex for all these processes involves the Levi-Civita symbol $\varepsilon_{\mu\nu\lambda\rho}$. A typical process of this kind is $K^+ K^- \to \pi^+ \pi^- \pi^0$, for which the effective vertex involves the term $\pi^0 \varepsilon^{\mu\nu\alpha\beta} \partial_\mu K^+ \partial_\nu K^- \partial_\alpha \pi^+ \partial_\beta \pi^-$. There is no Lorentz invariant function of $L_\mu = U^\dagger \partial_\mu U$ consistent with chiral symmetry which involves the ε symbol and only four derivatives. For example, $\varepsilon^{\mu\nu\lambda\rho} \, \mathrm{Tr} \, L_\mu L_\nu L_\lambda L_\rho$ is identically zero as we showed when we demonstrated the conservation of the topological current J_μ in Chapter 13. Thus, any action of the standard form forbids such vertices.

Witten [118] has pointed out that there is an extra conservation law in chiral models based on such actions. It is due to the presence of a discrete symmetry which is absent in QCD. Thus, if we parametrize $U(\vec{x}, t)$ in terms of the pseudoscalar meson field $\phi_\alpha = \phi_\alpha(\vec{x}, t)$, $\alpha = 1, 2, \ldots, 8$, in the form

$$U(\vec{x}, t) = e^{\frac{2i}{F_\pi} \phi_\alpha \lambda_\alpha} \ , \tag{16.2}$$

where λ_α are the Gell-Mann matrices, then since $\phi_\alpha(\vec{x}, t)$ becomes $-\phi_\alpha(-\vec{x}, t)$ under parity P, the action of P on U is

$$P: \quad U(\vec{x}, t) \to U^\dagger(-\vec{x}, t) \ . \tag{16.3}$$

We can also define two other discrete operations P' and P'' on U:

$$\begin{aligned} P': \quad & U(\vec{x}, t) \to U(-\vec{x}, t) \\ P'': \quad & U(\vec{x}, t) \to U^\dagger(\vec{x}, t) \ . \end{aligned} \tag{16.4}$$

Then

$$P = P' \, P'' \ . \tag{16.5}$$

There are no symmetries in QCD which correspond to P' or P''. Nonetheless, chiral models with actions of the form (16.1) (and with only four derivatives) do have P' and P'' as symmetries. The reason is simple: It is made up of L_μ's, and all vector indices have to appear in the contracted form in $\mathcal{L}$. Thus, $\mathcal{L}$ contains an even number of L_μ's. Further, $\mathcal{L}$ contains no ε symbols. It follows that P' for example is a symmetry. One can verify these statements explicitly in the model Lagrangian of Skyrme.

The selection rule implied by the symmetry P'' can now be simply stated. All Green's functions have to be invariant under the substitution $\phi_\alpha(x) \rightarrow -\phi_\alpha(x)$ and thus vanish for an odd number of ϕ''s. That is , all reactions with an odd number of pseudoscalars vanish.

We have stated the selection rule in the chiral models with only pseudoscalars. It can be generalized to chiral models with vector, pseudovector and higher spin mesons [123]. In any event, this selection rule is unwelcome so that the action S has to be modified. Since the extra term S_{WZ} to be added to S to get rid of this selection rule can not be the four dimensional integral of a local Lagrangian, it is clear that such a modifying term is characterized by unusual features. This term is the Wess-Zumino term. As we have seen, the anologue of the Wess-Zumino term occurs in magnetic monopole theory. Here we discuss this term for the $1 + 1$ dimensional chiral model in Section 2 and the $3 + 1$ dimensional chiral model in Section 3. (The latter was briefly introduced in Chapter 6.) The treatment of the Wess-Zumino term in the $3+1$ model is taken up in Section 4 and it is shown to lead to Eq.(15.28). Finally, in Section 5, we show how the inclusion of the Wess-Zumino term leads to the correct Gell-Mann-Nishijima formula [117] for the $B = 1$ states.

16.2 THE CHIRAL MODEL IN $1 + 1$ SPACE-TIME

There are terms similar to the monopole interaction S_{WZ} [Cf. Eq. (6.4)] that we can construct in some field theories. In the latter context, they are generically called Wess-Zumino terms. We shall now illustrate how they are constructed in field theory models in one space and one time where the fields have values in the group $SU(N)$ [$N \geq 2$]. The necessity for introducing these terms in certain contexts will also be discussed.

Thus, we consider a field g valued in the group $SU(N)$ which we represent concretely by $N \times N$ unitary matrices of determinant unity. The configuration space Q is composed of fields on one dimensional space. We also impose the boundary condition that the fields go to the unit matrix at spatial infinity ($x_1 = \pm\infty$). With this condition, the physical space at a given time may be thought of as a circle S^1, and Q as the set of maps from S^1 to $SU(N)$:

$$Q: \; S^1 \to SU(N) \quad ,$$

$$x_1 \to g(x_1) \in SU(N) \quad . \tag{16.6}$$

Following the monopole example in Chapter 6, we consider a disc Δ in Q. It is parametrized by two variables σ and x_0, where x_0 has the significance of time. The field associated with (σ, x_0) is $g(\sigma, x_0, .)$ where the dot indicates that it is a map from S^1 to $SU(N)$. The boundary $\partial\Delta$ of Δ is parametrized as follows:

$$\partial\Delta = \partial\Delta_1 \cup \partial\Delta_2 \cup \partial\Delta_3 \quad ,$$

$$\partial\Delta_1 = \{(\sigma, x_0^{(1)})|0 \le \sigma \le 1\} \quad , \quad \partial\Delta_2 = \{(\sigma, x_0^{(2)})|0 \le \sigma \le 1\} \quad ,$$

$$\partial\Delta_3 = \{(1, x_0)|x_0^{(1)} \le x_0 \le x_0^{(2)}\} \quad . \tag{16.7}$$

Thus, $x_0^{(1)}$ and $x_0^{(2)}$ denote the initial and final times in the action principle. Further, the field at $\sigma = 0$ is independent of x_0 and is fixed once and for all,

$$g(0, x_0, x_1) = g_0(x_1) \tag{16.8}$$

$g_0(\cdot)$ being the reference point (field) used to define the path space $\mathcal{P}Q$.

Given such a disc Δ, our task is to associate a real number S_{WZ} to Δ which has the following properties:

i) When the interior Int Δ of Δ is varied infinitesimally, S_{WZ} does not change.

ii) Let Δ' be another disc in Q such that $\partial\Delta' = \partial\Delta$ and Δ' can not be continuously deformed to Δ. The change of S_{WZ} when Δ is replaced by Δ' should be 2π times an integer.

Let us now show that the expression

$$S_{WZ} = \frac{n}{12\pi} \int_0^1 d\sigma \int_{x_0^{(1)}}^{x_0^{(2)}} dx_0 \int_{-\infty}^{\infty} dx_1 \; \varepsilon_{ijk} \; \mathrm{Tr} \; [g^{-1}\partial_i g \; g^{-1}\partial_j g \; g^{-1}\partial_k g] \tag{16.9}$$

fulfills both these requirements. Here n is an integer, the indices take the values 0, 1, 2 and $x_2 = \sigma$.

Proof of i) consists in observing that S_{WZ} is the integral of a closed three form. A physical way of showing this result is as follows. Think of g as a function of four coordinates $\vec{x}, \sigma$ and x_4, where x_4 is a fictitious time. The integral above defining S_{WZ} may be regarded as involving g at some fixed x_4. The deformation of Int Δ may be regarded as evolution of g in x_4. Since we have seen before that the current

$$J^\mu = \varepsilon^{\mu\nu\lambda\rho} \frac{1}{24\pi^2} \text{Tr}[g^{-1}\partial_\nu g g^{-1}\partial_\lambda g g^{-1}\partial_\rho g] \qquad (16.10)$$

is conserved,

$$\partial_\mu J^\mu = 0 \quad , \qquad (16.11)$$

the corresponding charge $S_{WZ}/2\pi n$ is a constant of motion for such x_4 evolution. This proves i).

As regards ii), we note that when Δ is replaced by Δ', their union $\Delta \cup \Delta'$ can be thought of as a two sphere S^2. Since the x_1 integration is over a circle S^1, ii) is the condition

$$(2\pi n) \times \frac{1}{24\pi^2} \int_{S^2 \times S^1} \text{Tr} \left(g^{-1}dg g^{-1}dg g^{-1}dg\right) = 2\pi \times \text{integer} \qquad (16.12)$$

for every choice of S^2. The coefficient of $2\pi n$ is integer valued. We have seen this in Chapter 12 when integration is over S^3, it is also true for integration over $S^2 \times S^1$. Thus, ii) is fulfilled when n is an integer.

Is it possible to motivate the addition of such terms to the action? In two dimensions, there is no fundamental distinction between fermions and bosons, and it is possible to transform a theory of fermions to a theory of bosons. This process is called "bosonization". Already, when the free fermion theory is bosonized, it is found that the bosonic action contains the term S_{WZ} [Cf. Witten in Ref. 87 and Section 16.6]. It is possible to understand why such a term should appear using anomaly arguments. Thus, consider the theory of N species of massless free left handed fermions. It has the symmetry of $U(N)$.

If this $U(N)$ or any of its subgroups is gauged, however, the corresponding currents are not in general covariantly conserved in the quantum theory, these covariant divergences are instead proportional to certain "anomalies" involving the gauge potentials. When the bosonic version of the free fermion model is gauged with gauge group $U(N)$ or one of its subgroups, the bosonized currents should also acquire these anomalies. It is possible to argue that such anomalies can be reproduced only by the gauged version of a Wess-Zumino term.

There are similar motivations for introducing S_{WZ} in the bosonized versions of interacting fermionic theories as well [124].

16.3 THE CHIRAL MODEL IN $3 + 1$ SPACE-TIME

Upon generalizing these considerations to the model of our greatest interest where the field U is valued in $SU(N_f)$ and space-time is four dimensional, we obtain the term (6.29). We shall, as usual, regard three dimensional space at a given time as the three sphere S^3 in view of our boundary condition on U. The generalization of the Wess-Zumino term of the two dimensional chiral model is then

$$S_{WZ} = -\frac{iN_c}{240\pi^2} \int_0^1 d\sigma \int_{x_0^{(1)}}^{x_0^{(2)}} dx_0 \int_{S^3} d^3x \varepsilon^{\alpha\beta\gamma\delta\rho} \operatorname{Tr} V_\alpha V_\beta V_\gamma V_\delta V_\rho \quad ,$$

$$V_\alpha = U^\dagger \partial_\alpha U \quad , \tag{16.13}$$

where the indices take values from 0 to 4 and $x_4 = \sigma$.

We can prove, as for the "baryon" current J^μ [Cf. Chapter 13], that the current

$$\varepsilon^{\mu\nu\lambda\rho\sigma\tau} \operatorname{Tr} V_\nu V_\lambda V_\rho V_\sigma V_\tau \tag{16.14}$$

is conserved regardless of the dynamics of U. Hence, if we define Δ in the usual way, we see that infinitesimal variations of U in Int Δ does not change S_{WZ}. On the other hand, the requirement that S_{WZ} changes by $2\pi \times$ integer when

Δ is replaced by any Δ' with the same boundary as Δ leads to the conclusion that N_c is an integer. If the chiral model is the effective Lagrangian for QCD, then N_c can also be identified with the number of colours by comparing the anomalies in the two models when the chiral flavour group $SU(N_f)_L \times SU(N_f)_R$ is gauged.

It is a simple exercise to show that S_{WZ} leads to reactions like $K^+ K^- \to \pi^+ \pi^- \pi^0$ [118,125]. As we discussed earlier, one may regard this feature as one of the motivations for the introduction of such a term in the chiral description. We refer the reader to the literature for a detailed treatment of such processes.

16.4 REDUCTION OF THE WESS-ZUMINO TERM

We have seen that in many calculations, it is customary to approximate U by the expression $A(t) U_c A(t)^\dagger$, where $A(t)$ is the collective coordinate and U_c is the classical static solution. We shall see below that in such an approximation, S_{WZ} can be written as follows:

$$S_{WZ}(U) = S_{WZ}(U_c) + \int_\Delta \int_{S^3} d\omega \quad . \tag{16.15}$$

Here $\omega = \omega(A, U_c)$ is a four form. Using Stokes' theorem, we can then write:

$$\int_\Delta \int_{S^3} d\omega = \left\{ \int_{\partial\Delta_1} - \int_{\partial\Delta_2} \right\} \int_{S^3} d\omega + \int_{x_0^{(1)}}^{x_0^{(2)}} \int_{S^3} \omega|_{\sigma=1} \quad , \tag{16.16}$$

where the boundaries $\partial\Delta_{1,2}$ are defined as in (16.7). [The minus sign in front of the $\partial\Delta_2$ integral arises because we assume that the integrals over $\partial\Delta_1$ are done in the increasing σ direction.] In obtaining the Euler-Lagrange equations, we do not vary $S_{WZ}(U_c)$ or the integrals over $\partial\Delta_{1,2}$. Thus, these terms can be discarded, and S_{WZ} can be replaced by a simplified action with an associated Lagrangian L_{WZ} where

$$\int_{x_0^{(1)}}^{x_0^{(2)}} dt \, L_{WZ} = \int_{x_0^{(1)}}^{x_0^{(2)}} \int_{S^3} \omega|_{\sigma=1} \quad . \tag{16.17}$$

The properties of L_{WZ} were already discussed in detail in the previous Chapter.

Let us now show how to derive the result (16.17). The Wess-Zumino term can be written in terms of the differential form

$$V = U^\dagger dU \quad , \tag{16.18}$$

as follows

$$S_{WZ}(U) = -\frac{iN_c}{240\pi^2} \int_\Delta \text{Tr } V^5 \quad , \qquad V^5 \equiv V \wedge V \wedge V \wedge V \wedge V \quad . \tag{16.19}$$

Introducing also the forms

$$\alpha = A^\dagger dA \quad , \qquad w = U_c^\dagger dU_c \quad , \tag{16.20}$$

we can easily verify that

$$\begin{aligned}
\text{Tr } V^5 &= \text{Tr } w^5 + 5d \text{ Tr}\{w^3\alpha - \alpha^3 w - \frac{1}{2}(w\alpha)^2 \\
&\quad + U_c(w - \alpha)^3 U_c^\dagger \alpha - \alpha^3 U_c(w - \alpha)U_c^\dagger \\
&\quad - \frac{1}{2}[\alpha U_c(w - \alpha)U_c^\dagger]^2\} \quad ,
\end{aligned} \tag{16.21}$$

where for example $\alpha^3 w = \alpha \wedge \alpha \wedge \alpha \wedge w$. This gives the simplified action in (16.16) involving only a four dimensional integral:

$$\begin{aligned}
S_{WZ}(U) &= S_{WZ}(U_c) - \frac{iN_c}{48\pi^2} \int_{x_0^{(1)}}^{x_0^{(2)}} \int_{S^3} \text{Tr } \{w^3\alpha - \alpha^3 w - \frac{1}{2}(w\alpha)^2 \\
&\quad + U_c(w - \alpha)^3 U_c^\dagger \alpha - \alpha^3 U_c(w - \alpha)U_c^\dagger \\
&\quad - \frac{1}{2}[\alpha U_c(w - \alpha)U_c^\dagger]^2\} \quad .
\end{aligned} \tag{16.22}$$

This result is valid even if A has a general spatial dependence. When A depends only on t, and U_c depends only on $\vec{x}$. There is a further drastic simplification since $\alpha^2 = (A^{-1}\dot{A})^2(dt)^2$, $\alpha^3 = (A^{-1}\dot{A})^3(dt)^3$ and $S_{WZ}(U_c)$ are zero. Thus, in this case, the simplified effective action is

$$\int_{x_0^{(1)}}^{x_0^{(2)}} dt \, L_{WZ} = -\frac{iN_c}{48\pi^2} \int_{x_0^{(1)}}^{x_0^{(2)}} \int_{S^3} \text{Tr } w^3(\alpha + U_c^\dagger \alpha U_c) \tag{16.23}$$

206

and the effective Lagrangian is

$$L_{WZ} = -\frac{iN_c}{48\pi^2} \int_{S^3} \text{Tr } w^3(A^{-1}\dot{A} + U_c^\dagger A^{-1}\dot{A}U_c) \quad . \tag{16.24}$$

For the Skyrme soliton, for $N_f \geq 3$, this expression can be further simplified. For definiteness, let $N_f = 3$. Since

$$U_c = \begin{pmatrix} \cos\theta(r)\mathbf{1}_{2\times2} + i\vec{\tau}\cdot\hat{x}\sin\theta(r) & \begin{matrix}0\\0\end{matrix} \\ \begin{matrix}0 & 0\end{matrix} & 1 \end{pmatrix} \quad , \tag{16.25}$$

$w = U_c^\dagger dU_c$ has vanishing entries along the third row and column. Further, if h is an element of the isospin subgroup,

$$h = \begin{pmatrix} \tilde{h} & \begin{matrix}0\\0\end{matrix} \\ \begin{matrix}0 & 0\end{matrix} & 1 \end{pmatrix} \quad , \quad \tilde{h} \in SU(2) \quad , \tag{16.26}$$

then since

$$\tilde{h}\tau_i\tilde{h}^\dagger = \tau_j R_{ji}(\tilde{h}) \quad , \quad R(\tilde{h}) \in SO(3) \quad , \tag{16.27}$$

we have

$$h\left[\int w^3(\vec{x})\right] h^\dagger = \int w^3[R(\tilde{h})x] = \int w^3(\vec{x}) \quad . \tag{16.28}$$

Here we have used the rotational invariance of the measure d^3x. Thus, the nonvanishing 2×2 block of the integral of w^3 commutes with $\tilde{h}$. By Schur's lemma, it is, therefore, a multiple of the identity:

$$\int w^3 = \text{ constant } \times \begin{pmatrix} 1 & & \\ & 1 & \\ & & 0 \end{pmatrix} \quad . \tag{16.29}$$

It is possible to evaluate the constant in terms of the baryon number $B(U_c)$ by taking the trace to finally find

$$\int w^3 = 12\pi^3 B(U_c)\begin{pmatrix} 1 & & \\ & 1 & \\ & & 0 \end{pmatrix} \quad . \tag{16.30}$$

The second term in L_{WZ} involves the integral

$$\int_{S^3} U_c w^3 U_c^\dagger = \int_{S^3}(dU_c U_c^\dagger)^3 \quad . \tag{16.31}$$

A calculation similar to the preceding one shows that it is equal to

$$\int w^3 \quad . \tag{16.32}$$

Thus, for the three flavour Skyrmion,

$$L_{WZ} = -\frac{iN_c}{2} \, \mathrm{Tr} \begin{pmatrix} 1 & & \\ & 1 & \\ & & 0 \end{pmatrix} A^{-1}\dot{A} \quad . \tag{16.33}$$

Writing

$$\begin{pmatrix} 1 & & \\ & 1 & \\ & & 0 \end{pmatrix} = \frac{2}{3}\, \mathbf{1}_{3\times3} + Y \quad , \quad Y = \frac{1}{3}\begin{pmatrix} 1 & & \\ & 1 & \\ & & -2 \end{pmatrix} \quad , \tag{16.34}$$

and using $\mathrm{Tr}\,[A^{-1}\dot{A}] = 0$, we find

$$L_{WZ} = -\frac{iN_c B(U_c)}{2} \, \mathrm{Tr}\, Y A^{-1}\dot{A} \tag{16.35}$$

which is the expression we considered earlier.

16.5 ELECTRIC CHARGES AND THE WESS-ZUMINO TERM

In Section 14.3, we pointed out a difficulty with the electric charges of the nucleons in the two flavour model. Unlike in the case of the two flavour $SU(2)$, the $U(1)$ of electromagnetism is a subgroup of the three flavour $SU(3)$. Its generator can be found by canonical methods following, for example, the methods of this Chapter. Here we verify that when this is done, the Gell-Mann-Nishijima formula [117] is correctly reproduced.

We begin by examining the contribution of the Wess-Zumino term to the total electric current $\tilde{J}_\mu^{em}$. Again we denote the matrix representation of the charge operator $\hat{Q}$ by Q. It induces the transformation

$$U(x) \to e^{ie_0\Lambda Q}U(x)e^{-ie_0\Lambda Q} \tag{16.36}$$

on the field $U(x)$. By making this a local transformation $\Lambda = \Lambda(x)$, the following Noether current results

$$\tilde{J}_\mu^{em}(x) = J_\mu^{em}(x) + J_\mu^{WZ}(x) \quad , \tag{16.37}$$

where $J_\mu^{em}(x)$ is given in Eq. (14.7) and

$$J_\mu^{WZ}(x) = \frac{e_0 N_c}{48\pi^2}\varepsilon_{\mu\nu\lambda\sigma} \, \mathrm{Tr} \, V^\nu V^\lambda V^\sigma (Q + U^\dagger Q U) \quad . \tag{16.38}$$

The contribution Eq.(16.38) to the electric current comes solely from the Wess-Zumino term. We note that even though the Wess-Zumino term vanishes for two flavours, its resulting contribution to the electric current does not. Thus, the two flavour reduction of the three flavour model has an interaction with electromagnetism not present in the original Skyrme model. This interaction is known to exist experimentally and is predicted by QCD. Its absence was the reason why we obtained the wrong values for the charges of the neutron and proton in Section 14.3.

To evaluate the extra contribution to the charge resulting from the new interaction in the two flavour model, we substitute (14.5) and the identity

$$\varepsilon_{\mu\nu\lambda\sigma} V^\nu V^\lambda V^\sigma = 12\pi^2 J_\mu(x) \, \mathbf{1}_{2\times 2} \tag{16.39}$$

into (16.38). $J_\mu(x)$ is the baryon current defined in Eq. (13.10). The result is

$$J_\mu^{WZ}(x) = \frac{e_0}{2}(q_1 + q_2)N_c J_\mu(x) \quad . \tag{16.40}$$

The correction to the electric charges is thus

$$\frac{e_0 N_c}{2}(q_1 + q_2) \int J_0 d^3 x \quad . \tag{16.41}$$

Upon substituting $U(\vec{x},t) = A(t)U_c(\vec{x})A(t)^\dagger$ into (16.41), this becomes

$$\frac{e_0}{2}(q_1 + q_2)N_c B(U_c) \quad . \tag{16.42}$$

With $N_c = 3$, the correct charges for the N and Δ states are recovered if we now set

$$q_1 + q_2 = \frac{1}{3} \quad . \tag{16.43}$$

This along with (14.10) fixes the charges q_1 and q_2 to be

$$q_1 = \frac{2}{3} \quad , \quad q_2 = -\frac{1}{3} \quad , \tag{16.44}$$

which is in agreement with the up and down quark charges.

In the case of the three flavour model, we replace the 2×2 charge matrix (14.5) by

$$Q = \begin{pmatrix} q_1 & & \\ & q_2 & \\ & & q_3 \end{pmatrix} \quad , \tag{16.45}$$

since the charge operator can be simultaneously diagonalized along with the third component of isospin and hypercharge. Assuming the usual electric charges for the pseudoscalar octet mesons, we find

$$q_1 - q_2 = 1 \quad , \quad q_2 = q_3 \quad . \tag{16.46}$$

Then using (16.46), we can write (16.45) according to

$$Q = (q_2 + \frac{1}{3})\mathbf{1}_{3\times3} + \frac{1}{2}\lambda_3 + \frac{1}{2\sqrt{3}}\lambda_8 \quad . \tag{16.47}$$

From the Skyrme model Lagrangian (without the Wess-Zumino term), we get the following contribution to the electric charge of the $B = 1$ soliton:

$$\int d^3x J_0^{em}(\vec{x}, t) = i\left\{ a(U_c)(A^\dagger Q A)_i (A^\dagger \dot{A})_i + b(U_c)(A^\dagger Q A)_a (A^\dagger \dot{A})_a \right\} \quad , \tag{16.48}$$

where again the index i runs over $1, 2, 3$ and a runs over $4, 5, 6, 7$. Using the constraint (15.53), we can write (16.48) in terms of the left handed generators L_α:

$$\frac{1}{e_0} \int d^3x J_0^{em}(\vec{x}, t) = \frac{1}{2}\left(L_3 - (A^\dagger \lambda_3 A)_8 \frac{N_c B(U_c)}{\sqrt{3}} \right)$$

$$+ \frac{1}{2\sqrt{3}}\left(L_8 - (A^\dagger \lambda_8 A)_8 \frac{N_c B(U_c)}{\sqrt{3}} \right) . \tag{16.49}$$

The contribution to the electric charge from the Wess-Zumino term is

$$\frac{1}{e_0} \int d^3x J_0^{WZ}(\vec{x}, t) = \frac{N_c}{48\pi^2} \int \text{Tr } A^\dagger Q A(w^3 + U_c w^3 U_c^\dagger) \quad , \tag{16.50}$$

where w is defined in Eq. (16.20). Using (16.30-32), we get

$$\frac{1}{e_0} \int d^3x J_0^{WZ}(\vec{x}, t) = N_c B(U_c) \left(q_2 + \frac{1}{3} + \frac{1}{2\sqrt{3}}(A^\dagger \lambda_3 A)_8 + \frac{1}{6}(A^\dagger \lambda_8 A)_8 \right).$$

$$(16.51)$$

Now upon adding (16.49) and (16.51), we find the total electric charge

$$\frac{Q}{e_0} = \frac{1}{2}L_3 + \frac{1}{2\sqrt{3}}L_8 + (q_2 + \frac{1}{3})N_c B(U_c) \quad . \tag{16.52}$$

The last term vanishes once we take the down quark charge q_2 to be $-1/3$, and we are left with the Gell-Mann-Nishijima formula [113]

$$\frac{Q}{e_0} = I_3 + \frac{1}{2}Y \quad . \tag{16.53}$$

Note that once the assignment $q_2 = -1/3$ is made, the remaining quark charges are determined through the Eqs. (16.46).

16.6 THE WESS-ZUMINO-NOVIKOV-WITTEN MODEL

We temporarily change our focus from the Skyrme model, and return to a discussion of lower dimensional chiral models. In Section 2 of this Chapter we introduced the Wess-Zumino [Cf. Eq. (16.9)] term for the $1+1$ dimensional chiral model. Upon adding the standard kinetic energy term

$$S_0 = -\frac{1}{2\lambda^2} \int d^2x \ \mathrm{Tr}[g^{-1}\partial_\mu g g^{-1}\partial^\mu g] \quad , \tag{16.54}$$

we obtain the actions for the "Wess-Zumino-Novikov-Witten (WZNW) model" [87,88]. This model is of current interest because (i) it is an example of a conformal field theory and (ii) it is equivalent to a theory of free massless fermions in $1+1$ dimensions. Here we will have nothing to say about (i) but rather refer the reader to the extensive literature on the subject [see e.g. Refs. [85]].

The fermion-boson equivalence is based on an identification of the conserved currents of the two theories. In the fermionic theory the conserved currents are the two-dimensional analogues of the vector and axial vector currents of QCD discussed in Chapter 12. [Here we will not consider the coupling to gauge fields and gravity. Thus there are no anomalies in the current conservation laws.] Conserved currents occur in the WZNW model as well. To see this we vary g, $\delta g = i\varepsilon g$, in the total action $S = S_0 + S_{WZ}$, where S_{WZ} is given by Eq. (16.9). Then $\delta S = 0$ implies

$$\partial_\mu J_L^\mu = 0 \quad , \quad J_L^\mu = \frac{1}{\lambda^2}\partial^\mu gg^{-1} + \frac{n}{4\pi}\varepsilon^{\mu\nu}\partial_\nu gg^{-1} \quad , \qquad (16.55)$$

where $\varepsilon^{\mu\nu} = -\varepsilon^{\nu\mu}$ and $\varepsilon^{01} = 1$. Alternatively, (16.55) can be written as

$$\partial_\mu J_R^\mu = 0 \quad , \quad J_R^\mu = \frac{1}{\lambda^2}g^{-1}\partial^\mu g - \frac{n}{4\pi}\varepsilon^{\mu\nu}g^{-1}\partial_\nu g \quad . \qquad (16.56)$$

The equivalence with the theory of fermions occurs for special values of the coupling constants λ^2 and n in S; namely, when

$$\lambda^2 = \pm 4\pi/n \quad . \qquad (16.57)$$

For the plus sign in (16.57), equation (16.55) reduces to $\partial_-(\partial_+ gg^{-1}) = 0$, where $\partial_\pm = \dfrac{\partial}{\partial x^\pm}$ and $x^\pm = \dfrac{1}{\sqrt{2}}(x^0 \pm x^1)$. This has the general solution $g = A(x^+)B(x^-)$ [A and B taking values in the group G], implying that left and right moving waves do not interfere. For the minus sign in (16.57), equation (16.55) reduces to $\partial_+(\partial_- gg^{-1}) = 0$ implying that $g = A(x^-)B(x^+)$.

We further note that when conditions (16.57) are satisfied, the total action $S = S_0 + S_{WZ}$ is invariant under a large group of symmetries. For the plus sign in (16.57), S is invariant under $g \to C(x^+)gD(x^-)$, whereas for the minus sign in (16.57), S is invariant under $g \to C(x^-)gD(x^+)$ [C and D take values in the group G]. In the Hamiltonian formulation of the theory, these symmetries are seen to generated by the current densities J_L^0 and J_R^0.

In the fermionic theory, the analogue of $(J_L^0)_i$ and $(J_R^0)_i$, [$i = 1, 2, \ldots$ $\dim(G)$], which we denote by $(\mathcal{J}_L)_i$ and $(\mathcal{J}_R)_i$, satisfy Kac-Moody algebras

[126,85] in the quantum theory. This is shown by the equal time commutation relations of $(\mathcal{J}_L)_i$ and $(\mathcal{J}_R)_i$:

$$
\begin{aligned}
[(\mathcal{J}_L)_i(x^1), (\mathcal{J}_L)_j(x'^1)] &= ic_{ij}^k (\mathcal{J}_L)_k(x^1)\delta(x^1 - x'^1) \\
&\quad + \frac{ki}{4\pi}\delta_{ij}\left(\frac{\partial}{\partial x^1} - \frac{\partial}{\partial x'^1}\right)\delta(x^1 - x'^1)\ , \quad (16.58)
\end{aligned}
$$

$$
\begin{aligned}
[(\mathcal{J}_R)_i(x^1), (\mathcal{J}_R)_j(x'^1)] &= ic_{ij}^k (\mathcal{J}_R)_k(x^1)\delta(x^1 - x'^1) \\
&\quad - \frac{ki}{4\pi}\delta_{ij}\left(\frac{\partial}{\partial x^1} - \frac{\partial}{\partial x'^1}\right)\delta(x^1 - x'^1)\ ,
\end{aligned}
$$

$$(16.59)$$

$$[(\mathcal{J}_R)_i(x^1), (\mathcal{J}_L)_j(x'^1)] = 0 \quad , \tag{16.60}$$

where c_{ij}^k are structure constants for the Lie algebra $\underline{G}$ associated with the group G. The c-number terms or "Schwinger terms" in (16.58) and (16.59) are quantum mechanical in origin with the coefficient k being an integer denoting the number of "flavours". In what follows we use the canonical quantization methods of Chapter 5 to show that identical commutation relations arise for J_R^0 and J_L^0 with k and n being equal. Here n is an integer due to the arguments in Section 2.

Before proceeding with the canonical quantization we note that since S_{WZ} is the integral of a closed but not exact three-form on G, we cannot write down a global Lagrangian density $\mathcal{L}_{WZ} = \mathcal{L}_{WZ}(g, \partial_0 g)$, such that $S_{WZ} = \int_{\partial\Delta_3} d^2x \mathcal{L}_{WZ}(g, \partial_0 g)$. [The domain $\partial\Delta_3$ is defined in Eq. (16.7).]

At best, we can define a Lagrangian density $\mathcal{L}_{WZ} = \mathcal{L}_{WZ}^{(u)}$ associated with patches u of the group manifold G. Thus locally, S_{WZ} is the integral of an exact three-form and hence using Stokes' theorem, we can also write it as the integral of a two form $A^{(u)}$ on coordinate patch u and then sum appropiately over all coordinate patches [20]. Let group elements g be parametrized by $\xi^i, i = 1, 2, \ldots \dim(G)$. Then

$$A^{(u)} = \frac{1}{2}A_{ij}^{(u)}(\xi)d\xi^i \wedge d\xi^j \quad , \tag{16.61}$$

where $A_{ij}(\xi) = -A_{ji}(\xi)$ can be regarded as antisymmetric potentials on the group space. Although A_{ij} cannot be expressed globally in terms of g (and its derivatives), the associated curvature tensor

$$F_{ijk} = \frac{\partial}{\partial \xi^i} A_{jk} + \frac{\partial}{\partial \xi^j} A_{ki} + \frac{\partial}{\partial \xi^k} A_{ij} \tag{16.62}$$

can be given explicitly for all fields $g(x)$. The latter, as well as the Wess-Zumino action

$$S_{WZ} = \sum_u \int_u \mathcal{L}^{(u)}_{WZ} d^2 x = \sum_u \int_u A^{(u)} \quad, \tag{16.63}$$

is invariant under the gauge transformation

$$A_{ij} \to A_{ij} + \frac{\partial}{\partial \xi^i} \Lambda_j - \frac{\partial}{\partial \xi^j} \Lambda_i \quad, \tag{16.64}$$

where Λ is a function of ξ_i. [In Eq. [16.63] the definition of the sum strictly speaking involves partitions of unity and transition functions. These technical details are, however, not important for the present discussion.] To determine F_{ijk} we compare variations of g in (16.9) and (16.63). For the former we obtain

$$\delta S_{WZ} = \frac{n}{4\pi} \int_\Delta \mathrm{Tr}(dgg^{-1})^2 d(\delta gg^{-1})$$

$$= \frac{n}{4\pi} \int_{\partial \Delta_3} \mathrm{Tr}(dgg^{-1}) d(\delta gg^{-1})$$

$$= \frac{n}{4\pi} \int_{\partial \Delta_3} \mathrm{Tr}\left(\frac{\partial g}{\partial \xi^i} g^{-1} \frac{\partial g}{\partial \xi^j} g^{-1} \frac{\partial g}{\partial \xi^k} g^{-1} \right) d\xi^i \wedge d\xi^j \delta\xi^k , \tag{16.65}$$

while for the latter

$$\delta S_{WZ} = \sum_u \frac{1}{2} \int_u F_{ijk} d\xi^i \wedge d\xi^j \delta\xi^k \quad. \tag{16.66}$$

In (16.65) we have assumed that the variations vanish on the boundaries $\partial\Delta_1$, and $\partial\Delta_2$ [Cf. Eq. (16.7)]. Further the integrand in (16.66) does not depend on the choice of coordinate patches of $\partial\Delta_3$. Hence we can make the identification

$$F_{ijk} = \frac{n}{4\pi} \mathrm{Tr}\left(\left[\frac{\partial g}{\partial \xi^i} g^{-1} , \frac{\partial g}{\partial \xi^j} g^{-1} \right] \frac{\partial g}{\partial \xi^k} g^{-1} \right) \quad. \tag{16.67}$$

214

We now proceed with the canonical quantization. The total Lagrangian density on coordinate patch u is

$$\mathcal{L}^{(u)} = \mathcal{L}_0 + \mathcal{L}_{WZ}^{(u)} \quad ,$$

$$\mathcal{L}_0 = -\frac{1}{2\lambda^2}\text{Tr}[g^{-1}\partial_\mu g g^{-1}\partial^\mu g] \quad ,$$

$$\mathcal{L}_{WZ}^{(u)} = \frac{1}{2}A_{ij}^{(u)}(\xi)\varepsilon^{\alpha\beta}\partial_\alpha\xi^i\partial_\beta\xi^j \quad . \tag{16.68}$$

We denote the canonical momentum density by $\pi_i = \pi_i(x)$,

$$\pi_i = \frac{\delta\mathcal{L}^{(u)}}{\delta\partial_0\xi^i} = \frac{1}{\lambda^2}\text{Tr}\left(g^{-1}\partial_0 g g^{-1}\frac{\delta g}{\delta\xi^i}\right) + A_{ij}^{(u)}\partial_1\xi^j \quad . \tag{16.69}$$

The system discussed here has some similarity to the system of an electric charge in a magnetic monopole field discussed in Chapter 4. Just as the canonical momenta for the electric charge are not invariant under $U(1)$ gauge transformations, the variables π_i in (16.69) are not invariant under gauge transformations (16.64). On the other hand,

$$v_i \equiv \pi_i - A_{ij}^{(u)}\partial_1\xi^j \tag{16.70}$$

are gauge invariant. These variables are the analogues of the charged particle velocities [Cf. Section 4.3]. The analogues of Poisson bracket relations Eqs. (4.8-4.10) are

$$\{\xi^i(x^1), \xi^j(x'^1)\} = 0 \quad , \tag{16.71}$$

$$\{\xi^i(x^1), v_j(x'^1)\} = \delta_j^i\delta(x^1 - x'^1) \quad , \tag{16.72}$$

$$\{v_i(x^1), v_j(x'^1)\} = F_{ijk}\partial_1\xi^k\delta(x^1 - x'^1) \quad . \tag{16.73}$$

These Poisson brackets are evaluated at equal times x^0.

As in Chapter 5 (also see Ref. [127]) we make a change of variables from the set $\{\xi^i(x), v_i(x), i = 1, 2, \ldots \dim(G)\}$. This is facilitated by the field theoretic generalization of the identity (5.3), i.e.

$$iT(i)g(\xi) = \frac{\delta g}{\delta \xi^j} N_{ji}(\xi) \quad , \tag{16.74}$$

where $T(i)$'s form a basis for the Lie algebra $\underline{G}$ associated with the group G. Thus

$$[T(i), T(j)] = ic_{ijk}T(k) \quad . \tag{16.75}$$

As in Chapter 5, $N = [N_{ij}]$ is nonsingular. We can therefore define the variables

$$t_i = -\pi_j N_{ji} \quad . \tag{16.76}$$

The field theoretic analogues of Poisson brackets relations (5.7) and (5.9) are

$$\{t_i(x^1) , g(x'^1)\} = iT(i)g(x'^1)\delta(x^1 - x'^1) \tag{16.77}$$

$$\{t_i(x^1) , t_j(x'^1)\} = c_{ijk}t_k(x^1)\delta(x^1 - x'^1) \quad . \tag{16.78}$$

The variables t_i are not invariant under (16.64). Let us instead define

$$L_i = -v_j N_{ji} + \frac{in}{4\pi}\text{Tr } T(i)\partial_1 g g^{-1} \quad , \tag{16.79}$$

$$R_i = v_j M_{ji} + \frac{in}{4\pi}\text{Tr } T(i)g^{-1}\partial_1 g \quad , \tag{16.80}$$

where the matrix $M = [M_{ij}]$ is defined by $N_{ij}^{-1}T(i) = M_{ij}^{-1}gT(i)g^{-1}$. The variables defined in (16.79) and (16.80) are gauge invariant and identical to the current densities $(J_L^0)_i$ and $(J_R^0)_i$ respectively. Using (16.77) and (16.78) we can now compute the Poisson brackets of the current densities:

$$\{L_i(x^1) , L_j(x'^1)\} = c_{ij}^k L_k(x^1)\delta(x^1 - x'^1) + \frac{n}{4\pi}\delta_{ij}\left(\frac{\partial}{\partial x^1} - \frac{\partial}{\partial x'^1}\right)\delta(x^1 - x'^1), \tag{16.81}$$

$$\{R_i(x^1), R_j(x'^1)\} = c_{ij}^k R_k(x^1)\delta(x^1 - x'^1) - \frac{n}{4\pi}\delta_{ij}\left(\frac{\partial}{\partial x^1} - \frac{\partial}{\partial x'^1}\right)\delta(x^1 - x'^1), \tag{16.82}$$

$$\{R_i(x^1) , L_j(x'^1)\} = 0 \quad . \tag{16.83}$$

Now upon replacing Poisson brackets { , } by $-i[$, $]$ we obtain commutation relations identical to (16.58-60) where k is identified with n. Also, although in (16.58) and (16.59) the central terms (i.e. the last terms) are quantum mechanical in nature, their analogues are already present in the classical Poisson brackets in (16.81) and (16.82). In the bosonic theory the central terms arise from the Wess-Zumino action S_{WZ}.

Chapter 17

A HIERARCHY OF "SPHERICALLY SYMMETRIC" ANSÄTZE

17.1 INTRODUCTION

When N_f=2, the "spherically symmetric" ansatz has the form (13.29). Skyrme conjectured that the corresponding quantum states represent ordinary baryons. When $N_f > 2$, one may define "spherically symmetric" ansätze which are not of the form (13.29). These ansätze along with the physical interpretation of the corresponding quantum states are the main topics of this Chapter. We also discuss new quantum ambiguities which can occur if $N_f > 2$. They are the generalizations of the two flavour ambiguites discussed in Chapter 13.

17.2 A GENERAL DEFINITION OF SPHERICAL SYMMETRY

Let W denote any static solution for N_f flavours. We expect such a solution to be characterized by a high degree of symmetry. When the baryon number $B(W) \neq 0$, experience with the standard Skyrmion suggests that such a solution, if it exists, will be characterized by the following generalized "spherical symmetry":

$$-i(\vec{x} \times \vec{\nabla})_i \, W(\vec{x}) + [\Lambda_i, W(\vec{x})] = 0 \quad . \tag{17.1}$$

Here Λ_i generate any $SU(2)$ or $SO(3)$ subgroup of the flavour $SU(N_f)$:

$$[\Lambda_i, \Lambda_j] = i\varepsilon_{ijk}\Lambda_k \quad . \tag{17.2}$$

As we are interested in static solutions, we have suppressed time from the argument of W.

The Skyrme soliton for three flavours corresponds to the following special choice of Λ_i:

$$\Lambda_i = \frac{1}{2} \begin{pmatrix} & & 0 \\ & \tau_i & 0 \\ 0 & 0 & 1 \end{pmatrix} \quad . \tag{17.3}$$

There are at least $N_f - 1$ physically inequivalent solutions for (17.1). This is because there are at least $N_f - 1$ $SU(2)$ subalgebras of $SU(N_f)$ which cannot be transformed into each other by an $SU(N_f)$ flavour rotation. Thus, any "spin" S representation of $SU(2)$ with $2S + 1 \leq N_f$ is a subgroup of the flavour $SU(N_f)$. (Here "spin" is not the physical spin of the soliton, but rather is just a label for the $SU(2)$ representation.) We choose this representation to act on the first $2S + 1$ flavours. The corresponding generators Λ_i (with any one of the S values $1/2, 1, \dots (N_f - 1)/2$) can be inserted in the defining equation (17.1) for W. For N_f flavours, if D is the dimension $2S + 1$ of the $SU(2)$ representation which goes into the definition of W, we shall call the corresponding ansatz and the corresponding static solution (assuming that such a solution exists) as an (N_f, D) (classical) soliton. In this notation, a Skyrme soliton for any B is an $(N_f, 2)$ soliton.

Note that under these $(N_f - 1)$ $SU(2)$ subalgebras, the N_f dimensional flavour multiplet splits up into a "spin S" multiplet and $(N_f - 2S - 1)$ singlets. There are also ansätze based on $SU(2)$ subalgebras under which the flavour multiplet splits up into several spin S_i multiplets such that $\sum_i (2S_i + 1) = N_f$.

In the next Section, we set $N_f = 3$ and examine the (3,3) classical soliton. It is believed to correspond to a "dibaryon" [128]. We then proceed with the standard semiclassical quantization of the soliton and show that an ambiguity results which is analogous to that discussed in Chapter 13 for the two flavoured Skyrme model [Balachandran, Lizzi, Rodgers and Stern in Ref. [128] and Ref. [129]]. These quantization ambiguities are discussed for any number of flavours in Section 4.

17.3 THE DIBARYON

In the preceding hierarchy, the (3,3) ansatz based on the $SO(3)$ subgroup of the flavour $SU(3)$ are particularly interesting phenomenologically. We shall see below that for the "$SO(3)$" ansatz, baryon number can assume only even values. The ansatz for $B = 2$ seems to lead to a quasistable set of states which appear to be the same as the H-baryon states found by Jaffe in the quark model [130].

In this Section, the method of construction of the ansatz and of the calculation of its baryon number are indicated. The Lagrangian and Hamiltonian formalisms in the collective coordinate approximation are also explained. It turns out that because of this approximation, the two fold quantization ambiguity of the (2,2) soliton reappears as a three fold ambiguity. The discussion of this ambiguity is postponed to a later Section, here we content ourselves merely by choosing the physically correct quantization. A detailed analysis of the phenomenology of the H-baryon states in the quark and chiral models is important and has been carried out in the literature [130,131]. Except for some brief remarks in the following Chapter, we shall not reproduce this analysis in this book, however.

The "angular momentum" which defines the spherical symmetry of the (3,3) soliton V_c is $-i(\vec{x} \times \vec{\nabla})_i + \theta_i$, where

$$\theta_1 = \lambda_7 \ , \quad \theta_2 = -\lambda_5 \ , \quad \theta_3 = \lambda_2 \ , \quad (17.4)$$

where λ_α are the Gell-Mann $SU(3)$ Lie algebra matrices. [We denote the Λ_i for $(3,3)$ soliton as θ_i. They are identical to the θ_i defined in Chapter 13.] Thus

$$-i(\vec{x} \times \vec{\nabla})_i \, V_c + [\theta_i, V_c] = 0 \quad . \tag{17.5}$$

The solution of this equation is a function of $\vec{\theta} \cdot \hat{x}$ and r. Now it is easy to verify the identity $(\theta_3)^3 = \theta_3$. By rotating it one also has the identity $(\vec{\theta} \cdot \hat{x})^3 = \vec{\theta} \cdot \hat{x}$. Thus, we can write

$$V_c = \alpha \mathbf{1}_{3\times 3} + \beta \, \vec{\theta} \cdot \hat{x} + \gamma (\vec{\theta} \cdot \hat{x})^2 \;, \quad \alpha = \alpha(r) \;, \quad \beta = \beta(r) \;, \quad \gamma = \gamma(r) \;. \tag{17.6}$$

The requirement that V_c is $SU(3)$ valued forces this expression to have the form

$$V_c = e^{i\psi} \mathbf{1}_{3\times 3} + i \, \sin \chi e^{-i\psi/2}(\vec{\theta} \cdot \hat{x}) + [\cos \chi e^{-i\psi/2} - e^{i\psi}](\vec{\theta} \cdot \hat{x})^2 \;. \tag{17.7}$$

Here ψ and χ are (real) functions of r only. [Evaluate V_c after rotating $\hat{x}$ to $(0,0,1)$ to verify that $V_c \in SU(3)$.] The boundary condition $V_c \to \mathbf{1}$ as $r \to \infty$ requires that either

$$\cos \chi(\infty) = 1 \quad , \quad e^{-i\psi(\infty)/2} = 1 \tag{17.8}$$

or

$$\cos \chi(\infty) = -1 \quad , \quad e^{-i\psi(\infty)/2} = -1 \quad . \tag{17.9}$$

Since we can reduce (17.9) to (17.8) by the substitutions

$$\begin{aligned}
\chi(\infty) &\rightarrow \chi(\infty) + \pi \quad , \\
\psi(\infty) &\rightarrow \psi(\infty) + 2\pi \quad ,
\end{aligned} \tag{17.10}$$

which together preserve the form of V_c, (17.9) can be ignored.

If V_c is to be well defined at the origin, the coefficients of $\vec{\theta} \cdot \hat{x}$ and $(\vec{\theta} \cdot \hat{x})^2$ must vanish there. Thus, we require also that either

$$\cos \chi(0) = -1 \quad , \quad e^{-i\psi(0)/2} = -\left[e^{\frac{2}{3}\pi i}\right]^k \tag{17.11}$$

or

$$\cos \chi(0) = 1 \quad , \quad e^{-i\psi(0)/2} = \left[e^{\frac{2}{3}\pi i}\right]^{k'} \quad , \tag{17.12}$$

where k and k' are integers mod three.

It can be proved that the baryon number for the first set of boundary conditions is restricted to odd multiples of 2, and for the second set of boundary conditions to even multiples of 2. Let us first prove the second result. Consider

$$R = \mathbf{1}_{3\times3} + i\ \sin\chi(\vec{\theta}\cdot\hat{x}) + (\cos\chi - 1)(\vec{\theta}\cdot\hat{x})^2 \ . \qquad (17.13)$$

It is well defined as $r \to 0$ in view of (17.12). Further, it is $SO(3)$ valued. [Rotate $\hat{x}$ to $(0, 0, 1)$ to simplify the verification of this statement.] In fact, it is the image of the $SU(2)$ matrix

$$h = \mathbf{1}_{2\times2}\cos\frac{\chi}{2} + i\ \vec{\tau}\cdot\hat{x}\sin\frac{\chi}{2} \qquad (17.14)$$

under the canonical $SU(2) \to SO(3)$ homomorphism defined by

$$h\ \tau_i h^\dagger = \tau_j R_{ji}(h) \ . \qquad (17.15)$$

We have also the familiar identities

$$h^{-1}\partial_i h = \frac{\tau_j}{2}\ \mathrm{Tr}\ (\tau_j h^{-1}\partial_i h) \qquad (17.16)$$

and

$$R^{-1}\partial_i R = \theta_j\ \mathrm{Tr}\ (\tau_j h^{-1}\partial_i h) \ . \qquad (17.17)$$

Writing $B(h)$ and $B(R)$ in the forms

$$\begin{aligned}
B(h) &= \frac{1}{48\pi^2}\varepsilon_{ijk}\int d^3x\ \mathrm{Tr}\ \left(h^{-1}\partial_i h[h^{-1}\partial_j h,\ h^{-1}\partial_k h]\right) \ , \\
B(R) &= \frac{1}{48\pi^2}\varepsilon_{ijk}\int d^3x\ \mathrm{Tr}\ \left(R^{-1}\partial_i R[R^{-1}\partial_j R,\ R^{-1}\partial_k R]\right) \ ,
\end{aligned} \qquad (17.18)$$

it follows that

$$B(R) = \frac{\mathrm{Tr}\ \theta_i^2}{\mathrm{Tr}\ (\tau_i/2)^2}B(h) = 4B(h) = \frac{2}{\pi}[\chi(0) - \chi(\infty)] \ . \qquad (17.19)$$

Because of the boundary conditions, the values of $B(R)$ are thus restricted to even multiples of 2.

Next consider

$$S = V_c R^\dagger = e^{i\psi} \mathbf{1}_{3\times 3} + [e^{-i\psi/2} - e^{i\psi}](\vec{\theta} \cdot \hat{x})^2 \quad . \qquad (17.20)$$

This S is symmetric. Using also the reality of $B(S)$, we have

$$
\begin{aligned}
B(S) &= B(S^T) = [B(S^T)]^\star = B(S^{-1}) \\
&= \frac{1}{24\pi^2} \varepsilon_{ijk} \int d^3x \; \mathrm{Tr}\; (S\partial_i S^{-1} S\partial_j S^{-1} S\partial_k S^{-1}) \\
&= -B(S) \quad , \qquad (17.21)
\end{aligned}
$$

where in obtaining the last line, we used $S\partial_i S^{-1} = -\partial_i SS^{-1}$ and the cyclic symmetry of the trace. Thus, $B(S) = 0$. Now a little algebra shows the identities $B(S) = B(V_c) + B(R^\dagger) = B(V_c) - B(R)$. It follows that

$$B(V_c) = B(R) = \frac{2}{\pi}[\chi(0) - \chi(\infty)] = 4n \quad ,$$

$$n = 0, \pm 1, \pm 2, \ldots \qquad (17.22)$$

This argument has to be modified for the first set of boundary conditions since R is not then well defined at $r = 0$. Let us look at V_c^2:

$$V_c^2 = e^{i\psi'}\mathbf{1} + i\sin\chi' \; e^{-i\psi'/2}(\vec{\theta} \cdot \hat{x}) + (\cos\chi' \; e^{-i\psi'/2} - e^{i\psi'})(\vec{\theta} \cdot \hat{x})^2 \quad ,$$

$$\chi' = 2\chi, \; \psi' = 2\psi \quad . \qquad (17.23)$$

Thus, V_c^2 fulfills the second set of boundary conditions. It follows that

$$B(V_c^2) = 2B(V_c) = \frac{2}{\pi}[\chi'(0) - \chi'(\infty)] \quad , \qquad (17.24)$$

and that

$$B(V_c) = \frac{2}{\pi}[\chi(0) - \chi(\infty)] = 2(2n+1) \quad ,$$
$$n = 0, \pm 1, \pm 2, \ldots \quad . \qquad (17.25)$$

The integers k and k' which appear in the ansatz are unaffected by deformations which respect spherical symmetry at every step. However, it is possible to construct deformations through nonspherical configurations which change their value. This suggests that k and k' may be approximately conserved quantum numbers in the chiral model.

Assuming that V_c is classically a static configuration, its classical energy for any B, k, and k' is found after some calculation to be

$$
\begin{aligned}
E(V_c) \;=\; & \frac{\pi F_\pi^2}{2} \int_0^\infty dr \; r^2 \left\{ \chi'^2 + \frac{3}{4}\psi'^2 + \frac{4}{r^2}\left(1 - \cos\frac{3}{2}\psi\cos\chi\right) \right. \\
& + \frac{4}{e^2 F_\pi^2 r^2}[(1 - \cos\frac{3\psi}{2}\cos\chi)(\chi'^2 + \frac{9}{4}\psi'^2) \\
& + 3\sin\frac{3\psi}{2}\sin\chi \; \chi'\psi' + \frac{3}{r^2}\sin^2\chi\sin^2\frac{3\psi}{2} \\
& \left. + \frac{1}{r^2}(1 - \cos\frac{3\psi}{2}\cos\chi)^2] \right\} \quad ,
\end{aligned}
\tag{17.26}
$$

where here prime denotes differentiation with respect to r.

For the $B = 2$ soliton, $\chi(\infty) = 0$ and $\chi(0) = \pi$. Further, the minimum of energy is expected to be realized when the magnitude $|\Delta\psi| = |\psi(\infty) - \psi(0)|$ of the variation of ψ is least. Numerical work shows that this expectation is correct. Let us, therefore, choose $\psi(0) = 2\pi/3$ and $\psi(\infty) = 0$. Variational calculations show that with this constraint, the minimum energy is

$$
E(V_c) \cong 70 \, \frac{F_\pi}{e} \cong 1.92 \, E(U_c) \quad ,
\tag{17.27}
$$

where $E(U_c)$ is the energy of the static configuration for the three flavour Skyrme soliton [the $(3,2)$ soliton].

The action we are working with is of course the usual one:

$$
S(U) = \int d^4x \left\{ \frac{F_\pi^2}{16}\mathrm{Tr} \, V_\mu^2 + \frac{1}{32e^2}\mathrm{Tr}[V_\mu, V_\nu]^2 \right\} + S_{WZ}(U) \quad ,
\tag{17.28}
$$

$$
S_{WZ}(U) = -\frac{iN_c}{240\pi^2}\int_\Delta \mathrm{Tr} \, V^5 \quad , \quad V = V_\mu dx^\mu = U^\dagger dU \quad .
$$

In the collective coordinate approximation

$$
U(\vec{x}, t) = A(t)U_c(\vec{x})A(t)^\dagger
\tag{17.29}
$$

we have already remarked in Chapter 15 that the first two terms in $S(U)$ can be written as follows:

$$S_{SK}(U) = S_{SK}(U_c) + \int dt \ L_{SK}(A, \dot{A}) \quad , \tag{17.30}$$

$$L_{SK}(A, \dot{A}) = I_{\alpha\beta}(U_c)(A^\dagger \dot{A})_\alpha (A^\dagger \dot{A})_\beta \quad . \tag{17.31}$$

The expressions for the moment of inertia tensors are given in (15.24). For the ansatz appropriate to the $(3,3)$ soliton, the moment of inertia tensors can be explicitly computed by replacing U_c by V_c in (15.24). They are diagonal. Their nonvanishing diagonal entries for any choice of B, k, and k' can be found, after a long but straightforward calculation. The literature may be consulted for the results.

We next show that L_{WZ} vanishes for the $SO(3)$ soliton. For this, it is sufficient to show that

$$\int w^3 \quad \text{and} \quad \int V_c w^3 V_c^\dagger \tag{17.32}$$

are proportional to the unit matrix. Here $w = V_c^\dagger dV_c$. For then from (16.24) (which is valid with the replacements $U_c \to V_c$ and $w \to V_c^\dagger dV_c$), L_{WZ} becomes proportional to $\text{Tr } \alpha = \text{Tr } A^\dagger dA$ which is zero since $A^\dagger dA$ is valued in the Lie algebra of $SU(3)$.

Let $\{R\}$ be the group generated by θ_i. It is isomorphic to $SO(3)$. Because of (17.5),

$$R V_c(\vec{x}) R^\dagger = V_c(\vec{R}x) \quad . \tag{17.33}$$

Using also the identity $d^3 x = d^3(\vec{R}x)$ we find that

$$R \left[\int w^3 \right] R^\dagger = \int w^3 \quad , \tag{17.34}$$

or that $\int w^3$ commutes with R. Since the representation $\{R\}$ is irreducible on three flavours, the first part of (17.32) follows from Schur's lemma. The second part follows similarly.

The effective Lagrangian for the collective coordinate A of the $SO(3)$ soliton can be calculated from (17.31). It is

$$L(A, \dot{A}) = -\frac{1}{2}\alpha(V_c)(A^\dagger \dot{A})_i (A^\dagger \dot{A})_i - \frac{1}{2}\beta(V_c)(A^\dagger \dot{A})_a (A^\dagger \dot{A})_a - E(V_c) \quad ,$$

$$\alpha(V_c) \cong \frac{243}{F_\pi e^3} \quad , \quad \beta(V_c) \cong \frac{364}{F_\pi e^3} \quad . \tag{17.35}$$

Here i runs over $2, 5, 7$ and a runs over $1, 3, 4, 6, 8$. α and β have been estimated numerically for the configuration which minimizes (17.26). [Cf. Balachandran, Lizzi, Rodgers and Stern in Ref. [128] for more details of this calculation.]

The quantization of this system uses the same method as for the $B = 1$ soliton. The calculations are, in fact, slightly simpler because (17.35) leads to no constraint. As before, we find the commutation relations of the observables to be

$$
\begin{aligned}
&[\hat{A}_{\alpha\beta}, \hat{A}_{\gamma\delta}] = 0 \quad , \\
&[\hat{R}_\alpha, \hat{A}_{\beta\gamma}] = -[\hat{A}\lambda_\alpha]_{\beta\gamma} \quad , \\
&[\hat{R}_\alpha, \hat{R}_\beta] = -2i f_{\alpha\beta\gamma} \hat{R}_\gamma \quad ,
\end{aligned} \tag{17.36}
$$

and the Hamiltonian to be

$$\hat{H} = E(V_c) + \frac{\hat{R}_i \hat{R}_i}{2\alpha(V_c)} + \frac{\hat{R}_a \hat{R}_a}{2\beta(V_c)} \quad . \tag{17.37}$$

We shall see later that for three flavours there are quantization ambiguities for the $SO(3)$ soliton in the approximations that we have made. A way to solve this problem is to embed the $SO(3)$ soliton in the four (or more) flavour model. This is analogous to Witten's imbedding of the $SU(2)$ soliton in the three (or more) flavour model. Then, as for the $SU(2)$ soliton, the ambiguity is resolved, and it is found that the $SO(3)$ soliton must be quantized using triality zero representations of $SU(3)$. The proof of this last statement will be left as an exercise. Thus, the Hilbert space $\mathcal{H}_{SO(3)}$ for the model at hand is spanned by the functions

$$D^{(p,q)}_{(I,I_3,Y)(I',I'_3,Y')}(A) \quad , \tag{17.38}$$

where (p, q) is the triality zero irreducible representation (IRR) of $SU(3)$. The scalar product for $\mathcal{H}_{SO(3)}$ is defined by (15.57) while the operators $\hat{A}$ and $\hat{R}$ are realized by imitating the rules (15.58) and (15.59).

The transformation $A \to GA, G \in SU(3)$ is (as before) the action of the flavour $SU(3)$. Thus, (I, I_3, Y) are flavour indices. The transformation $A \to AR, R \in SO(3) \subset SU(3)$ is the action of spatial angular momentum. It transforms the indices (I', I'_3, Y'). However, since $SO(3)$ is not the isospin $SU(2)$ subgroup, I', I'_3 do not have the significance of spin and its third component [as was the case for the (3,2) soliton]. To find the spin content, we have to reduce the IRR (p, q) with respect to the $SO(3)$ subgroup.

Let us consider a few examples. For $(p, q) = (0, 0)$, we get a flavour singlet with spin zero. For $(p, q) = (1, 1)$, since $(1, 1)$ reduces under $SO(3)$ to a spin 1 and a spin 2 IRR, we get an octet of spin 1 and another octet of spin 2 states. For $(p, q) = (3, 0)$, in a similar way, we get decuplets of spin 1 and 3 states.

The energy eigenvalues can be obtained by writing $\hat{R}_a \hat{R}_a$ as

$$\sum_{\alpha=1}^{8} \hat{R}_\alpha \hat{R}_\alpha - \sum_{i=2,5,7} \hat{R}_i \hat{R}_i \qquad (17.39)$$

and inserting the values of the Casimir invariants for $SU(3)$ and $SO(3)$. For the spin J system corresponding to (p, q), it reads

$$\begin{aligned}
M_{(p,q)J} &= E(V_c) + \left[\frac{1}{\alpha(V_c)} - \frac{1}{\beta(V_c)} \right] \frac{J(J+1)}{2} \\
&\quad + (p^2 + q^2 + 3(p+q) + pq) \frac{2}{3\beta(V_c)} \quad . \qquad (17.40)
\end{aligned}$$

For the singlet and the two spin values of the octet and decuplet, it becomes

$$\begin{aligned}
M_{(0,0)0} &= m_{B=2} = E(V_c) \quad . \\
M_{(1,1)1} &= m_{B=2} + \frac{1}{\alpha(V_c)} + \frac{5}{\beta(V_c)} \quad , \\
M_{(1,1)2} &= m_{B=2} + \frac{3}{\alpha(V_c)} + \frac{3}{\beta(V_c)} \quad , \\
M_{(3,0)1} &= m_{B=2} + \frac{1}{\alpha(V_c)} + \frac{11}{\beta(V_c)} \quad , \\
M_{(3,0)3} &= m_{B=2} + \frac{6}{\alpha(V_c)} + \frac{6}{\beta(V_c)} \quad . \qquad (17.41)
\end{aligned}$$

There is an important aspect in which this quantization scheme is deficient and that is the following. The classical ansatz V_c is not invariant under the parity operation

$$V_c(\vec{x}) \rightarrow V_c(-\vec{x})^\dagger$$

or

$$\psi(r) \rightarrow -\psi(r) \quad , \quad \chi(r) \rightarrow \chi(r) \quad . \tag{17.42}$$

[Note that if $\psi(0) = 2\pi/3$, the parity transformed function $\tilde{\psi}$ fulfills the boundary condition $\tilde{\psi}(0) = -2\pi/3$. The classical energy $E(V_c)$ is, of course, parity invariant.] Thus, all degenerate states are doubled, one corresponds to V_c and the other to its parity transform. Further, since there will definitely be effects in a full quantum treatment which will mix such a state with its parity transform, solitons of definite energy will not be associated with V_c or its parity transform, but with states of definite parity. One does not see these effects in the preceding approximate treatment. We shall, therefore, simply assume that the soliton energies we compute correspond to the even parity states and bear in mind the possibility of odd parity states of somewhat higher energies.

17.4 QUANTIZATION AMBIGUITIES FOR ANY NUMBER OF FLAVOURS

For any finite number N_f of flavours, consider the ansatz W with spherical symmetry based on the "spin" $S = (N_f - 1)/2$ subgroup of $SU(N_f)$. Thus, Λ_i in this case are the angular momentum matrices for spin $(N_f - 1)/2$. Then we will show in this Section that there are *at least* $N_f + 1$ ways to quantize the associated zero mode in the approximation where other modes are ignored.

There is reason to believe that this problem arises for $N_f \geq 3$ because of the approximation of retaining only collective coordinates and neglecting other modes when we pass to quantum theory. [As regards the two flavour case, see the comments after (13.77).] Thus, consider the dibaryon state for $N_f = 3$.

228

Its collective coordinates can be quantized in at least three different ways. [Actually, there are four ways. We ignore the fourth way in these preliminary remarks.] The corresponding Hilbert spaces $\mathcal{H}^{(\nu)}$ ($\nu = 0, 1, 2$) are characterized by triality quantum numbers z^ν, $z = \exp(i2\pi/3)$ as we shall see below. The $B = 1$ Skyrmion has triality zero for $N_f = 3$. Thus, *in the collective coordinate approximation*, if we quantize the dibaryon using $\mathcal{H}^{(1)}$ or $\mathcal{H}^{(2)}$, there will be a selection rule forbidding the decay of the dibaryon into two $B = 1$ Skyrmions for $N_f = 3$. However, without this approximation there are certainly histories with finite action $S(U)$ where the field U is continuously evolved from the dibaryon configuration into two widely separated $B = 1$ states. Thus, in the full quantum theory of the U's , there will be no such selection rule (unless there are unexpected cancellations between such histories). Consistency with the unapproximated quantum field theory therefore seems to require that one should quantize the collective coordinates using $\mathcal{H}^{(0)}$.

The topological reason for this quantization ambiguity seems to be the following. For the dibaryon, the collective coordinate approximation does not correctly reproduce the relevant topology of the space of U's. For the three ways mentioned above, the relevant configuration space Q of the collective coordinates is $SU(3)/\mathbf{Z}_3$. For the fourth way (see below), Q is $SU(3)$. The topological properties important for quantization are $H^2(Q, \mathbf{Z})$ and $\pi_1(Q)$, $H^2(Q, \mathbf{Z})$ being the second cohomology group with coefficients in integers. [$H^2(Q, \mathbf{Z})$ classifies $U(1)$ bundles over Q. The importance of these bundles and of $\pi_1(Q)$ for quantization have been discussed at length in earlier Chapters. The role of $\pi_1(Q)$ in the quantization of the system under consideration was examined in Chapter 8.] They can be shown to be different for any of these Q's and the space of U's. This suggests that when we restrict ourselves to three flavours, we should somehow augment the collective coordinates for the dibaryon by introducing new modes so that the relevant topology is correctly reproduced. The remarks in the preceding paragraph suggest that these modes are likely to be associated with the tunneling process from the dibaryon to two $B = 1$ states.

Similar remarks can be made for any N_f larger then three.

Note that the ambiguity in quantization disappears when the state characterized by "spin" $S = (N_f - 1)/2$ is embedded in $N_f + 1$ or more flavours. (Again, "spin" S is not the physical spin of the soliton, but rather it just serves to label the representation of the $SU(2)$ subgroup of $SU(N_f)$ which is associated with Eq. (17.1).) For example, there is a unique quantization of the usual $B = 1$ Skyrmion when it is embedded in 3 or more flavours.

If the number of flavours is infinite, then the ambiguity in question, at least formally, disappears altogether. This is because it makes no sense to talk about states with "infinite spin" S. Therefore, the state responsible for the ambiguity ceases to exist.

The quantization ambiguity is resolved for "non-maximal" solitons like the $B = 1$ soliton embedded in $SU(3)$ because a certain *gauge* symmetry under a continuous group appears in $L(A, \dot{A})$. For "maximal" solitons, for which Λ_i describe maximal angular momentum for the given N_f, this gauge symmetry under a continuous group disappears from $L(A, \dot{A})$.

We now discuss these quantization ambiguities in detail. There is first the uncertainty in determining what is the configuration space Q associated with the Lagrangian $L(A, \dot{A})$. We can regard it as the space of A's, that is as $SU(N_f)$. On the other hand, L is obtained by substituting $A(t)WA(t)^\dagger$ for U in the original Lagrangian so that U is invariant under the action of the center $\mathbf{Z}_{N_f}$ of $SU(N_f)$ on A:

$$U \to U, \quad \text{when} \quad A \to Az \quad , \quad z = e^{2\pi i/N_f}\mathbf{1} \in \mathbf{Z}_{N_f} \quad . \tag{17.43}$$

Therefore, L as well is invariant under this action:

$$L(Az, \dot{A}z) = L(A, \dot{A}) \quad . \tag{17.44}$$

It follows that if D is a discrete subgroup of $\mathbf{Z}_{N_f}$, L can be thought of as a function on (the tangent bundle of) $SU(N_f)/D$, and the configuration space Q can be equally well identified with $SU(N_f)/D$. The algebra of observables and hence its IRR's are in general different for this new configuration space. We shall illustrate this fact explicitly below for the case $D = \mathbf{Z}_{N_f}$.

When $Q = SU(N_f)$, the operator algebra can be realized on the Hilbert space $\mathcal{H} = \{\psi\}$ of functions on $SU(N_f)$ with the scalar product defined as usual, i.e.

$$(\psi, \phi) = \int d\mu(A) \, \psi(A)^* \, \phi(A) \quad , \tag{17.45}$$

where $d\mu(A)$ is the invariant $SU(N_f)$ measure. On this space, the operator $\hat{A}$, which represents the classical collective coordinate A, is realized by the rule

$$[\hat{A}_{ij}\psi](A) = A_{ij}\psi(A) \quad . \tag{17.46}$$

[The operators conjugate to $\hat{A}$ have a natural action on ψ which we need not go into here (Cf. (15.59)).] The irreducibility of this representation is a simple consequence of the Peter-Weyl theorem [54].

Consider next the case where $D = \mathbf{Z}_{N_f}$ and $Q = SU(N_f)/\mathbf{Z}_{N_f}$ is the adjoint representation $Ad\ SU(N_f)$ of $SU(N_f)$. Let us denote the classical observables of this Q [which are the coordinates of Q] by $Ad\ A$ and the corresponding quantum operators by $Ad\ \hat{A}$. Recall first that the well known homomorphism $SU(N_f) \to Ad\ SU(N_f)$ associates to an element A of $SU(N_f)$ an element $Ad\ A$ of $Ad\ SU(N_f)$. [$Ad\ A$ is defined by the equation $A\lambda_\alpha A^{-1} = \lambda_\beta (Ad\ A)_{\beta\alpha}$, λ_α being appropriate generalizations of Gell-Mann matrices.] With this knowledge, we can realize the operators $Ad\ \hat{A}$ on the Hilbert space $\mathcal{H}$ introduced above:

$$[(Ad\ \hat{A})_{ij}\psi](A) = (Ad\ A)_{ij}\psi(A) \quad . \tag{17.47}$$

However, such a representation of $Ad\ \hat{A}$ is not irreducible. Let $\mathbf{Z}_{N_f} = \{e, z, z^2, z^3, \ldots, z^{N_f - 1}\}$ where e is the identity. We can define a representation of $\mathbf{Z}_{N_f}$ on $\mathcal{H}$ by the association $z^k \to U(z)^k$ where

$$[U(\hat{z})\psi](A) = \psi(Az) \quad . \tag{17.48}$$

The operator $U(\hat{z})$ commutes with $Ad\ \hat{A}$:

$$
\begin{aligned}
[U(\hat{z})Ad\ \hat{A}_{ij}\psi](A) &= [Ad\ \hat{A}_{ij}\psi](Az) \\
&= [Ad(Az)]_{ij}\psi(Az) \\
&= [Ad\ A]_{ij}\psi(Az) \\
&= [Ad\ \hat{A}_{ij}U(\hat{z})\psi](A) \quad .
\end{aligned}
\tag{17.49}
$$

The existence of such $U(\hat{z})$ (which are not necessarily multiples of the identity) shows that the representation defined by (17.49) is reducible. [Here we have not discussed the operators $\hat{R}_\alpha$ which play the role of momenta. It is not difficult to show that they too commute with $U(\hat{z})$.]

The group $\mathbf{Z}_{N_f}$ has N_f inequivalent IRR's. We can decompose $\mathcal{H}$ into the direct sum

$$
\mathcal{H} = \overset{N_f-1}{\underset{\nu=1}{\oplus}}\ \mathcal{H}^{(\nu)} \quad ,
\tag{17.50}
$$

where $\mathcal{H}^{(\nu)}$ carries all the infinite number of copies of the same IRR of Z_{N_f}. [On $\mathcal{H}^{(\nu)}$, $U(\hat{z})$ is represented by z^ν.] Each of these subspaces $\mathcal{H}^{(\nu)}$ is invariant under $Ad\ \hat{A}$. Thus, there are N_f inequivalent IRR's of the operator algebra generated by $(Ad\ \hat{A})_{ij}$.

The decomposition (17.50) corresponds to the decomposition of $\mathcal{H}$ according to different N_f-alities [trialities, quadralities, ...]. The remark then is that we have the freedom to choose any one of N_f-alities in the process of quantization. If we choose any one of these Hilbert spaces $\mathcal{H}^{(\nu)}$, the chiral model agrees with QCD only if the number of colours in QCD is restricted to a discrete set of values. For example, for $N_f = 2$, let us consider the Skyrme soliton with $B(U_c) = 1$. In the decomposition (17.50), we can take $U(\hat{z})$ to be unity on $\mathcal{H}^{(0)}$ and to be -1 on $\mathcal{H}^{(1)}$. In the quark model, the baryon is composed of N_c quarks and each quark flips sign under the action of the center of the flavour $SU(2)$. Thus, the baryon acquires a phase $(-1)^{N_c}$ under the action of the center of $SU(2)$. If this baryon is in some sense the same as the soliton state in $\mathcal{H}^{(0)}$, then N_c is restricted to even values. Similarly, if $\mathcal{H}^{(1)}$ is used to quantize the Skyrme soliton for $B = 1$, then the chiral model and QCD agree only if N_c is odd. Let us illustrate this point for the $SO(3)$ dibaryon for $N_f = 3$. On $\mathcal{H}^{(\nu)}(\nu = 0, 1, 2)$, let $U(\hat{z})$ be represented by z^ν. The quark states which acquire this phase under the center of the flavour $SU(3)$ are composed of $\nu, \nu + 3, \nu + 6, \ldots$ quarks since each quark acquires the phase z under this center. Since we also need $2N_c$ quarks to make a dibaryon, we find that $2N_c$ is restricted to one of the values $\nu, \nu + 3, \nu + 6, \ldots$ if the QCD states are to resemble the states in $\mathcal{H}^{(\nu)}$. [The choice of the solution for N_c is, of course, further constrained by the requirement that N_c is an integer. Also note that in this example the center of *colour* $SU(3)$ too induces the transformation $q \to zq$ of quarks. Thus, with colour confinement, the two models agree only if $\nu = 0 \bmod 3$.]

For the Skyrme soliton and $N_f = 2$, the two methods of quantization end up by describing the soliton either as a boson or as a fermion. However, in general the quantization ambiguity is not such a boson-fermion ambiguity. For example, in the preceding example of the $SO(3)$ soliton for $N_f = 3$, the defining triplet representation of $SU(3)$ carries spin one under this $SO(3)$. Since all IRR's of $SU(3)$ can be built up by considering tensor products of triplets and each triplet has spin one, the physical angular momenta contained in these IRR's are all integral regardless of the values of ν.

We shall now explain how a gauge symmetry appears in $L(A, \dot{A})$ for non-maximal solitons and its role in resolving quantization ambiguities. For maximal solitons, this gauge symmetry is absent and there are quantization ambiguities associated with $L(A, \dot{A})$. We shall also show that L_{WZ} is zero for maximal solitons.

The Lagrangian $L(A, \dot{A})$ is obtained by the substitution $U = A(t)WA(t)^\dagger$ in the action for U. It is U that is our dynamical variable and not $A(t)$ and W separately. It is U that appears in the original action. A consequence is that if U is unchanged under a certain transformation $A \to A'$, then the equation of motion for A derived from $L(A, \dot{A})$ should have the symmetry $A \to A'$. Therefore, under such a transformation, L changes at most by a total derivative:

$$L(A', \dot{A}') = L(A, \dot{A}) + \frac{d\phi}{dt} \quad . \tag{17.51}$$

Suppose now that such transformations form a continuous time dependent set of symmetries of the form $A(t) \to A(t)h(t)$. Since L is changed only by total derivatives by these transformations, they form a group of gauge symmetries of $L(A, \dot{A})$.

For the $B = 1$ soliton, for $N_f = 3$, such a symmetry group is $U(1)$ with the action $A(t) \to A(t)e^{i\phi(t)\lambda_8}$. For non-maximal solitons, we generally expect such gauge symmetries simply because Λ_i's do not act irreducibly of N_f flavours so that there are Lie subgroups of $SU(N_f)$ which commute with Λ_i. The static solutions based on these Λ_i are expected to be invariant under such subgroups as well.

For the maximal (N_f, N_f) solitons, the situation is different. For instance, for the $SO(3)$ soliton for $N_f = 3$, direct inspection of V_c shows that no continuous subgroup of $SU(3)$ leave V_c invariant. A more general argument to demonstrate this result is as follows. The gauge symmetries on the collective coordinates are of the form $A(t) \to A(t)h(t)$ where $h(t)$ commutes with the static solution W. The generalized spherical symmetry of W implies that it is a polynomial in $\vec{\Lambda} \cdot \hat{x}$ of the type $\sum_n \alpha_n(r)(\vec{\Lambda} \cdot \hat{x})^n$. Thus, by considering different values of $\hat{x}$, we infer that $[h(t), \Lambda_i] = 0$. For maximal solitons, Λ_i's generate an irreducible representation of $SO(3)$. Hence, $h(t)$ is a multiple of the identity. Since $h(t) \in SU(N_f)$, it follows that $h(t) \in \mathbf{Z}_{N_f}$ and that $h(t)$ is in fact independent of t since $\mathbf{Z}_{N_f}$ is discrete. Thus, there is no gauge (that is, time dependent) symmetry for maximal solitons.

Consider again the case of non-maximal solitons with the gauge symmetry

$$A(t) \to A(t)e^{i\theta_\alpha(t)T_\alpha(t)} \quad . \tag{17.52}$$

Let K_α be the quantum mechanical operators for the infinitesimal generators of the symmetry. Then K_α can be calculated from $L(A, \dot{A})$ and are constrained by equations of the form

$$K_\alpha - \xi_\alpha \approx 0 \quad , \tag{17.53}$$

where ξ_α's are numbers determined by the fact that under the change (17.52) for small θ_α, the variation δL of L looks like $d[\theta_\alpha(t)\xi_\alpha]/dt$. In the absence of L_{WZ}, ξ_α actually vanish. In any event, we know from the general theory of constraint dynamics (cf. Chapter 2) that we can interpret this equation as a constraint on the states:

$$(K_\alpha - \xi_\alpha)\psi = 0 \quad . \tag{17.54}$$

This condition singles out the allowed family of states and hence the method of quantization, as we saw in detail for the $B = 1$ soliton. For maximal solitons, gauge symmetries do not exist, and hence this condition does not hold.

We shall finally explain why L_{WZ} is zero for maximal solitons for which Λ_i carries "spin" $S = (N_f - 1)/2$ for N_f flavours. The structure of L_{WZ} for any number of flavours is of the form (16.24). Now the condition of spherical symmetry says that a spatial rotation of the ansatz W can be compensated by a rotation generated by Λ_i. It follows, as in the derivation of (16.28), that

$$[\Lambda_i \ , \ \int \omega^3] = 0 \quad . \tag{17.55}$$

For the maximal soliton, for which Λ_i act irreducibly on N_f flavours, this equation implies that

$$\int \omega^3 = \ c \times \mathbf{1} \quad , \tag{17.56}$$

where c is a constant. But then in (16.24), we get zero for $\text{Tr}[\omega^3 A^{-1}\dot{A}]$ since $\text{Tr}[A^{-1}\dot{A}] = 0$, $A^{-1}\dot{A}$ being valued in the Lie algebra of $SU(N_f)$. Similarly, the remaining term in L_{WZ}, and hence L_{WZ} itself, are seen to vanish for maximal solitons.

The quantization ambiguities we have discussed in this Chapter are in a large part the ambiguities which appear when the configuration space is multiply connected. A systematic discussion of such ambiguities is contained in Chapter 8.

Chapter 18

SKYRMION PHENOMENOLOGY

18.1 INTRODUCTION

There has been extensive work on the phenomenological aspects of the Skyrme model, including several reviews [58,122]. Since the aim of our discussion of the Skyrme model is primarily to discuss its mathematical aspects, we will not make any attempt at a comprehensive review of the phenomenological work which has been done. Rather, we refer the reader to the literature.

On the other hand, it is important to make some contact with experiment. With this in the mind, we shall discuss some of the most basic phenomenological results of the Skyrme model. Aspects which will not be covered here are the many modifications and extensions of the model which have been pursued. For instance, models have been constructed (for a review see Ref. [123]) where more meson fields and more interaction terms have been included in the Lagrangian. Although they often yield phenomenological improvements over the "minimal" Skyrme model [that is, the system given by the Lagrangian density (12.15) and the Wess-Zumino term (16.31)], the results tend to be inconclusive, since new parameters must be introduced into the theory. A systematic method of approximation whereby one could compute the values of these parameters (starting from QCD) has not yet been established for the Skyrme model.

In Chapter 14, we gave a brief discussion of baryon masses and charges in the two-flavoured Skyrme model. In Section 2 of this Chapter, we add a discussion of certain static properties of baryons like radii and magnetic moments in the "minimal" Skyrme model for two flavours. This discussion closely follows the work of Adkins, Nappi and Witten [116,122].

In Section 3 of this Chapter, we discuss baryon masses in the three-flavoured Skyrme model. Formulae for octet and decouplet masses were already found in Chapter 15. Here we also add $SU(3)$ flavour symmetry breaking effects to these formulae in order to determine the individual baryon masses. This discussion closely follows the work of Guadagnini [132].

18.2 STATIC PROPERTIES IN THE TWO-FLAVOURED MODEL

Following Adkins, Nappi and Witten [116], we shall compute mean square radii and magnetic moments for the Skyrmion. The radii we shall compute utilize the zeroth component of baryon current $J_\mu(x)$ and the electromagnetic current $\tilde{J}_\mu^{em}(x)$. The radius associated with the baryon current is denoted by $< r^2 >_{I=0}^{1/2}$. It is obtained from

$$< r^2 >_{I=0} = \int d^3x \; \vec{x}^2 J_0(\vec{x}) \quad , \tag{18.1}$$

where $J_0(\vec{x})$ is defined in (13.10). We denote the radius associated with the electromagnetic current by $< r^2 >_{em}^{1/2}$. It will be defined below in Eq.(18.4).

Equation (18.1) is a functional of $U(\vec{x},t)$, for which we make the usual semiclassical ansatz

$$U(\vec{x},t) = A(t)U_c(\vec{x})A(t)^\dagger \quad . \tag{18.2}$$

Upon substituting (18.2) in (18.1), the variable $A(t)$ drops out, indicating that for this approximation the value of $< r^2 >_{I=0}$ is the same for the nucleon and delta. Its numerical value can be easily computed:

$$< r^2 >_{I=0} = \frac{2}{\pi} \int_0^\infty dr \; r^2 \frac{d\theta}{dr} \sin^2 \theta$$

$$\approx \frac{4.47}{e^2 F_\pi^2} \quad . \tag{18.3}$$

With the values of F_π and e determined from the experimental values of the N and Δ masses [Cf. Eq. (14.4)], we arrive at the result $< r^2 >_{I=0}^{1/2} = 0.58$ fm. This is to compared with .72 fm, which is the measured $I = 0$ rms radius of the nucleon.

The semiclassical approximation yields the same radius for N and Δ. But unfortunately this prediction can not be compared with experiment, the rms radius of Δ being unknown.

The mean square "electric" radius is obtained from

$$< r^2 >_{em} = \int d^3x \; \vec{x}^2 \tilde{J}_0^{em}(\vec{x}) \quad . \tag{18.4}$$

From Section 16.5, we know that $\tilde{J}_\mu^{em}(x)$ has contributions $J_\mu^{em}(x)$ from the original Skyrme model Lagrangian, as well as $J_\mu^{WZ}(x)$ coming from the Wess-Zumino term. The latter is given by $\frac{1}{2}J_\mu(x)$ if we substitute $q_1 + q_2 = \frac{1}{3}$ and $N_C = 3$ in (16.40). Thus

$$< r^2 >_{em} = \int d^3x \; \vec{x}^2 J_0^{em}(\vec{x}) + \frac{1}{2} < r^2 >_{I=0} \quad . \tag{18.5}$$

Now upon substituting the semiclassical approximation in (18.5), we get

$$< r^2 >_{em} = \frac{i}{2}\tilde{a}[U_c] \; \text{Tr} \; \dot{A}A^\dagger Q + \frac{1}{2} < r^2 >_{I=0}$$

$$= \frac{\tilde{a}[U_c]}{a[U_c]}L_3 + \frac{1}{2} < r^2 >_{I=0} \quad . \tag{18.6}$$

Here

$$\tilde{a}[U_c] = \frac{4\pi}{3} \int_0^\infty dr \; r^4 \sin^2\theta \left[F_\pi^2 + \frac{4}{e^2}\left(\theta'^2 + \frac{1}{r^2}\sin^2\theta \right) \right] \quad , \tag{18.7}$$

while $a[U_c]$ was defined in (13.42) and L_3 is as in Eq. (16.49). It is easy to check that $\tilde{a}[U_c]$ is linearly divergent when U_c is the classical solution. This is so since

$$\theta \approx \pi - c/r^2 \quad \text{as} \quad r \to \infty \quad . \tag{18.8}$$

Consequently, we are unable to give a numerical value for $< r^2 >^{1/2}_{em}$. Note in any case that the answer depends on the third component of isospin.

The $1/r^2$ behavior in Eq. (18.8) and consequently the divergence in (18.7) is due to the assumed masslessness of the pion fields. The situation is cured with the addition of a mass term such as (12.19) to the Lagrangian. This addition was first performed by Adkins and Nappi [133]. [We refer the reader to the reference for full details.] Using the nucleon, delta and pion masses as inputs, they arrived at the result

$$\tilde{a}[U_c]/a[U_c] \approx (1.04 \text{ fm})^2 \quad . \tag{18.9}$$

From Eq. (18.6), the value of $\tilde{a}[U_c]/a[U_c]$ is also obtained by taking the difference of the experimental values for $< r^2 >_{em}$ for the proton and neutron. The result is $\approx (.88 \text{ fm})^2$.

One can obtain the magnetic moment from the space components of J_μ and $\tilde{J}^{em}_\mu$. They are

$$\vec{\mu} = \int d^3x \ \vec{\rho}(\vec{x}) \quad \text{and} \quad \vec{\mu}^{em} = \int d^3x \ \vec{\rho}^{\ em}(\vec{x}) \quad , \tag{18.10}$$

respectively, where

$$\vec{\rho}(\vec{x}) = \frac{1}{2}\vec{x} \times \vec{J}(\vec{x}) \quad \text{and} \quad \vec{\rho}^{\ em}(\vec{x}) = \frac{1}{2}\vec{x} \times \vec{J}^{em}(\vec{x}) \quad . \tag{18.11}$$

We first evaluate $\vec{\mu}(\vec{x})$. In terms of collective coordinates $A(t)$,

$$J_i(\vec{x}) = \frac{1}{18\pi^2}\varepsilon_{ijk} \text{Tr } A^\dagger \dot{A}\{[U_c^\dagger \partial_j U_c, U_c^\dagger \partial_k U_c] - [\partial_j U_c U_c^\dagger, \partial_k U_c U_c^\dagger]\} \tag{18.12}$$

while $\rho_i(\vec{x})$ takes the form

$$\rho_i(\vec{x}) = i\nu_{ij}[U_c] \text{ Tr } \tau_j A^\dagger \dot{A} \quad . \tag{18.13}$$

For the spherically symmetric ansatz,

$$\nu_{ij}[U_c] = -\frac{1}{4\pi^2}\theta' \sin^2\theta(\hat{x}_i\hat{x}_j - \delta_{ij}) \quad . \tag{18.14}$$

Upon substituting (8.13) and (8.14) in the expression for $\vec{\mu}$ in (8.10), we find

$$\mu_i = \frac{i}{3} < r^2 >_{I=0} \text{Tr } \tau_i A^\dagger \dot{A} = \frac{4}{3}\frac{< r^2 >_{I=0}}{a[U_c]}R_i \quad . \tag{18.15}$$

In quantum theory, we replace R_i by the operator $\hat{R}_i$. $\hat{R}_i$ is the generator of right translations on A. The former was shown to be the spin $\vec{S}$ in Chapter 13. We thus recover the known result that the magnetic moment is proportional to spin. We can now identify the g-factor from its definition:

$$\vec{\mu} = \frac{g_N^{I=0}}{2m_N}\vec{S} \quad . \tag{18.16}$$

Thus,

$$g_N^{I=0} = \frac{8m_N < r^2 >_{I=0}}{3a[U_c]} \quad . \tag{18.17}$$

Upon substituting the values (14.4) for F_π and e and the experimental value for m_N, we get $g_N^{I=0} \approx 1.06$. This is to be compared with the known value of 1.76.

Now we evaluate μ_i^{em}. Once again we use

$$\tilde{J}_\mu^{em}(\vec{x}) = J_\mu^{em}(\vec{x}) + \frac{1}{2}J_\mu(\vec{x}) \quad . \tag{18.18}$$

In the semiclassical approximation,

$$\begin{aligned}
\tilde{J}_i^{em} &= i(A^\dagger Q A)_\alpha \Big\{ (\partial_i U_c U_c^\dagger)_\beta \, \rho_{\alpha\beta}[U_c] \\
&\quad - \frac{1}{8e^2}(A^\dagger \dot{A})_\beta (A^\dagger \dot{A})_\gamma \, \sigma_{\alpha\beta\gamma}[U_c] \Big\} + \frac{1}{2}J_i \quad ,
\end{aligned} \tag{18.19}$$

where we have used the notation $(N)_\alpha = \frac{1}{2}\text{Tr } \tau_\alpha N$ for any 2×2 matrix N and

$$\begin{aligned}
\rho_{\alpha\beta}[U_c] &= \frac{F_\pi^2}{8}(2\delta_{\alpha\beta} - \text{Tr } U_c^\dagger \tau_\beta U_c \tau_\alpha) \\
&\quad - \frac{1}{8e^2}\text{Tr}[\tau_\alpha, \partial_j U_c U_c^\dagger][\tau_\beta - U_c^\dagger \tau_\beta U_c, \partial_j U_c U_c^\dagger] \quad ,
\end{aligned}$$

$$\sigma_{\alpha\beta\gamma}[U_c] = Tr[\tau_\gamma - U_c\tau_\gamma U_c^\dagger , \tau_\alpha] \times$$
$$\times [\partial_i U_c U_c^\dagger - U_c^\dagger \partial_i U_c , \tau_\beta - U_c\tau_\beta U_c^\dagger] \quad . \tag{18.20}$$

This yields

$$\mu_i^{em} = \frac{1}{4}a[U_c](A^\dagger Q A)_i + d[U_c]\{Tr(A^\dagger \dot{A})^2(A^\dagger Q A)_i$$
$$- Tr(\dot{A}A^\dagger Q)(A^\dagger \dot{A})_i\} + \frac{1}{2}\mu_i \quad , \tag{18.21}$$

$$d[U_c] = \frac{2\pi}{3e^3}\int_0^\infty dr \ r^2 \ \sin^4\theta(r) \quad . \tag{18.22}$$

For obtaining (18.22), we once again have assumed the spherically symmetric form for U_c. It can be argued that the second term in (18.21) is negligible compared to the other terms. This is because after writing it in terms of the left- and right-handed generators L_α and R_α, it is seen to be suppressed by a factor of $d[U_c]/(a[U_c])^3 \sim (e/158)^3$ in comparison to the first term in (8.21). Thus, we have the approximate expression

$$\mu_3^{em} \approx \frac{1}{8}a[U_c](\hat{A}^\dagger \tau_3 \hat{A})_3 + \frac{1}{2}\mu_3 \quad . \tag{18.23}$$

To complete the calculation, we need to evaluate matrix elements of the operator $(\hat{A}\tau_3\hat{A})^\dagger)_3$ between physical states of the form $|J, I_3, J_3 >$, where J is total isospin or angular momentum and I_3 and J_3 are third components of isospin and angular momentum. Such a matrix element is thus

$$< J, I_3, J_3|(\hat{A}\tau_3\hat{A}^\dagger)_3|J', I_3', J_3' > .$$

We shall utilize the functions $\{D_{I_3,J_3}^J\}$ introduced in Chapter 13 to evaluate this expression. These functions are normalized according to

$$(D_{I_3,J_3}^J, D_{I_3',J_3'}^{J'}) = \int_{SU(2)} d\mu(A)D_{I_3,J_3}^J(A)^\star D_{I_3',J_3'}^{J'}(A)$$
$$= \frac{1}{2J+1}\delta_{JJ'} \ \delta_{I_3'I_3} \ \delta_{J_3J_3'}$$
$$\equiv \frac{1}{2J+1} < J, I_3, J_3|J', I_3', J_3' > \quad . \tag{18.24}$$

The appropriate matrix elements for the nucleon are

$$< \frac{1}{2}, I_3, J_3 | (\hat{A}\tau_3 \hat{A}^\dagger)_3 | \frac{1}{2}, I_3, J_3 >$$

$$= 2 \int_{SU(2)} d\mu(A) |D^{1/2}_{I_3,J_3}(A)|^2 (A^\dagger \tau_3 A)_3$$

$$= 2 \int_{SU(2)} d\mu(A) |D^{1/2}_{I_3,J_3}(A)|^2 D^1_{0,0}(A) \quad . \tag{18.25}$$

The third line in (18.25) follows because (i) $R_{ij} = \frac{1}{2} Tr \, \tau_j (A^\dagger \tau_i A) \equiv (A^\dagger \tau_i A)_j$ defines a two to one map from $SU(2)$ to $SO(3)$ or, equivalently, the matrix R defines the $J = 1$ representation of $SU(2)$ and (ii) $(A^\dagger \tau_3 A)_3$ is invariant when A undergoes left and right multiplication by $\exp(i\alpha\tau_3)$, i.e.

$$(e^{-i\alpha\tau_3} A^\dagger \tau_3 A e^{i\alpha\tau_3})_3 = (A^\dagger e^{-i\alpha\tau_3} \tau_3 \, e^{i\alpha\tau_3} A)_3$$

$$= (A^\dagger \tau_3 A)_3 \tag{18.26}$$

for any number α. Hence $(A^\dagger \tau_3 A)_3$ is a $J_3 = I_3 = 0$ eigenstate.

Finally, using (18.24) and

$$D^J_{I_3 J_3} D^{J'}_{I'_3 J'_3} = \sum_{J'', J''_3, I''_3} C^{I_3 I'_3 I''_3}_{J J' J''} \, C^{J_3 J'_3 J''_3}_{J J' J''} \, D^{J''}_{I''_3 J''_3} \quad , \tag{18.27}$$

where the C's are Clebsch-Gordan coefficients, we get

$$< \frac{1}{2}, I_3, J_3 | (\hat{A}\tau_3 \hat{A}^\dagger)_3 | \frac{1}{2}, I_3, J_3 > = \frac{4}{3} I_3 J_3 \tag{18.28}$$

or more generally,

$$< \frac{1}{2}, I_3, J_3 | (\hat{A}\tau_i \hat{A}^\dagger)_j | \frac{1}{2}, I'_3, J'_3 > = \frac{4}{3} < \frac{1}{2}, I_3, J_3 | \hat{L}_j \hat{R}_i | \frac{1}{2}, I'_3, J'_3 > \quad . \tag{18.29}$$

The magnetic moment for the nucleon is thus

$$\mu^{em}_3 \approx \frac{1}{6} a[U_c] \hat{I}_3 \hat{J}_3 + \frac{1}{2} \mu_3$$

$$\approx (6.40 \, \hat{I}_3 + .53) \frac{\hat{J}_3}{2m_N} \quad . \tag{18.30}$$

In computing $a[U_c]$, we have used the values of F_π and e given in (14.4). For the proton and neutron, we get 1.86 and -1.33 respectively in units of nuclear magnetons $(\frac{1}{2m_N})$. This is to be compared with the experimental values of 2.79 and -1.91.

Since we have an expression for the magnetic moment operator (18.23), we can also compute the off-diagonal matrix elements

$$\mu_{N\Delta} = <\frac{3}{2}, I_3, J_3 | \mu_3^{em} | \frac{1}{2}, I_3, J_3 > \quad . \tag{18.31}$$

Such matrix elements are known experimentally from electromagnetic decays of the Δ. For computing (18.31), we use

$$< \frac{3}{2}, I_3, J_3 | (\hat{A}^\dagger \tau_3 \hat{A})_3 | \frac{1}{2}, I_3, J_3 >$$

$$= 2\sqrt{2} \int_{SU(2)} d\mu(A) D_{I_3 J_3}^{3/2}(A)^\star D_{00}^1(A) D_{I_3 J_3}^{1/2}(A) = \sqrt{2}/3 \tag{18.32}$$

and find

$$\mu_{N\Delta} = \frac{a[U_c]}{12\sqrt{2}} \quad . \tag{18.33}$$

Upon substituting the values (14.4) for F_π and e, we have $\mu_{N\Delta} \approx 2.26$ (nuclear magnetons) as opposed to the experimental value of 3.3. On the other hand, from (18.30) and (18.33), we get the relation

$$\frac{\mu_{N\Delta}}{\mu_{proton} - \mu_{neutron}} = \frac{1}{\sqrt{2}} \quad , \tag{18.34}$$

independent of the parameters F_π and e. This relation agrees very well with experiment, as the experimental value for the ratio in (18.34) is $.70 \pm 0.1$. In Ref. [116], other such parameter independent relations were found.

In the previous discussion, we have computed the parameter dependent quantities using the values for F_π and e determined from the N and Δ masses. This was the procedure followed in Ref. [116]. Alternatively, we can perform a least square fit of the static properties to determine F_π and e. The results are

	Theory		Experiment	
m_N =	953 MeV		939	MeV
m_Δ =	1153	-"-	1232	-"-
$< r^2 >_{I=0}^{1/2}$ =	.686 fm		.72	fm
μ_{proton} =	2.64 nuclear magnetons		2.79	nuclear magnetons
$\mu_{neutron}$ =	−2.13 -"-	-"-	−1.91 -"-	"
$\mu_{N\Delta}$ =	3.37 -"-	-"-	3.3 "	-"-
$g_{\pi NN}$ =	11.5 -"-	-"-	13.5 -"-	-"-

$$(18.35)$$

The theoretical value for the pion-nucleon coupling constant $g_{\pi NN}$ in (18.35) is obtained from the expression

$$g_{\pi NN} \approx 12.8 \, \frac{m_N}{F_\pi e^2} \quad , \tag{18.36}$$

derived in Ref. [116]. We shall not repeat this derivation here, but instead refer the reader to the reference.

The results of (18.35) are obtained with the parameters

$$F_\pi \approx 123 \text{ MeV} \quad , \quad e \approx 4.95 \quad . \tag{18.37}$$

Based on just these two parameters, we thus obtain a fit of the static properties which is, on the average, better than 10%. However, the value for F_π is once again quite far from the experimental value of 186.4 MeV.

18.3 BARYON MASSES IN THE THREE-FLAVOURED MODEL

In Chapter 15, we computed the masses associated with the $SU(3)$ octet and $SU(3)$ decouplet of baryons in the absence of $SU(3)$ breaking [see (15.64)].

As stated in Chapter 12 this is unrealistic in light of the C quark "current algebra" mass. We now suitably extend the "minimal" Skyrme model by including explicit $SU(3)$ breaking effects in order to obtain the individual baryon masses.

A natural term to add to the Lagrangian density to account for $SU(3)$ breaking is

$$\mathcal{L}_I = \frac{F_\pi^2}{16} \text{Tr}\{MU + M^\dagger U^\dagger - M - M^\dagger\} \quad , \tag{18.38}$$

where M is a diagonal matrix whose elements are associated with the masses of the pseudoscalar octet mesons. For our purposes, it suffices to consider $SU(3)$ flavour symmetry breaking to $SU(2)$. Thus for M, we take

$$M = (\mu_1, \mu_1, \mu_2)_{diagonal} \quad , \tag{18.39}$$

where the constants μ_i can be expressed in terms of the pion and kaon masses

$$\mu_1 = \frac{1}{2}m_\pi^2 \quad , \qquad \mu_2 = \frac{1}{2}(m_K^2 - m_\pi^2) \quad . \tag{18.40}$$

Eq. (18.40) can be verified by expanding the total action to second order in the pseudoscalar octet meson fields. In this way, one also recovers the familiar Gell-Mann-Okubo mass formula [130] for the pseudoscalar meson octet:

$$3m_\eta^2 = 4m_K^2 - m_\pi^2 \quad . \tag{18.41}$$

The contribution H_I of $\mathcal{L}_I$ to the effective Hamiltonian of the solitonic configuration $U_c(\vec{x})$ is, as usual, found by substituting $U = A(t)U_c(\vec{x})A(t)^\dagger$ in $\mathcal{L}_I$, and integrating over $\vec{x}$. The spectrum for the octet and decouplet baryons can now be computed treating H_I as a perturbation of the Hamiltonian H [Cf. Eq. (15.55)]. The results, however, are known to agree rather poorly with experiment [134].

In order to improve agreement with experiment, in addition to the term H_I, Guadagnini [132] included the term

$$H_I' = -K(U_c)\hat{Y} \tag{18.42}$$

in the total Hamiltonian. Here $\hat{Y}$ is the hypercharge operator and $K(U_c)$ is a functional of U_c. The numerical value for $K(U_c)$ for the configuration that minimizes the classical energy is to be determined phenomenologically. Like H_I, H'_I serves to break the flavour $SU(3)$ group to $SU(2)$.

The Lagrangian associated with H'_I will necessarily involve derivatives of U. Its origin is not as clear as the origin of H_I. A justification for including the interaction (18.42) in the total Hamiltonian is that not all of the $SU(3)$ symmetry breaking effects are due to differences in the pseudoscalar meson masses. Rather $SU(3)$ symmetry breaking involving derivatives of U are also in general permitted in the effective Lagrangian.

We now proceed with the semiclassical approximation. After substituting $U(\vec{x},t) = A(t)U_c(\vec{x})A(t)^\dagger$ in $\mathcal{L}_I$ and performing a spatial integration, we find that H_I has the following form:

$$H_I = -\frac{1}{3}C(U_c)\{(\mu_2 - \mu_1)\,\mathrm{Tr}\,(\lambda_8 \hat{A}\lambda_8 \hat{A}^\dagger) - 2(2\mu_1 + \mu_2)\}\ . \qquad (18.43)$$

For the Skyrme solution, we find

$$C(U_c) = \frac{\pi F_\pi^2}{2}\int_0^\infty dr\ r^2[1 - \cos\theta(r)] \approx \frac{26.7}{F_\pi e^3}\ \ . \qquad (18.44)$$

The eigenstates of the Hamiltonian in the absence of $H_I + H'_I$ are the D-functions discussed in Chapter 15. They are normalized according to

$$\begin{aligned}
&(D^{(p,q)}_{(I,I_3,Y)(J,J_3,Y')}\ ,\ D^{(p',q')}_{(I',I'_3,Y'')(J',J'_3,Y''')})\\
&= \frac{1}{d_{p,q}}\delta_{(p,q)(p',q')}\delta_{(I,I_3,Y)(I',I'_3,Y'')}\delta_{(J,J_3,Y')(J',J'_3,Y''')}\ , \qquad (18.45)
\end{aligned}$$

where $d_{p,q}$ is the dimension of the representation $D^{(p,q)}$. In first order perturbation theory, the contribution of H'_I to the level shift is $-K(U_c)$ times the hypercharge of the state, while the contribution of H_I is governed by the matrix element

$$\begin{aligned}
&\int_{SU(3)} d\mu(A)\,|D^{(p,q)}_{(I,I_3,Y)(J,J_3,1)}(A)|^2\,Tr(\lambda_8 A\lambda_8 A^\dagger)\\
&= 2\int_{SU(3)} d\mu(A)\,|D^{(p,q)}_{(I,I_3,Y)(J,J_3,1)}|^2\,D^{(1,1)}_{(0,0,0)(0,0,0)}(A) \qquad (18.46)
\end{aligned}$$

which can be evaluated using known Clebsch-Gordan coefficients. The results for the masses of the octet and decouplet states are

$$M_N = M_8' - \frac{1}{5}C(U_c)(\mu_2 - \mu_1) - K(U_0) ,$$

$$M_\Lambda = M_8' - \frac{1}{15}C(U_c)(\mu_2 - \mu_1) ,$$

$$M_\Sigma = M_8' + \frac{1}{15}C(U_c)(\mu_2 - \mu_1) ,$$

$$M_\Xi = M_8' + \frac{2}{15}C(U_c)(\mu_2 - \mu_1) + K(U_0) ,$$

$$M_\Delta = M_{10}' - \frac{1}{12}C(U_c)(\mu_2 - \mu_1) - K(U_0) ,$$

$$M_{\Sigma^\star} = M_{10}' ,$$

$$M_{\Xi^\star} = M_{10}' + \frac{1}{12}C(U_c)(\mu_2 - \mu_1) + K(U_0) ,$$

$$M_\Omega = M_{10}' + \frac{1}{6}C(U_c)(\mu_2 - \mu_1) + 2K(U_0) , \qquad (18.47)$$

where

$$M_8' = M_8 + \frac{2}{3}C(U_c)(2\mu_1 + \mu_2) ,$$

$$M_{10}' = M_{10} + \frac{2}{3}C(U_c)(2\mu_1 + \mu_2) \qquad (18.48)$$

and M_8 and M_{10} are given by (15.64). Good agreement with experiment was found for the following values of the parameters:

$$M_8' = 1151 \text{ MeV}, \qquad M_{10}' = 1382 \text{ MeV} ,$$

$$C(U_c)(\mu_2 - \mu_1) = 581 \text{ MeV}, \quad K(U_c) = 96 \text{ MeV} . \qquad (18.49)$$

With these parameters, the following results are obtained for the baryon masses:

	Theory	Experiment	
$M_N =$	938.8 MeV	938.9 MeV	
$M_\Lambda =$	1112.3 MeV	1115.6 MeV	
$M_\Sigma =$	1189.7 MeV	1193.1 MeV	
$M_\Xi =$	1324.5 MeV	1318.1 MeV	(18.50)
$M_\Delta =$	1237.6 MeV	1232 MeV	
$M_{\Sigma^\star} =$	1382 MeV	1383.9 MeV	
$M_{\Xi^\star} =$	1526.4 MeV	1533.4 MeV	
$M_\Omega =$	1670.8 MeV	1672.2 MeV	

Let us fix the pion mass m_π at the experimental value of 135 MeV. Then since M_8, M_{10} and $C(U_c)$ are known in terms of F_π and e (see (15.64), (15.25) and (18.44)), and since μ_i are given in terms of m_π and m_K (see (18.40)), the two equations in (18.48) and the values of M_8', M_{10}' and of $C(U_c)(\mu_2 - \mu_1)$ in (18.49) give three equations for the three unknowns m_K, F_π and e. We find

$$m_K \approx 622 \text{ MeV} \quad , \quad F_\pi \approx 62.2 \text{ MeV} \quad , \quad e \approx 6.49 \quad . \tag{18.51}$$

The value for m_K here is approximately 25% off from the experimental value of 494 MeV while the value obtained here for F_π is considerably worse than the two-flavour result. This is due to the fact that additional (positive) ground state energy corrections have been included in the octet and decouplet masses M_8' and M_{10}'. Specifically, in going from the two- to the three-flavour case, the spin 1/2 and spin 3/2 multiplets receive an additional contribution 3/b [Cf. Eq. (15.64)]. Further positive contributions also entered when we included meson masses through the term H_I. The problem seems to be due to the addition of zero-point corrections in the estimation of the total energy. Note also that for the values for F_π and e found in (18.51), the classical energy for the $B = 1$ soliton is found to be ≈ 349 MeV which represents only a small percentage of the $B = 1$ octet and decouplet masses. The quantum corrections in (15.64), (18.47) and (18.48) are therefore making the main contributions to the $B = 1$ masses. The validity of the semiclassical approximation is thus in doubt. Furthermore, it is unreasonable to expect that zero point corrections due to modes which have been neglected in the above treatment are small.

The above discussion indicates that it may be too ambitious to attempt to compute absolute masses of baryons, since for instance zero-point corrections can in general be large. It may thus be more reasonable to compute mass differences.

As an illustration of this, we will see what happens if we try to use the values for m_K, F_π and e found in (18.51) to determine the absolute mass of the singlet dibaryon discussed in Chapter 17. We then compare this procedure with the one where we compute the dibaryon mass relative to baryon masses.

The expression for the singlet $SO(3)$ classical soliton mass (uncorrected for $SU(3)$ breaking effects) is given in (17.27) and (18.43). It remains to evaluate the expression for H_I for the $SO(3)$ soliton and its contribution to the singlet mass. (It is not at all clear what the analogue of H_I' is for the dibaryon and we thus choose to ignore it.)

Upon substituting $U(\vec{x}, t) = A(t)V_c(\vec{x})A(t)^\dagger$ into $\mathcal{L}_I$ integrating over $\vec{x}$, it is found that the dependence on $A(t)$ drops out because, as a simple calculation shows,

$$\int d^3x V_c(\vec{x})$$ (18.52)

is proportional to the unit matrix. The result for H_I is

$$H_I = (2\mu_1 + \mu_2)\gamma(V_c) \quad,$$

$$\gamma(V_c) = \frac{\pi F_\pi^2}{6} \int_0^\infty dr \, r^2(3 - \cos\psi - 2\cos\chi\cos\psi/2) \quad, \quad (18.53)$$

where we used the spherically symmetric form (17.7) for V_c. For the configuration V_c which minimizes the classical static energy, we find

$$\gamma(V_c) \approx \frac{26.9}{F_\pi e^3} \quad. \quad (18.54)$$

We can now calculate the singlet dibaryon mass (including the $SU(3)$ breaking effects) when the values of the parameters are as in (18.51). The answer is 1030 MeV. Thus, not only are the results (18.51) in poor agreement with experiment, but they also lead to an unreasonable value for the singlet dibaryon mass.

Now, instead of using (17.27) and (18.54) to determine the absolute mass of the singlet dibaryon, let us compute the mass differences of the dibaryon with the $B = 1$ Skyrmion states. In other words, we shall add an overall constant X to both the $B = 1$ and the $B = 2$ masses. Physically, X is supposed to represent zero point energy corrections from all neglected modes. It will be determined phenomenologically from the $B = 1$ masses. [In general, zero point corrections for the $B = 1$ and $B = 2$ states will differ, but we must ignore this fact since we wish to make a prediction for the dibaryon mass.]

Using the values for M'_8 and M'_{10} found in (18.49) and the experimental value of 132 MeV for F_π, we obtain

$$X = -1194 \text{ MeV} \quad , \quad e = 4.47 \quad , \quad m_K = 622 \text{ MeV} \quad , \qquad (18.55)$$

where again we have used the experimental value $m_\pi = 135$ MeV. (In the case of some soliton examples in two-space time dimensions, Rajaraman [70] has shown that zero-point corrections can be negative in the presence of solitons.) With the values in (18.55), the classical energy $m_{B=1}$ of the $B = 1$ baryons is now 1544 MeV.

Next, we substitute the value for e from (18.55) into the sum of (17.27) and (18.53), using the experimental values of F_π and m_K. To this we add the value of X obtained in (18.55) to arrive at the following value for the H-baryon mass:

$$M_H = 2100 \text{ MeV} \quad . \qquad (18.56)$$

This is to be compared with the value of 2150 MeV originally found by Jaffe [130] in the quark model.

Chapter 19

ELECTROWEAK SKYRMIONS

19.1 INTRODUCTION

In previous Chapters, the chiral group was implemented as a global symmetry group; that is the action was invariant under a global $SU(N)_L \times SU(N)_R$. In this Chapter, we shall consider systems where part of the chiral group is implemented as a local symmetry. Specifically, we shall study the two-flavour chiral model possessing the additional feature that the $SU(2)_L$ group is "gauged". We shall refer to it as the "gauged chiral model".

We shall examine certain soliton solutions in the gauged chiral model. They will be the analogues of Skyrmions in the chiral model. Although the two models appear quite similar, the corresponding solitons are distinct in several respects. For example, the solitons of the gauged chiral model are not topological. Thus, this Chapter is in contrast with the previous Chapters on solitons. Nevertheless, due to the similarities of the gauged chiral model with the usual chiral model, and the possible physical relevance of the former, we feel that the discussion which follows is appropriate.

We now briefly discuss the possible physical relevance of the gauged chiral model. The fundamental fields of this model will not be interpreted as pions, and the solitons will not be interpreted as known baryons or multibaryons. Instead, we are interested in a possible connection of this model with the Glashow-Salam-Weinberg (GSW) theory of weak interactions [135]. [For a review of the latter, see e.g. Ref. [136].] The connection is that in the limit of an infinite Higgs self-coupling constant λ and Higgs mass μ, the bosonic fields of the GSW theory are described by a gauged chiral model with two flavours. The experimental values of λ and μ are not as yet known. Furthermore, the strict limit $\lambda = \mu = \infty$ cannot yet be ruled out. In any case, if λ and μ are large (but finite), the above model should serve as an approximation to the GSW theory. [In order that the model approximates the GSW theory it is also necessary that the parameter $\sin^2 \theta_W = 0$ in the GSW theory is zero or "small". This choice for the value of $\sin^2 \theta_W$ is convenient in constructing the model because it allows us to decouple the $U(1)_Y$ gauge field of the GSW theory, and as a result, only the $SU(2)_L$ subgroup of the chiral group is "gauged". Since, in nature, $\sin^2 \theta_W$ is small (≈ 0.227), the resulting model should give a reasonable first approximation to the true theory].

The possibility of finding Skyrmion-like solutions to the GSW theory was first pointed out by Gipson and Tze [115,121]. They, however, did not take into account the effects due to the gauge coupling. The latter is responsible for the nontopological nature of the soliton and was investigated in Refs. [137] and [138,139].

In the next section we review the strong coupling limit of the GSW theory while the the classical soliton solutions are exhibited in Section 3. A collective coordinate quantization is carried out in Section 4 while some phenomenological properties of the resulting quantum states are discussed in Section 5.

19.2 THE GSW THEORY AND THE STRONG COUPLING LIMIT

We begin with a review of the bosonic sector of the GSW theory. The bosonic fields are $SU(2)_L \times U(1)_Y$ gauge bosons (A_μ, B_μ) and a complex Higgs doublet $\Phi = (\phi_1, \phi_2)$. The Lagrangian density for the bosonic fields is

$$
\begin{aligned}
L &= L_A + L_B + L_\Phi \quad , \\
L_A &= -\frac{1}{8} \, Tr \, F_{\mu\nu} F^{\mu\nu} \quad , \qquad L_B = -\frac{1}{4} B_{\mu\nu} B^{\mu\nu} \quad , \\
L_\Phi &= (\mathcal{D}_\mu \Phi)^\dagger (\mathcal{D}^\mu \Phi) + \mu^2 \Phi^\dagger \Phi - \lambda (\Phi^\dagger \Phi)^2 \quad ,
\end{aligned}
\tag{19.1}
$$

where the covariant derivative and the field strengths are defined by

$$
\mathcal{D}_\mu \equiv \partial_\mu - \frac{ig}{2} A_\mu - \frac{ig'}{2} B_\mu \quad , \qquad A_\mu \equiv A_\mu^a \tau^a
$$

and

$$
\begin{aligned}
F_{\mu\nu} &= \partial_\mu A_\nu - \partial_\nu A_\mu - \frac{ig}{2} [A_\mu, A_\nu] \quad , \\
B_{\mu\nu} &= \partial_\mu B_\nu - \partial_\nu B_\mu \quad .
\end{aligned}
$$

Here g and g' are the $SU(2)_L$ and $U(1)_Y$ coupling constants respectively and $\mu^2, \lambda > 0$.

For our purposes, it is convenient to define the matrix

$$
M = \begin{pmatrix} \phi_2^\star & \phi_1 \\ -\phi_1 & \phi_2 \end{pmatrix} \quad ,
\tag{19.2}
$$

We can then make a polar decomposition $M = hU$, expressing the four real degrees of freedom in Φ in terms of a real field h and a field U which takes values in $SU(2)$. $h = \sqrt{\Phi^\dagger \Phi}$ is the field for the physical Higgs particle, while U describes the (unphysical) Goldstone bosons. Now, using (19.2), we can rewrite the Higgs Lagrangian density in (19.1) as follows:

$$
L_\Phi = -\frac{1}{2} h^2 \, Tr \, R_\mu R^\mu + \partial_\mu h \partial^\mu h + \mu^2 h^2 - \lambda h^4 \quad ,
$$

$$R_\mu = D_\mu U U^\dagger \quad .$$ (19.3)

Here when acting on U, the covariant derivative has the form

$$D_\mu U = \partial_\mu U - \frac{ig}{2} A_\mu U + \frac{ig'}{2} U \tau_3 B_\mu \quad .$$ (19.4)

As indicated earlier, we wish to examine the theory in the limit (i) $\sin^2 \theta_W \to 0$ and (ii) $\mu^2, \lambda \to \infty$, with μ^2/λ fixed. Limit (i) is equivalent to setting $g' \to 0$ in the covariant derivative. In addition to decoupling B_μ from the system, this limit will allow us to write down a consistent "spherically symmetric" ansatz for the remaining fields. Such an ansatz is not suitable when $g' \neq 0$, essentially due to the presence of τ_3 in the covariant derivative (19.4). We shall demonstrate this explicitly later. (ii) is the strong coupling limit. It "freezes out" the h degree of freedom. The latter then assumes its (finite) vacuum value $<h>$, where $<h>^2 = \mu^2/2\lambda$. In Section 5 , we will consider the effects of dropping the limits (i) and (ii).

With the limits (i) and (ii), the bosonic fields of the theory reduce to just A_μ and U. Their dynamics is described by the Lagrangian

$$L_\Phi + L_A = -\frac{1}{2} <h>^2 \text{ Tr } R_\mu R^\mu - \frac{1}{8} \text{ Tr } F_{\mu\nu} F^{\mu\nu} \quad ,$$ (19.5)

where g' in R_μ is zero. The first term in the Lagrangian is the gauged analogue of the nonlinear chiral model. Here $\sqrt{8} <h>$ plays the role of the pion decay constant F_π in the chiral model. In nature, the two scales differ by a factor of $\sqrt{8} <h> / F_\pi \cong 2610$. The second term in (19.5) gives dynamics to A_μ.

Under the action of the chiral group $G = SU(2)_L \times SU(2)_R$, U and A_μ transform according to

$$U \to V_L U V_R^\dagger$$
$$A_\mu \to V_L A_\mu V_L^\dagger \quad ,$$ (19.6)

where $(V_L, V_R) \in SU(2)_L \times SU(2)_R$. G has the subgroup $H = SU(2)_{diagonal} = \{(V, V)\}$ similar to the isospin group in the chiral model. We shall refer to it as "vector isospin" in the gauged chiral model [so as to distinguish it from weak isospin].

In addition to the global symmetry group G, there is, of course, a local symmetry group $\mathcal{G}$. The Lagrangian is invariant under the following action of $\mathcal{G} = \{v(x)\}$, $x = (\vec{x}, t)$:

$$U \to vU \quad ,$$
$$A_\mu \to vA_\mu v^\dagger - \frac{2i}{g}\partial_\mu vv^\dagger \quad . \tag{19.7}$$

Here $v(x)$ takes values in $SU(2)$. The transformation (19.7) indicates that all the degrees of freedom in U are indeed unphysical since we can choose $v = U^\dagger$ [corresponding to the unitary gauge]. Then the remaining physical fields are,

$$W_\mu \equiv U^\dagger A_\mu U + \frac{2i}{g}U^\dagger \partial_\mu U \quad . \tag{19.8}$$

The physical fields W_μ of the gauged chiral model take values in the Lie algebra of $SU(2)$. We denote this set of physical fields by M. Since M is topologically a trivial space, we expect that no topological solitons, and more generally no topological conservation laws, occur in this theory. (This is in contrast to the case of the [ungauged] chiral model.) Thus, any soliton we may find in this theory must be nontopological in character.

To elaborate on this point further, recall that in the two-flavour chiral model, the field U was required to go to a (direction independent) constant value at spatial infinity. This was due to the requirement of finite energy. One could then identify all points at spatial infinity which converted Euclidean three-space $\mathbf{R}^3 = \{\vec{x}\}$ to a three-sphere S^3. Further, the physical field U took values in $SU(2)$. Since $\pi_3(SU(2))$ is nontrivial, topologically disconnected field configurations for U resulted. These configurations were labelled by an integer $B[U]$. An explicit formula for $B[U]$ was given in (13.15).

In the "gauged" chiral model, the above is no longer the case. Finite energy now does not imply that U becomes a constant field at $|\vec{x}| \to \infty$. Instead, we have

$$D_i U \to 0 \qquad \text{as} \quad |\vec{x}| \to \infty \quad . \tag{19.9}$$

Since U need not go a constant at spatial infinity, the domain of U cannot be regarded as S^3: it must be regarded as $\mathbf{R}^3$. The space of mappings from $\mathbf{R}^3$

to $SU(2)$ is topologically trivial. Furthermore, $B[U]$ is no longer restricted to the set of integers, and any two configurations U_1 and U_2, with $B[U_1] = t_1$ and $B[U_2] = t_2$, are homotopic.

If we like, we can choose $U \to 1$ as $|x| \to \infty$ in the gauged chiral model, thereby restricting $B[U]$ to the set of integers. The value of this integer, however, is arbitrary since by the application of a suitable gauge transformation $U \to U' = gU$, we can change the integer $n = B[U]$ to any other integer $n' = B[U'] = B[U] + B[g]$. (For this we need $B[g] = B[U'] - B[U] \neq 0$.) The gauge group here can be taken to be $\{g\}$ where g is a function of $\vec{x}$ with values in $SU(2)$ and fulfills the condition $g(x) \to 1$ as $|\vec{x}| \to \infty$. Then $B[g] \neq 0$ means that the desired gauge transformation is not connected to the identity. Thus, the index $B[U]$ has no gauge invariant meaning.

Instead of $B[U]$, Gipson and Tze [111] introduced a gauge invariant quantity

$$\tilde{B}[U, A_\mu] = \frac{1}{24\pi^2}\varepsilon_{ijk} \int \mathrm{Tr}\,\left(R_i R_j R_k + \frac{3i}{4}gR_i F_{jk}\right)d^3x \quad . \tag{19.10}$$

In the limit $g \to 0$, $\tilde{B}[U, A^\mu] \to B[U]$. Although (19.10) is gauge invariant, it is not a constant of the motion. Note that $\tilde{B}[U, A^\mu]$ can be identified with the charge associated with the current

$$\tilde{k}^\mu[U, A_\mu] = \frac{1}{24\pi^2}\varepsilon^{\mu\nu\lambda\sigma} \,\mathrm{Tr}(R_\nu R_\lambda R_\sigma + \frac{3i}{4}gR_\nu F_{\lambda\sigma}) \quad . \tag{19.11}$$

This current is not conserved

$$\partial_\mu \tilde{k}^\mu[U, A_\mu] = -\frac{g^2}{128\pi^2}\varepsilon^{\mu\nu\lambda\sigma} \,\mathrm{Tr}\, F_{\mu\nu}F_{\lambda\sigma} \quad . \tag{19.12}$$

In addition to $\tilde{B}[U, A^\mu]$ not being a constant in time, its values are not discrete. Rather, $\tilde{B}[U, A^\mu]$ can take any real value and clearly can have no topological meaning. Thus, to conclude, for any two pairs of field $(U^1, A_\mu^{(1)})$ and $(U^2, A_\mu^{(2)})$, (whether or not $\tilde{B}[U^1, A_\mu^{(1)}] \neq \tilde{B}[U^2, A_\mu^{(2)}]$), $(U^1, A_\mu^{(1)})$ is homotopic to $(U^2, A_\mu^{(2)})$.

The field theory of the Lagrangian (19.5) is nonrenormalizable. Loop corrections induce interactions not present at the tree level, and the associated interaction terms have divergent coefficients. These terms have been computed at the one loop level by Appelquist and Bernard [140]. It was shown that away from the strong coupling limit these coefficients can be related to the mass of the physical Higgs field in the original renormalizable theory. When the strong coupling limit was taken, the tree level Higgs mass became singular, and so did the coefficients. Thus, the Higgs mass can be regarded as a cutoff in a low energy theory where λ and μ^2 had large but finite values. In general, the corresponding regularized Lagrangian contains an infinite number of terms which are of higher order in derivatives.

Now instead of finding soliton solutions to the classical theory, it may be more appropriate to look for soliton solutions in an "effective" theory where higher order terms are included in the Lagrangian. This approach is similar to the one taken in the Skyrme model and shall be adopted here as well.

As mentioned above, the general one loop correction terms have been computed in the literature. In part for reasons of simplicity, we shall not deal with the general one loop expression for the Lagrangian, but rather only consider the consequences of having the sample one loop correction

$$ L_{SK} = -\frac{1}{32e^2} \text{ Tr } [R_\mu, R_\nu]^2 \qquad (19.13) $$

to the Lagrangian, where e is treated as an unknown phenomenological coefficient. Equation (19.13) reduces to the "Skyrme term" in the limit $g \to 0$.

In the Skyrme model, the "Skyrme term" was needed to ensure the existence of a classical soliton solution. This was seen in Section 13.2. The gauged chiral model however does not require the presence of a term like (19.13) to ensure the existence of static solutions. This is so since the term $-\frac{1}{8} \text{ Tr } F_{\mu\nu} F^{\mu\nu}$ in (19.5) already scales like (19.13).

The details of the construction of the soliton solution now follow.

19.3 THE "SPHERICALLY SYMMETRIC" ANSATZ

Our definition of spherical symmetry is that in the $A_0 = 0$ gauge U and A_i satisfy the conditions:

$$-i(\vec{x} \wedge \nabla)_i U + [\tau_i, U] = 0 \quad,$$

$$-i(\vec{x} \wedge \nabla)_i A_j - i\varepsilon_{ijk} A_k + [\tau_i, A_j] = 0 \quad, \quad (i, j = 1, 2, 3) \quad.$$

$$(19.14)$$

Eqs. (19.14) imply that the fields are invariant under simultaneous space rotations and vector "isospin" transformations [that is , the action $U \to hUh^\dagger$ of the group $H = \{h\} = SU(2)_{diagonal}$]. This property was also shared by the $B[U] = 1$ soliton in the Skyrme model.

The general solution to (19.14) is $(U, A_i) = (U^c, A_i^c)$ where

$$U^c = \mathbf{1}_{2 \times 2} \cos\theta + i\tau \cdot \hat{x} \sin\theta \quad,$$

$$-\frac{g}{2} A_i^c = \frac{\alpha - \frac{1}{2}}{r} (\hat{x} \wedge \tau)_i + \frac{\beta}{r} \tau_i + \frac{\delta - \beta}{r} (\tau \cdot \hat{x}) \hat{x}_i \quad. \quad (19.15)$$

Here θ, α, β and δ are functions of the radial coordinate $r = |\vec{x}|$, and $\hat{x}_i = x_i / r$. The ansatz (19.15) still contains a remaining $U(1)$ gauge degree of freedom. It corresponds to

$$U^c \to v U^c \quad,$$

$$A_i^c \to v A_i^c v^\dagger - \frac{2i}{g} \partial_i v v^\dagger \quad,$$

$$v = \mathbf{1}_{2 \times 2} \cos\chi - i\tau \cdot \hat{x} \sin\chi \quad, \quad (19.16)$$

where $\chi = \chi(r)$. This $U(1)$ gauge transformation is realized on the functions θ, α, β and δ as follows:

$$\theta \to \theta - \chi \quad,$$

$$\alpha \to \alpha \cos 2\chi - \beta \sin 2\chi \quad,$$

$$\beta \to \alpha \sin 2\chi + \beta \cos 2\chi \quad,$$

$$\delta \to \delta + r\frac{d\chi}{dr} \quad. \quad (19.17)$$

We now eliminate the gauge freedom (19.16) by choosing the gauge

$$\beta = 0 \quad . \tag{19.18}$$

Eq. (19.18) is an allowed gauge provided α does not pass through a zero for $0 < r < \infty$. [Note that $\alpha^2 + \beta^2$ is gauge invariant. When $\alpha^2 + \beta^2 = 0$, the gauge transformation which takes the functions θ, α, β, and δ to the $\beta = 0$ gauge is thus ill defined.] We are, therefore, restricted to solutions with this property.

Next, we substitute the ansatz (19.15), with the gauge condition (19.18), into the expression for the static energy $E[U, A_i]$ associated with the Lagrangian (19.5) plus (19.13). The result is

$$
\begin{aligned}
E[U^c, A_i^c] \;=\; & \frac{8\pi <h>}{g} \int_0^\infty d\rho \left\{ \frac{2}{\rho^2}(\alpha^2 - \frac{1}{4})^2 + \alpha'^2 + 4\sigma^2\alpha^2 \right. \\
& \left. + \rho^2(\theta' + \sigma)^2 + F \cdot [1 + \frac{g^2}{32e^2\rho^2}\left(F + 4\rho^2(\theta' + \sigma)^2\right)] \right\} ,
\end{aligned}
\tag{19.19}
$$

where

$$F \equiv 2(\alpha + \sin^2\theta - \frac{1}{2})^2 + \frac{1}{2}\sin^2 2\theta \quad . \tag{19.20}$$

Here $\rho = g <h> r/2$ is a dimensionless variable, the prime denotes differentiation with respect to ρ and $\sigma \equiv \delta/\rho$. The variable σ appears in (19.19) without any derivatives. The equation which follows from its variation is, therefore, an algebraic equation:

$$\sigma = -\theta' \left\{ \frac{\rho^2 + \frac{g^2 F}{8e^2}}{\rho^2 + 4\alpha^2 + \frac{g^2 F}{8e^2}} \right\} \quad . \tag{19.21}$$

This equation can be substituted back into the energy functional (19.19) thereby eliminating σ from that functional. The equations following from variations of the remaining two variables θ and α are

$$\frac{d}{d\rho}\left[(\rho^2 + \frac{g^2 F}{8e^2})(\theta' + \sigma)\right] = 2\alpha \sin 2\theta \left[1 + \frac{g^2}{16e^2\rho^2}\left(F + 2\rho^2(\theta' + \sigma)^2\right)\right]$$

$$\alpha'' = \frac{4\alpha}{\rho^2}(\alpha^2 - 1/4) + 4\sigma^2\alpha + (F - \frac{1}{2}\sin^2 2\theta)^{1/2} \times$$

$$\times \left[1 + \frac{g^2}{16e^2\rho^2}\left(F + 2\rho^2(\theta' + \sigma)^2\right)\right] \quad . \tag{19.22}$$

The boundary conditions for α and θ result from the requirement of finite energy as well as from the requirement that the ansatz (19.15) be well defined when $r \to 0$. The former means that

$$\theta(\infty) = m\pi/2 \quad , \quad \alpha(\infty) = (-1)^m/2 \tag{19.23}$$

while the latter gives

$$\theta(0) = n\pi \quad , \quad \alpha(0) = 1/2 \quad , \tag{19.24}$$

where m and n are both integers. The integer n can be set to zero by a suitable redefinition of θ. One integer m then remains, which could possibly serve to classify solutions. In the Skyrme model, an analogous integer appeared from the boundary conditions. It was related to the topological index. Here, however, the appearance of the integer m is only an artifact of the $\beta = 0$ gauge. For instance, had we gone to the unitary gauge, that is $\theta = 0$, as was done by Ambjørn and Rubakov [137], no such index would have appeared.

One may expect that the static solution with $|m| = 1$ has the lowest energy. For such a case, however, α must pass through a zero for some r between 0 and ∞, and we have excluded such α while making the gauge choice. We therefore look instead for an $|m| = 2$ solution. The boundary conditions for θ are then identical to those of the $B = 1$ solution in the Skyrme model. For solitons with $|m| > 2$ see Ref. [141].

The solutions are functions of two parameters g and e and can be obtained numerically. Actually, from (19.19), we see that there is effectively only one parameter g/e.

Before discussing the general family of solutions, we examine two special limits. They are (i) $\frac{1}{e} = 0$, $g \neq 0$ and (ii) $\frac{1}{e} \neq 0$, $g = 0$. The first limit corresponds to the classical Lagrangian (19.5) for the strongly coupled theory with no higher order terms added. The second limit corresponds to the Skyrme model. We now discuss (i) in some detail.

Upon setting $1/e = 0$ (and $g \neq 0$), the equations (19.21) and (19.22) reduce to

$$\sigma = \frac{-\theta' \rho^2}{\rho^2 + 4\alpha} \quad , \tag{19.25}$$

$$\frac{d}{d\rho}[\rho^2(\theta' + \sigma)] = 2\alpha \sin 2\theta \quad , \tag{19.26}$$

$$\alpha'' = \frac{4\alpha}{\rho^2}\left(\alpha^2 - \frac{1}{4} + \rho^2\sigma^2\right) + 2\alpha - \cos 2\theta \quad . \tag{19.27}$$

The solution to these equations can be obtained numerically. The classical energy of the solution is

$$E \approx \frac{8\pi <h>}{g}(1.79) \quad . \tag{19.28}$$

Upon using the experimental values of $M_W = 83$ GeV and $g = .67$, and the relation $<h> = \sqrt{2}M_W/g \approx 175$ GeV, we obtain $E \approx 11.6$ TeV.

Upon substituting the ansatz (19.15) into $\tilde{B}[U, A_i]$, one obtains

$$\tilde{B}[U, A_i] = \frac{1}{\pi}(\theta + \frac{1}{2}\sin 2\theta)|_{r\to\infty} - \frac{4}{\pi}\int_0^\infty d\rho \, \sigma(\alpha^2 - \frac{1}{4}) \quad . \tag{19.29}$$

Due to the last term in (19.29), the value for $\tilde{B}[U, A^i]$ is not an integer as alluded to earlier. For the above solution where $1/e = 0$, we get $\tilde{B} \cong 0.63$.

Another solution to the equations of motion (19.21) and (19.22) is the Skyrme soliton solution. It corresponds to the limit (ii) $g = 0$, [and $\frac{1}{e} \neq 0$] and has $\tilde{B} = B = 1$. Note that equations of motion (19.21) and (19.22) are written in terms of the radial variable ρ, and that the definition of ρ is singular in the limit (ii). Thus, instead of using the variable ρ, it is more appropriate to define a new variable $\tilde{\rho} = e <h> r$. When written in terms of $\tilde{\rho}$, equations (19.21) and (19.22) reduce to Skyrme's equations for $g/e \to 0$. Of course in nature $g \neq 0$, so the limit (ii) is unphysical.

Next we discuss the numerical results for general values of g/e. For each value of $(g/e)^2$ in the range $0 \leq (g/e)^2 \leq .39$, we find two classical solutions to the equations of motion. In the limit $(g/e)^2 \to 0$, one of them tends to the solution in (i), while the other tends to the one in (ii). In the limit $(g/e)^2 \approx .39$, both solutions tend to the same field configuration. For $(g/e)^2 \geq .39$, we find no solutions.

Next we discuss the question of the classical stability of the static solutions. We can show that the solution in the limit (i) is classically unstable, that is it is not a local minimum of energy. For this it is sufficient to make the variation $\delta\theta$ of θ in E. This gives for the change in energy,

$$\delta E = \frac{32\pi <h>}{g} \int_0^\infty d\rho \left[\frac{\alpha^2 (\frac{d\delta\theta}{d\rho})^2}{1 + (\frac{2\alpha}{\rho})^2} + \alpha(\delta\theta)^2 \cos 2\theta + \theta(\delta\theta^3)) \right] . \qquad (19.30)$$

Although the first term in the integrand is positive, the second is negative for $\frac{\pi}{4} < \theta < \frac{3\pi}{4}$. It is not hard numerically to find variations $\delta\theta(\rho)$ such that the second term dominates the first and $\delta E < 0$. Thus, although there exists a static solution to the strongly coupled GSW theory with no higher order terms added, it is classically unstable.

On the other hand, if we take the limit (ii) corresponding to the Skyrme soliton, the result is generally believed to be classically stable. Consequently, in traversing the path in the space of solutions parametrized by e and g from (i) to (ii), we should encounter a transition from an unstable to a stable field configuration. There is some reason to believe that this transition occurs at $(g/e)^2 \cong 0.39$. Reference [139] may be consulted for arguments supporting this conjecture.

19.4 SEMICLASSICAL QUANTIZATION

We now discuss the semiclassical quantization of the electroweak Skyrmion. For this purpose, we can adopt the same procedure which was utilized to quantize the Skyrme solitons in Chapter 13. We first identify the zero frequency modes of the soliton and then quantize these modes.

The particular zero-modes of interest for the Skyrmion were associated with space (or isospace) rotations. The analogous modes for the electroweak Skyrmions are obtained from

$$U^c \to V U^c V^\dagger \quad ,$$
$$A_i^c \to V A_i^c V^\dagger \quad , \quad V \in SU(2) \quad . \tag{19.31}$$

Under (19.31), the form of the solution is changed, but the static energy is not. Transformation (19.31) corresponds to a vector isospin transformation. It may also be thought of as a spatial rotation since by the spherical symmetry (19.14),

$$V U^c(\vec{x}) V^\dagger = U^c(\vec{R}x) \quad ,$$
$$V A_k^c(\vec{x}) V^\dagger = R_{jk} A_j^c(\vec{R}x) \tag{19.32}$$

where

$$R_{ij} = \frac{1}{2} \operatorname{Tr} \left(\tau_i V \tau_j V^\dagger \right)$$

are the elements of an $SO(3)$ rotation matrix R.

We now wish to consider the collective motion induced by V. For this purpose we elevate V to a dynamical variable $V = V(t)$ and set

$$U(\vec{x}, t) = V(t) U^c(\vec{x}) V(t)^\dagger \quad ,$$
$$A_k(\vec{x}, t) = V(t) A_k^c(\vec{x}) V(t)^\dagger \tag{19.33}$$

and substitute (19.33) into the total Lagrangian density (19.5) plus (19.13). Upon integrating over $\vec{x}$, we obtain

$$L = \int d^3 x (L_\phi + L_A + L_{SK})$$
$$= -\frac{1}{2} a[U^c, A_i^c] \operatorname{Tr} (V^\dagger \dot{V})^2 - E[U^c, A_i^c] \quad . \tag{19.34}$$

Thus, L has the same form as in Eq. (13.42), only here

$$a[U^c, A_i^c] = \frac{256\pi}{3} \frac{1}{g^3 <h>} \int_0^\infty d\rho \rho^2 \left\{ \sin^2 \theta \left[1 + \frac{g^2}{8e^2} \left(\theta'^2 + \frac{\sin^2 \theta}{\rho^2} \right) \right] \right.$$
$$\left. + \left(\frac{\alpha - 1/2}{\rho} \right)^2 + \frac{\sigma^2}{2} \right\} \quad . \tag{19.35}$$

Since the Lagrangian (19.34) is identical in form to the Lagrangian describing rotational zero modes in the Skyrme model, we expect that the corresponding quantum spectra for the two systems are identical. However, we know that the spectrum associated with Lagrangian (19.34) is not unique due to the quantization ambiguities discussed in Chapter 13. As shown in Chapter 13, quantization can lead to three different cases: (a) All states in the theory are bosonic, (b) all states are fermionic and (c) both bosonic and fermionic states occur. In the Skyrme model, the ambiguity was resolved by going to a three flavored chiral model and including the Wess-Zumino term. The coefficient of the Wess-Zumino term contains the number of colours in the underliyng QCD theory. For an odd number of colours, quantization scheme (b) was uniquely selected.

If a similar resolution of the ambiguity in the electroweak theory is at all possible, it would appear to require a somewhat drastic change in the model. First, it is not clear how to write a Wess-Zumino term without enlarging the gauge group to, say, $SU(3)$. Second, it is necessary to justify the inclusion of such a term. Further unless the normalization of this term is known, the quantization ambiguity remains unresolved.

Actually, however, there is little doubt that we must quantize the model so that all states are tensorial (have integral spins). This is because this model is an approximation to the GSW theory with no fermions, and all the states in this theory are tensorial.

The results for the semiclassical quantization are essentially identical to those in Chapter 13. Once again, we can introduce the left and right $SU(2)$ generators L_i and R_i respectively. The former are associated with vector isospin transformations while the latter correspond to spatial rotations. As in that Chapter, we get a strong coupling spectrum, the masses being

$$m_\ell = E[U^c, A_i^c] + \frac{2\ell(\ell+1)}{a[U^c, A_i^c]} \quad , \quad \ell = 0, 1, 2, \ldots \quad . \tag{19.36}$$

According to (19.36), the ground state energy is just the classical energy.

19.5 PHENOMENOLOGICAL PROPERTIES

Here we compute electric charges for the above quantum states and also discuss questions of stability.

We now determine the electric charges for the soliton states found in the previous Section. Actually, for the theory we have been investigating, all electric charges are zero. This is because we have set $\sin^2 \theta_W = 0$. The theory we have been examining has no $U(1)$ electromagnetic subgroup, and the $SU(2)_L$ gauge symmetry of the Lagrangian was completely broken by the vacuum.

In a realistic theory, $\sin^2 \theta_W \neq 0$ and the charges of the soliton states are not zero. To determine the latter, we need to drop the condition that $g' \neq 0$, in order that the $U(1)$ gauge field B_μ couples to the other fields. The latter will be done perturbatively. We will also estimate the lowest order correction to the classical energy due to the introduction of the B_μ field.

For reinstating the $U(1)$ gauge field, we must consider the Lagrangian density

$$L = -\frac{1}{4} B_{\mu\nu} B^{\mu\nu} - \frac{<h>^2}{2} \text{ Tr } R_\mu R^\mu \quad , \tag{19.37}$$

where $R_\mu = D_\mu U U^\dagger$, and the covariant derivative is defined in (19.4). The static equation of motion for B_i is

$$\partial_i B_{ij} = j_j^B \equiv -\frac{ig'}{2} \text{ Tr } \tau_3 \left\{ <h>^2 U^\dagger D_j U \right.$$
$$\left. - \frac{1}{8e^2} [[U^\dagger D_i U, U^\dagger D_j U], U^\dagger D_i U] \right\} \quad . \tag{19.38}$$

Thus, to lowest order, B_i is linear in g'. Because of the presence of τ_3 in Eq. (19.38), the spherical symmetry of the classical solution is broken to, at best, an axial symmetry. Of course, it is not clear that with the additional field B_μ, which couples to A_μ and U, we can still find static solutions to the system of equations which tend to the previous ones when $g' \to 0$. Even if this is the case, we must also be concerned with the question of stability, specifically with regards to perturbations in the added field B_μ.

Since the above question is nontrivial and unsolved, we shall instead assume that classically stable soliton solutions can be found for the coupled system of fields A_μ, B_μ and U, and that these solutions tend to the previously found ones when $g' \to 0$.

We can now determine the energy of these "static solutions" to lowest order in g'. The static energy is a functional $E = E(U, A_i, B_i; g')$. If we vary E with respect to g' along the surface of solutions $(U, A_i, B_i) = (U^c, A_i^c, B_i^c)$ to the equations of motion, we have

$$\frac{dE}{dg'}\bigg|_{(U,A_i,B_i)=(U^c,A_i^c,B_i^c)} = \frac{\partial E}{\partial g'}\bigg|_{(U,A_i,B_i)=(U^c,A_i^c,B_i^c)} \qquad (19.39)$$

We have eliminated terms involving derivatives of fields in (19.39) by using the equations for U^c, A_i^c and B_i^c. Therefore,

$$
\begin{aligned}
\frac{dE}{dg'}\bigg|_{(U,A_i,B_i)=(U^c,A_i^c,B_i^c)} &= -\frac{i}{2} <h>^2 \int d^3x B_i^c \, \mathrm{Tr}\,(\tau_3 U^{c\dagger} D_i U^c) \\
&= -\frac{1}{2g'} \int d^3x (B_{ij}^c)^2 \quad .
\end{aligned}
\qquad (19.40)
$$

Since as (19.38) shows, B_i^c is linear in the coupling constant g', the quantity in (19.40) is also linear in g'. The change in the energy is thus

$$\delta E = E(U^c, A_i^c, B_i^c; g') - E(U^c, A_i^c, 0; 0) = -\frac{1}{4} \int d^3x (B_{ij}^c)^2 \quad , \qquad (19.41)$$

which is quadratic in g'. [Here $B_{ij}^c = \partial_i B_j^c - \partial_j B_i^c$.] This result indicates that the inclusion of the $U(1)$ field will reduce the soliton energy by order $\sim \sin^2 \theta_W$. The same result was obtained by Manton and Klinkhamer [142] in the context of a different classical solution known as the "spahleron".

We can now compute the electric charges using the semiclassical method. The electromagnetic current in the standard model is

$$j_\mu^{em} = \frac{g j_\mu^{(B)} + g' j_\mu^{(A^3)}}{\sqrt{g^2 + g'^2}} \quad , \qquad (19.42)$$

where $j_\mu^{(A^3)}$ is the current source for A_μ^3, that is $\partial^\nu(\partial_\nu A_\mu^3 - \partial_\mu A_\nu^3) = j_\mu^{(A^3)}$. In the strongly coupled theory, with the identification of $g \sin \theta_W$ with the unit of electric charge e_0, we find

$$
\begin{aligned}
j_\mu^{em} = \ & \frac{ie_0}{2} \, \mathrm{Tr} \, \tau_3 \Big\{ <h>^2 \, (U^\dagger D_\nu U - D_\nu U U^\dagger) \\
& + \frac{1}{2}(\partial^\nu [A_\mu, A_\nu] + [A^\mu, F_{\mu\nu}]) \\
& - \frac{1}{8e^2}[[U^\dagger D_\nu U, U^\dagger D_\mu U], U^\dagger D^\nu U] \\
& + \frac{1}{8e^2}[[D_\nu U U^\dagger, D_\mu U U^\dagger], D^\nu U U^\dagger] \Big\} \quad .
\end{aligned} \tag{19.43}
$$

After substituting the semiclassical approximation (19.33) into the corresponding expression for the electric charge $Q^{em} = \int d^3 x j_0^{em}(x)$, and performing the spatial integration, we get

$$
\begin{aligned}
Q^{em} &= \frac{ie_0}{2} a[U^c, A_i^c] \, \mathrm{Tr} \, (\dot{V} V^\dagger \tau_3) \\
&= -\frac{ie_0}{4} a[U^c, A_i^c] \, \mathrm{Tr} \, (\dot{R} R^T \theta_3) \\
&= e_0 L_3 \quad .
\end{aligned} \tag{19.44}
$$

Now L_i generates the vector isospin group H. Therefore in quantum theory Q^{em}/e_0 takes on the eigenvalues $-\ell, -\ell+1, \ldots \ell$ when acting on the basis D^ℓ. [D^ℓ is the same as D^L in Section 13.4 for $L = \ell$.] So, the ground state $\ell = 0$ is neutral, while the isotriplet $\ell = 1$ has charges 0 and ± 1. Note that had we allowed for fermionic states, the vector isospin and consequently the electric charges would have taken on half-odd integer values! This is also the case for solitons in the Skyrme model if one ignores the Wess-Zumino term.

In the above derivation, we have used perturbation theory to compute the charge spectrum of soliton states. The results, however, must be true nonperturbatively as well. Since the values of Q^{em}/e_0 are integers, they cannot change while continuously increasing the value of $\sin^2 \theta_W$.

Previously, we suggested the existence of classically stable static solutions. Even if this is the case, it is still necessary to check for quantum mechanical stability before concluding that these states are observables. [Of course, before saying with confidence whether or not such states can actually be observed in a future accelerator, much more work is necessary.] A quantum mechanically unstable state with $\tilde{B} \neq 0$ would readily tunnel to the vacuum (corresponding to $\tilde{B} = 0$) and the state may become unobservable if the decay rate by tunneling is too large.

To get an indication of the quantum mechanical stability of our states, one can compute an upper bound to the lowest order semiclassical tunneling probability. This was done in Ref. [139]. The latter is determined from $\exp\{-2S_E\}$, where S_E is the Euclidean action evaluated along a path connecting a $\tilde{B} \neq 0$ classical configuration to a configuration having $\tilde{B} = 0$. In the semiclassical approximation, the relevant path corresponds to a classical solution of the Euclidean field equations. Two relevant solutions are a) instantons and b) merons.

For case a), one finds the usual suppression factor for tunneling via instantons [143],

$$\exp(-2S_E) < \exp\left\{-\frac{16\pi^2|\tilde{B}_2 - \tilde{B}_1|}{g^2}\right\} \quad , \qquad (19.45)$$

where $\tilde{B}_1$ and $\tilde{B}_2$ are initial and final values for $\tilde{B}[U, A]$. This constrains $\exp(-2S_E)$ to be an extremely tiny number. It is known that it is this smallness of $\exp(-2S_E)$ which prevents the baryon from decaying too rapidly in electroweak theory. Here, as well, it prevents the electroweak Skyrmion from decaying too rapidly. Because of the similarity of these two decay processes, it is tempting to identify $\tilde{B}$ with baryon number. We believe, however, that this is incorrect since $\tilde{B}$ is not an integer. On the other hand, perhaps a change in $\tilde{B}$ can be identified with a change in baryon number.

On the other hand, the suppression factor for case b) is found to be quite large. For tunneling via merons, we must relax the strong coupling limit $\lambda, \mu^2 = \infty$. That is, we let λ and μ^2 to be large but finite numbers with the usual constraint that $\mu^2/2\lambda = <h>^2$. This relaxation means that we are now admitting the extra degree of freedom corresponding to the physical Higgs field h to be excited. [In principle, one should then check that there still exist classically stable static solutions to the equations of motion for A_μ, B_μ, U

and h, and that these solutions reduce to the ones we found previously upon taking the limit $\sin^2 \theta_W \to 0$ and $\lambda, \mu^2 \to \infty$. We will not however do that here. Rather, we will assume that such solutions exist when λ and μ^2 are large (and $\sin^2 \theta_W$ is small, but not precisely zero), and that they are to zeroth order in $\frac{1}{\lambda}$ and g' given by the solutions found for Eqs. (19.21) and (19.22).]

The upper bound on the suppression factor for semiclassical tunneling rate via merons is then found to be [135]

$$e^{-2S_E} < \exp\{-\frac{\pi^2}{e^2}|\tilde{B}_2 - \tilde{B}_1|\} \qquad (19.46)$$

Unlike the result for decay via instantons, this upper bound can be quite large. For example, for $(g/e)^2 \approx 0.39$, 0.2 and 0.1, we find the values 0.0016, 0.014, 0.11 for the bounds. For $(g/e)^2 \approx 0.05$ [and $E(U^c, A_i^c) \approx 6$ TeV], we get the bound $\approx 1/3$. Although we have not computed the corresponding width of such a state, we would expect it to be large. Consequently, the observation of states with $(g/e)^2 \leq 0.05$ (and $E(U^c, A_i^c) \leq 6$ TeV) may be difficult. The only conclusion we can make with some certainty is that as the value of $(g/e)^2$ and the energy $E(U^c, A_i^c)$ of the solutions go down, these states become more and more quantum mechanically unstable. Of course, if λ were actually infinite in nature, solitons in the electroweak theory would be stable under decay via merons.

PART IV

GAUGE, GRAVITY

AND

STRING THEORIES

Chapter 20

MULTIPLY CONNECTED CONFIGURATION SPACES IN GAUGE AND GRAVITY THEORIES

20.1 INTRODUCTION

Multiply connected configuration spaces occur in nonabelian gauge theories and in gravity so that our discussion in Chapter 8 about the construction of quantum theories on such spaces is relevant for these systems. The θ states of quantum chromodynamics (QCD) [for reviews see e.g. Refs. [31,31,145,146]] are consequences of such multiple connectivity, the fundamental group of the appropriate configuration space being the abelian group $\mathbf{Z}$ as we shall later indicate. [Also see e.g. Refs. [61,62,147].] As regards gravity, the topology of the configuration space can be very involved and the fundamental group can be abelian or nonabelian. The topological features in gravity that are of our interest are associated with certain excitations called geons [69] which resemble the topological solitons of the previous Chapters. The purpose of this Chapter is to discuss how multiply connected configuration spaces arise in gauge and gravity theories, while the properties of geons will be described in Chapter 21.

In Section 2, the canonical formalism for gauge and gravity theories will be briefly recalled and one approach to the θ states of QCD and the analogous gravity states will be outlined [147-149]. In this approach, the connectivity of the configuration space does not seem to play an obvious role for the existence of these states. The next three Sections develop an alternative approach to the θ states of gauge theories and their gravity analogues which emphasize the connectivity properties of the configuration spaces. The relation of this approach to the discussion of Section 2 is also explained. Section 6 is devoted to brief comments on the role of locality in the definition of observables in gauge theories.

20.2 THE CANONICAL FORMALISM AND AN APPROACH TO "θ STATES" IN GAUGE AND GRAVITY THEORIES

In nonabelian gauge theories such as chromodynamics, the canonically conjugate variables are the spatial components A_i and E_i of the vector potentials and electric fields. [We shall work in the $A_0 = 0$ gauge and ignore matter fields.] We now recall that the initial data for A_i and E_i (which are then evolved by the Hamiltonian) can not be prescribed freely, but are subject to the Gauss law constraint

$$D_i E_i \approx 0 \quad , \tag{20.1}$$

where D_i is covariant derivative and "$\approx$" denotes weak equality in the sense of Dirac [see Chapter 2]. The phase space function $D_i E_i$ is Lie algebra valued so that in chromodynamics there are actually eight constraints $(D_i E_i)_a \approx (a = 1, 2, \ldots, 8)$ implied by (20.1), one for each direction in the $SU(3)$ Lie algebra. These functions $(D_i E_i)_a$ are the generators of the $SU(3)$ gauge group $\mathcal{G}^\infty$ and form a set of first class constraints. The group $\mathcal{G}^\infty$ is the group of maps from R^3 into $SU(3)$ which become asymptotically trivial. Thus if $g \in \mathcal{G}^\infty$, $g(\vec{x}) \in SU(3)$ and

$$g(\vec{x}) \to 1 \quad \text{as} \quad |\vec{x}| \to \infty \ . \tag{20.2}$$

The reason for the boundary condition on g is that it is necessary to first smear the Gauss law constraints with test functions Λ^a which vanish at infinity before we obtain functions on the phase space which are well defined generators of canonical transformations. Thus the well defined generators of canonical transformations are

$$\int d^3x \, \Lambda^a (D_i E_i)_a \ , \quad \Lambda^a \to 0 \quad \text{as} \quad |\vec{x}| \to \infty \ . \tag{20.3}$$

Because of the behavior of Λ^a at infinity, these smeared functions are associated with gauge transformations fulfilling the condition (20.2). In view of (20.2), we may regard $\mathcal{G}^\infty$ as the group of maps of a three sphere S^3 to $SU(3)$ with the condition that a distinguished point ("infinity") is mapped to the identity of $SU(3)$ by all its elements.

Now the constraints $(D_i E_i)_a \approx 0$ form a first class set. They can therefore be imposed in quantum theory as conditions which pick out the physical states ψ. Using a hat to distinguish quantum operators, we can thus impose the conditions

$$(\widehat{D_i E_i})_a \psi = 0 \ . \tag{20.4}$$

[Here we should really smear the left hand side with test functions Λ^a as in (20.3). We shall omit exhibiting this smearing here and often in what follows below]. The states ψ may be regarded as functions of A_i. The operators $(\widehat{D_i E_i})_a$ are the generators of the Lie algebra of $\mathcal{G}^\infty$ so that on exponentiation they generate only the component $\mathcal{G}_0^\infty$ of $\mathcal{G}^\infty$ connected to identity. Thus (20.4) means that the states ψ are invariant under the connected group $\mathcal{G}_0^\infty$. It follows that the group $\mathcal{G}^\infty/\mathcal{G}_0^\infty$ can act nontrivially on the states. As is well known, this group is isomorphic to the additive group $\mathbf{Z}$ of integers. This is so because the maps g are similar to the chiral field U in the Skyrme model and hence each such map is characterized by a winding number n. [See Chapters

13 and 15.] The connected component of $\mathcal{G}_n^\infty$ of $\mathcal{G}^\infty$ consists of all maps g with the winding number n while $\mathcal{G}_n^\infty$ and $\mathcal{G}_m^\infty$ are not path connected if $n \neq m$. Hence $\mathcal{G}^\infty/\mathcal{G}_0^\infty = \mathbf{Z}$. If $T \in \mathcal{G}^\infty$ is any winding number 1 map, $\mathcal{G}^\infty/\mathcal{G}_0^\infty$ can be identified with $\{T^n\}_{n \in Z}$. The unitary irreducible representations of this group are given by $T \to e^{i\theta}$ (θ real) so that we can assume that the physical states fulfill

$$\hat{T}\psi = e^{i\theta}\psi \tag{20.5}$$

for some fixed θ. [$\hat{T}$ is the quantum operator analogue of T]. These angles are the famous θ angles of QCD [145].

Let us now see how to apply such considerations to quantum gravity.

We will first review the conventional canonical approach to gravity briefly emphasizing those results which are of relevance to our subsequent considerations on geons [62,151-154].

In the canonical approach, the four manifold ^4M is assumed to have the topology of the Cartesian product of a three manifold ^3M with the real line R^1, where the points of R^1 can be taken to label instants of time and ^3M is the spatial slice. The canonical variables of the theory are the components q_{ab} of the three metrics q on ^3M (with signature $+++$) and their conjugate momenta π^{ab}, where π^{ab} has an interpretation in terms of the extrinsic curvature of ^3M regarded as a submanifold of ^4M. The fields q_{ab}, π^{ab} are the gravity analogues of the components A_i of the QCD vector potentials and the components E_i of the QCD electric fields in the $A_0 = 0$ gauge. In addition to q_{ab} and π^{ab}, there may also be variables associated with sources of gravitational fields, but we shall ignore these degrees of freedom for simplicity.

We shall deal only with asymptotically flat space times with one asymptotic region. Thus there is a compact set $\mathcal{C}$ in ^3M, the complement $^3M \backslash \mathcal{C}$ of which is diffeomorphic to the complement of a ball in $\mathbf{R}^3$. Further,

$$q_{ab} \to \delta_{ab} \quad , \quad \pi^{ab} \to 0 \quad , \tag{20.6}$$

as spatial infinity is approached.

Now the action of any gravity theory is invariant under the diffeomorphism group of ^4M. This group is the analogue for gravity of the four dimensional $SU(3)$ gauge group in QCD. The existence of such an invariance group implies from general principles that there are four first class constraints in the Hamiltonian formulation of gravity for each point $\vec{x}$ of ^3M. These constraints may be written as

$$
\begin{aligned}
C_a &\approx 0 \quad, \quad a = 1, 2, 3 \quad, \\
C_0 &\approx 0 \quad.
\end{aligned}
\tag{20.7}
$$

$C_\lambda(\lambda = 0, 1, 2, 3)$ are functions of q_{ab} and π^{ab}, but their specific form will not concern us. The constraints C_a ($a = 1, 2, 3$) are the generators of the diffeomorphism group of ^3M while C_0 is associated with time reparametrizations. As the constraints C_λ are first class, they may be imposed in quantum theory as conditions on the physical states ψ:

$$
\hat{C}_\lambda \psi = 0 \quad.
\tag{20.8}
$$

Here ψ may be regarded as a function of the three metrics $q = (q_{ab})$ and $\hat{C}_\lambda$ is the quantum operator analogue of C_λ.

We will not have occasion to discuss the "scalar" constraint $C_0 \approx 0$ to any extent. Of course it does have a role in our discussion since classically the initial data we prescribe must be compatible with all the constraints, and not all ^3M admit initial data which are consistent with all the constraints. Thus classically certain three manifolds are excluded by the constraints. We shall comment on this issue briefly in Section 21.2. We shall also assume that if a certain three manifold with suitable initial data is compatible with all the constraints classically, then states annihilated by these constraints can be prepared in quantum theory as well on such a manifold.

The "vector" constraints C_a are generators of the diffeomorphism group of ^3M. They generate only those diffeomorphisms of ^3M which become asymptotically trivial. This is because just as in QCD we must first smear them with test functions N^a which vanish at ∞ before we obtain functions on the phase space which are well defined generators of canonical transformations. Thus the well defined generators of canonical transformations are

$$\int_{^3M} N^a C_a dV \quad , \quad N^a \to 0 \quad \text{asymptotically} . \quad (20.9)$$

In analogy with QCD, we may therefore infer that the group generated by the vector constraints is the identity component of asymptotically trivial diffeomorphisms. Let us denote the full group of asymptotically trivial diffeomorphisms by D^∞ and its identity component by D_0^∞. The constraints (20.8) require only that D_0^∞ acts trivially on the physical states so that just as in QCD the group $G = D^\infty / D_0^\infty$ can act nontrivially on the states. As discussed in detail by Friedman, Sorkin and Witt [62,69,151], this G can be quite complicated and interesting for suitable ^3M. In particular, unlike in QCD, G can be nonabelian. The possibility that it acts nontrivially on the physical states implies a rich variety of phenomena discovered by these and other authors, and one purpose of this Chapter and the next one is to review their work.

20.3 THE CONFIGURATION SPACE FOR NONABELIAN GAUGE THEORIES

When the gauge group $\mathcal{G}^\infty$ of a theory has many disconnected components, it seems generically to be true that the appropriate configuration space Q for this theory is multiply connected. If $\mathcal{G}_0^\infty$ is the connected subgroup of $\mathcal{G}^\infty$, it turns out in all the cases we encounter that $\pi_1(Q) = \mathcal{G}^\infty / \mathcal{G}_0^\infty$. The analogous result in gravity theories is that $\pi_1(Q) = D^\infty / D_0^\infty$. In the following Sections, we shall explain this connection for nonabelian gauge theories like QCD and for quantum gravity. We shall also see in this Chapter and the next one that the remarkable topological results like the existence of θ states of QCD and of spinorial states in quantum gravity can be interpreted within the general framework developed in Chapter 8.

We have seen in Section 20.2 that in the canonical approach to nonabelian gauge theories in $3+1$ dimensions, a group $\mathcal{G}^\infty$ of certain time independent gauge transformations plays a critical role. The elements of $\mathcal{G}^\infty$ are functions g on R^3 which take values in a compact semisimple Lie group $K : g(\vec{x}) \in K$ [Here we do not restrict K to be $SU(3)$.] They are also subjected to the boundary condition $g(\vec{x}) \to 1$ as $|\vec{x}| \to \infty$, where 1 is the identity of K. Because of this boundary condition, we can regard $\mathcal{G}^\infty$ as consisting of maps of the three sphere S^3 to the group K. We assume in the further discussion that K is simple since the treatment of the nonsimple cases follows easily from that of the simple case. The group $\mathcal{G}^\infty$ consists of several disconnected components labelled by a winding number, $\pi_0(\mathcal{G}^\infty)$ being $\mathbf{Z}$ for a simple K. [This result is true for any simple Lie group K and not just for $SU(N)(N \geq 2)$.] Let $\mathcal{G}_0^\infty$ be the component of $\mathcal{G}^\infty$ connected to the identity. Then $\pi_0(\mathcal{G}^\infty) = \mathcal{G}^\infty/\mathcal{G}_0^\infty = \mathbf{Z}$.

Let $\mathcal{A}$ be the set of vector potentials at a given time in the $A_0 = 0$ gauge. The observable functions on $\mathcal{A}$ are assumed to be invariant under $\mathcal{G}^\infty$. [See later for further remarks on this point.] They may be regarded as functions on $\mathcal{A}/\mathcal{G}^\infty$, the space which results when $\mathcal{A}$ is quotiented by the action of $\mathcal{G}^\infty$. The configuration space Q is thus $\mathcal{A}/\mathcal{G}^\infty$. [We ignore matter fields in this discussion.]

Let $\pi_1(\mathcal{A}/\mathcal{G}^\infty)$ denote the fundamental group of the configuration space $\mathcal{A}/\mathcal{G}^\infty$. We want to show that it is $\mathbf{Z}$. The following remarks are helpful for this demonstration: Let $A = A_i dx^i \in \mathcal{A}$ denote a gauge potential which we write as a one form, and let $^g A$ denote its gauge transform by an element $g \in \mathcal{G}^\infty$, $^g A \equiv gAg^{-1} + gdg^{-1}$. If the equation $^g A = A$ for any fixed A implies that g is the identity of $\mathcal{G}^\infty$, then following Section 3.5, we say that $\mathcal{G}^\infty$ acts freely on $\mathcal{A}$. In this case, $\mathcal{A}$ may be regarded as a principal fibre bundle on $\mathcal{A}/\mathcal{G}^\infty$ with structure group $\mathcal{G}^\infty$.

Let us assume for a moment that $\mathcal{G}^\infty$ does act freely on $\mathcal{A}$. Now $\mathcal{A}$ is topologically very simple. In particular $\pi_1(\mathcal{A}) = 0$. This is because if

$$\Gamma = \{A^{(s)} | 0 \leq s \leq 1 \; ; \; A^{(0)} = A^{(1)}\} \tag{20.10}$$

is a loop in $\mathcal{A}$, then the loops

$$\Gamma(t) = \{A^{(0)} + (1-t)(A^{(s)} - A^{(0)})\} \quad , \quad 0 \leq t \leq 1 \qquad (20.11)$$

provide a homotopy of Γ to the trivial loop $\Gamma(1) = \{A^{(0)}\}$. Thus Q is the quotient of the simply connected $\mathcal{A}$ by the free action of $\mathcal{G}^{(\infty)}$. In such a case, it is known that [62,69,151,153,154] $\pi_1[\mathcal{A}/\mathcal{G}^{(\infty)}]$ is equal to the group $\mathcal{G}^{(\infty)}/\mathcal{G}_0^{(\infty)} = \mathbf{Z}$ associated with the distinct components of $\mathcal{G}^{(\infty)}$.

It remains to prove that $\mathcal{G}^\infty$ acts freely on $\mathcal{A}$. For this purpose, consider the path ordered integral

$$U(A) = P \exp \int_{-\infty}^{x} dx' A_x(x', y, z) \quad , \qquad (20.12)$$

where we assume for simplicity that the integral exists. A standard property of $U(A)$ is

$$U(^g A) = g(-\infty, y, z) \, U(A) \, g(\vec{x})^{-1} \quad , \quad \vec{x} = (x, y, z) \quad . \qquad (20.13)$$

When $g \in \mathcal{G}^\infty$, the prefactor $g(-\infty, y, z)$ is the identity of K and

$$U(^g A) = U(A) \, g(\vec{x}). \qquad (20.14)$$

It follows immediately that $^g A = A$ for $g \in \mathcal{G}^\infty$ only if g is the identity of $\mathcal{G}^\infty$. Thus $\mathcal{G}^\infty$ acts freely on $\mathcal{A}$.

The general theory of quantization on multiply connected spaces developed in Chapter 8 now shows that the quantum theories for $3+1$ dimensional nonabelian gauge theories are classified by the unitary irreducible representations (UIR's) of $\pi_1(\mathcal{A}/G^\infty) = \mathbf{Z}$. These UIR's ρ_θ are labelled by an angle θ and defined by

$$\mathbf{Z} \ni n \to e^{in\theta} \quad . \qquad (20.15)$$

It is not difficult to show that this angle θ labelling a quantum theory is the same as the angle θ which occurred in Section 20.2. The equivalence of the method of quantization outlined in Section 20.2 and the one developed in Chapter 8 will be discussed further in Section 20.5.

20.4 THE CONFIGURATION SPACE FOR GRAVITY THEORIES

In the canonical approach to Einstein's theory of gravitation in $3 + 1$ dimensions, the roles of the groups $\mathcal{G}^\infty$ and $\mathcal{G}_0^\infty$ are assumed by the groups D^∞ and D_0^∞. The elements of D^∞ are diffeomorphisms of the three manifold ${}^3M$ which act asymptotically as identities on ${}^3M$. The group D_0^∞ is the subgroup of D^∞ which is connected to the identity. Instead of the space $\mathcal{A}$ of connections, we now have the space $\tilde{Q} \equiv$ Riem $({}^3M) = \{q\}$ of three metrics on ${}^3M$ which are asymptotically flat [see Eq. (20.6)]. The space $\tilde{Q}$ is convex and hence can be shown to be topologically trivial: $\pi_n(\tilde{Q}) = 0$. The convexity of $\tilde{Q}$ proved by observing that $q \in \tilde{Q}$ if and only if $q_{ab}(\vec{x})\xi^a\xi^b > 0$ for all nonzero real three vectors $\vec{\xi} = (\xi^1, \xi^2, \xi^3)$. Now if q and $q' \in \tilde{Q}$, then clearly $[tq_{ab}(\vec{x}) + (1 - t)q'_{ab}(\vec{x})]\xi^a\xi^b > 0$ for $0 \le t \le 1$. Therefore, $tq + (1 - t)q' \in \tilde{Q}$ and $\tilde{Q}$ is convex. Thus $\pi_n(\tilde{Q}) = 0$. If it can also be shown that D^∞ acts freely on $\tilde{Q}$, it will immediately follow as in the case of nonabelian gauge theories that

$$\pi_1[\tilde{Q}/D^\infty] = \pi_0[D^\infty] = D^\infty/D_0^\infty.$$

Since the observable functions on $\tilde{Q}$ are required to be invariant under D^∞ and hence may be regarded as functions on $\tilde{Q}/D^\infty$, the analogy to gauge theories will then be perfect, the appropriate configuration space Q for gravity being $\tilde{Q}/D^\infty$.

The proof that D^∞ acts freely on $\tilde{Q}$ is not so simple as in the gauge theory case. We will only indicate it briefly and refer the reader to the literature [62,69,151,153,154]. First we have to be a bit more precise in stating the asymptotic conditions on the elements of D^∞. The action of the elements of D^∞ is required to approach the action of identity as infinity is reached sufficiently rapidly so as to leave *both* the flat metric *and* an orthonormal frame at infinity invariant. Instead of trying to make this statement involving infinity rigorous, let us consider a different diffeomorphism group D^P which leaves a frame at a finite point P invariant. The latter implies that the vector fields at P are left invariant by D^P. Consider a subgroup R of D^P which acts as identity on a metric q. We want to show that R consists only of the identity diffeomorphism. Since every direction ξ at P and the metric q on ^3M are invariant under R, and ξ and q uniquely determine a geodesic starting at P (and tangent to ξ), all geodesics passing through P are (point-wise) invariant under R. If geodesics through P also go through every point of ^3M, then it follows that no point of ^3M is moved by the action of R. Therefore R then consists only of the identity and D^P acts freely on $\tilde{Q}$. Even if the geodesics through P do not fill up ^3M, it is still possible to show by further reasoning that D^P acts freely on $\tilde{Q}$. In a somewhat similar way it can be shown that D^∞ as well acts freely on $\tilde{Q}$.

The fundamental group $\pi_1(Q)$ for gravity theories depends on the three manifold ^3M. We will be classifying three manifolds in Chapter 21. The properties of quantum theories associated with different UIR's of $\pi_1(Q)$ have not yet been explored in depth in the literature. The possibility of spinorial states in quantum gravity on certain ^3M is a consequence of the properties of $\pi_1(Q)$. We will be studying this remarkable possibility as well in Chapter 21. There are other striking results as well caused by the properties of $\pi_1(Q)$ such as the invalidity of our central ideas about statistics for certain geons. Chapter 22 may be consulted for a discussion of these results.

20.5 THE DOMAIN OF WAVE FUNCTIONS IN GAUGE THEORIES AND GRAVITY

According to the discussion thus far to construct the physical states in gauge theories, we can start with functions on the space $\mathcal{A}$ of connections and subject them to the Gauss law constraint (20.4). The physical states can therefore be regarded as function of $\mathcal{A}/\mathcal{G}_0^\infty$. This space is simply connected as the line of argument in Section 20.3 shows:

$$\pi_1(\mathcal{A}/\mathcal{G}_0^\infty) = \pi_0(\mathcal{G}_0^\infty) = 0 . \tag{20.16}$$

On the other hand, the functions on $\mathcal{A}$ which are classical observables (and which become quantum observables on quantization) are invariant under $\mathcal{G}^\infty$ and hence may be regarded as functions on

$$Q = \mathcal{A}/\mathcal{G}^\infty \quad . \tag{20.17}$$

This space for nonabelian gauge theories is not simply connected since

$$\pi_1(Q) = \pi_0(\mathcal{G}^\infty) = \mathbf{Z} \quad . \tag{20.18}$$

The fundamental group of Q is in fact $G = \mathcal{G}^\infty/\mathcal{G}_0^\infty$:

$$\pi_1(Q) = G \quad . \tag{20.19}$$

What is the relation between $\mathcal{A}/\mathcal{G}_0^\infty$ and Q? It is not difficult to show that $\mathcal{A}/\mathcal{G}_0^\infty$ is the universal covering space of Q. For this purpose let us first repeat what we said about universal covering spaces in 8.3.

Consider a general configuration space Q and let $q = (q^1, q^2, \ldots q^k)$ be the coordinates of Q. When Q is not simply connected, there is always a universal covering space $\overline{Q}$ of Q which is distinct from Q and which has the following properties: Let $\overline{q}$ be the coordinates of $\overline{Q}$ and let t denote a generic element of the group $\pi_1(Q)$. Then $\pi_1(Q)$ has a free action on $\overline{Q}$ which we denote by $\overline{q} \to \overline{q}t$. The points which can be reached from $\overline{q}$ by the action of $\pi_1(Q)$ constitute the set $\overline{q}\pi_1(Q) \equiv \{\overline{q}t | t \in \pi_1(Q)\}$. When we identify such sets $\overline{q}\pi_1(Q)$ for every $\overline{q}$ as a single point, we get back the space Q. That is, Q can be thought of as the set of sets $\{\overline{q}\pi_1(Q)\}$. Our original coordinates q are the coordinates of the set $\overline{q}\pi_1(Q)$. We can also define a projection map $\pi : \overline{Q} \to Q$ as follows:

$$\pi : \overline{Q} \to Q \quad ,$$
$$\overline{q} \to \pi(\overline{q}) \equiv \overline{q}\pi_1(Q) \equiv q \quad . \tag{20.20}$$

Note that since $\pi_1(Q)$ is a group and $t \in \pi_1(Q)$, we have $t\pi_1(Q) = \pi_1(Q)$ and

$$\pi[\overline{q}t] = \pi(\overline{q}) \quad . \tag{20.21}$$

Thus $\overline{Q}$ is a principal fibre bundle over Q with structure group $\pi_1(Q)$.

Let us now show that $\mathcal{A}/\mathcal{G}_0^\infty$ is in fact the universal covering space of $\mathcal{A}/\mathcal{G}^\infty$. The elements of $\mathcal{A}/\mathcal{G}_0^\infty$, $\mathcal{A}/\mathcal{G}^\infty$ and $\mathcal{G}^\infty/\mathcal{G}_0^\infty$ are the sets $\mathcal{G}_0^\infty \circ A$, $\mathcal{G}^\infty \circ A$ and $\mathcal{G}_0^\infty g$, where $A \in \mathcal{A}$ and $g \in \mathcal{G}^\infty$. Here, for example if $g \in \mathcal{G}^\infty$, then $g \circ A$ is the gauge transform of A by g,

$$g \circ A = g\,A\,g^{-1} + g\,d\,g^{-1} \quad , \tag{20.22}$$

while $\mathcal{G}^\infty \circ A$ is the collection of all such transforms of A by elements of $\mathcal{G}^\infty$. Note that since $\mathcal{G}_0^\infty$ is an invariant subgroup of $\mathcal{G}^\infty$, the cosets $\mathcal{G}_0^\infty g$ and $g\mathcal{G}_0^\infty$ are the same:

$$\mathcal{G}_0^\infty g = g\mathcal{G}_0^\infty \quad . \tag{20.23}$$

Now the group $\mathcal{G}^\infty/\mathcal{G}_0^\infty$ acts on $\mathcal{A}/\mathcal{G}_0^\infty$ according to the rule

$$\mathcal{G}_0^\infty \circ A \rightarrow (g\mathcal{G}_0^\infty)(\mathcal{G}_0^\infty \circ A) = g\mathcal{G}_0^\infty \circ A \quad . \tag{20.24}$$

Since $\mathcal{G}^\infty$ acts freely on $\mathcal{A}$, this action of $\mathcal{G}^\infty/\mathcal{G}_0^\infty$ on $\mathcal{A}/\mathcal{G}_0^\infty$ is easily seen to be free. The quotient of $\mathcal{A}/\mathcal{G}_0^\infty$ by $\mathcal{G}^\infty/\mathcal{G}_0^\infty$ is the space with elements $(\mathcal{G}^\infty\mathcal{G}_0^\infty)\,(\mathcal{G}_0^\infty \circ A) = \mathcal{G}^\infty \circ A$, that is, it is $\mathcal{A}/\mathcal{G}^\infty$:

$$(\mathcal{A}/\mathcal{G}_0^\infty)/(\mathcal{G}^\infty/\mathcal{G}_0^\infty) = \mathcal{A}/\mathcal{G}^\infty \quad . \tag{20.25}$$

The space $\mathcal{A}/\mathcal{G}_0^\infty$ is also simply connected by (20.16). It is now clear that $\mathcal{A}/\mathcal{G}_0^\infty$ is in fact the universal covering space $\overline{Q}$ of $Q = \mathcal{A}/\mathcal{G}^\infty$.

Thus we find that in the approach to quantization of gauge theories outlined in Section 20.2, the wave functions need not be functions on the configuration space Q. Rather they are best regarded as functions on the universal covering space $\overline{Q}$. The approach to quantization on multiply connected configuration spaces developed in Chapter 8 also assumes that the wave functions are functions on the universal covering space. As a consequence, it is easy to see the equivalence of the two approaches.

Similar considerations apply to quantum gravity as well. Here the physical wave functions may be regarded as functions on $\tilde{Q}$ which are invariant under the action of D_0^∞ [see Eq. (20.8)]. They are thus functions on

$$\overline{Q} = \tilde{Q}/D_0^\infty \quad . \tag{20.26}$$

As is now easily shown, $\overline{Q}$ is simply connected:

$$\pi_1(\overline{Q}) = 0 \quad . \tag{20.27}$$

The observables functions on $\tilde{Q}$ are required to be invariant under D^∞. They are thus functions on

$$
\begin{aligned}
Q &= \tilde{Q}/D^\infty \\
&= \overline{Q}/G \quad , \\
G &= D^\infty/D_0^\infty \quad .
\end{aligned} \tag{20.28}
$$

We can identify Q as our configuration space. It is in general multiply connected. In fact

$$\pi_1(Q) = G \tag{20.29}$$

since the action of G on $\overline{Q}$ is free. Further $\overline{Q}$, being simply connected, is the universal covering space of Q. Thus analogously to nonabelian gauge theories and consistently with the discussion of Chapter 8, the domain of wave functions here as well is the universal covering space $\overline{Q}$ of Q.

20.6 COMMENTS ON OBSERVABLES

In the preceding discussion, we have assumed in the case of gauge theories that the observables are invariant under the gauge group $\mathcal{G}^\infty$, while in the case of gravity we assumed that they are invariant under D^∞. However, the validity of these assumptions is not so clear, especially in the case of gravity. It does not seem to be implied by the general principles of quantization of constrained systems as developed by Dirac, Bergmann and others as discussed in Chapter 2. These principles seem to require only that observables are invariant under

$\mathcal{G}_0^\infty$ for gauge theories and $\mathcal{D}_0^\infty$ in the case of gravity. We shall now discuss this point.

We recall that in the approach to the quantization of constrained systems, the observables or first class variables in the classical theory are those functions on the phase space which have (weakly) zero Poisson brackets (PB's) with first class constraints. This implies that they are functions on the constrained phase space which are invariant under $\mathcal{G}_0^\infty$ in gauge theories and under D_0^∞ in gravity. It is not immediately evident that they must be invariant under $\mathcal{G}^\infty$ or D^∞ as well.

In nonabelian gauge theories there are functions which are invariant under $\mathcal{G}_0^\infty$, but not under $\mathcal{G}^\infty$. One such function is the integral

$$I(A) = \int \text{Tr} \left[A \wedge dA + \frac{2}{3} A \wedge A \wedge A\right] \qquad (20.30)$$

of the Chern-Simons three-form which was discussed in Chapter 11. As is well known, $I(A)$ is invariant under the action of $\mathcal{G}_0^\infty$, that is under "small" gauge transformations. However under the action of an element of $\mathcal{G}^\infty$, it changes by a term proportional to the winding number of the transformation so that it is not invariant under $\mathcal{G}^\infty$.

There is the possibility of the existence of similar functions in gravity as well. We are thus confronted by the question as to why we exclude such functions as observables.

The reason appears to be related to the fact that such functions are intrinsically nonlocal and that it is difficult to imagine a reasonable experiment which will determine their values for a state. Note in this context that the integrand of $I(A)$ or its integral over any finite volume is not invariant under $\mathcal{G}_0^\infty$ and hence is not an observable.

Further clarification of these issues seems necessary before $I(A)$ and other functions with similar properties can be firmly excluded from the set of observables. In our discussion, we shall simply assume that all observables must be invariant under $\mathcal{G}^\infty$ in gauge theories and under D^∞ in gravity. We thereby rule out functions like $I(A)$ as observables by assumption.

Chapter 21

GEONS AND THEIR PROPERTIES

21.1 INTRODUCTION

We saw in the previous Chapter that the configuration space of gravity without matter in $3+1$ dimensional space-time, i.e. $Q = \mathrm{Riem}(^3M)/D^\infty$, has the fundamental group $\pi_1(Q) = G = D^\infty/D_0^\infty$, the general theory of multiply connected configuration spaces developed in Chapter 8 being applicable to such Q.

It has been shown by Friedman, Sorkin and Witt [62,69,151] that for suitable asymptotically flat three manifolds ^3M, the group G can have a rich structure with striking physical implications.

In what follows, we shall review the investigations of Friedman, Sorkin and Witt on the properties of asymptotically flat three manifolds ^3M. As shown by these authors, the study of such three manifolds naturally leads to the concept of gravitational geons. These geons may be thought of as the gravitational analogues of Skyrmions in the chiral model. They are constructed from certain elementary three manifolds called prime manifolds. These prime manifolds with their associated dynamical variables characterize the geons.

The original manifold can always be decomposed in a well defined sense in terms of these prime manifolds while the prime manifolds can not be so decomposed. After explaining this decomposition theorem and consequently the construction of geons in Section 2, we finally discuss in Section 3 the remarkable result of Friedman and Sorkin that many of these geons can be so quantized that their states change sign under a 2π rotation and are therefore spinorial. This is a consequence of the fact that for such geons, the 2π rotation is not in the identity component D_0^∞ of D^∞.

21.2 THE PRIME DECOMPOSITION THEOREM AND GRAVITATIONAL GEONS

An essential tool in the construction of gravitational geons is the decomposition of ^3M into certain "prime" manifolds. In this Section, we shall explain this decomposition theorem.

If we are given two N dimensional manifolds M_1 and M_2, the connected sum $M_1 \# M_2$ of M_1 and M_2 is defined as follows. We first remove two N dimensional balls or solid spheres [spheres with their interiors] B_1 and B_2 from M_1 and M_2. This leaves us with the manifolds $M_1 \backslash B_1$ and $M_2 \backslash B_2$. Assuming that M_1 and M_2 are boundaryless, the boundaries of $M_1 \backslash B_1$ and $M_2 \backslash B_2$ are $(N-1)$ spheres S_1^{N-1} and S_2^{N-1}. We then glue together $M_1 \backslash B_1$ and $M_2 \backslash B_2$ by identifying the points of S_1^{N-1} and S_2^{N-1}. The result is an N dimensional manifold $M_1 \# M_2$ with no boundary, called the "connected sum" of M_1 and M_2.

For $N = 1$, if we take M_1 and M_2 to be circles, $M_1 \backslash B_1$ and $M_2 \backslash B_2$ are two circles from which two intervals have been removed. The connected sum $M_1 \# M_2$ is thus again a circle:

$$S^1 \# S^1 = S^1 \quad .$$

(21.1)

If M_1 and M_2 are two oriented circles, there are actually two connected sums we can define as can be seen by drawing simple diagrams. In one, $M_1 \backslash B_1$ and $M_2 \backslash B_2$ are joined together in a way which is compatible with their orientations, while in the other they are joined together in a way which is incompatible with their orientations. This ambiguity arises in all dimensions. We shall hereafter assume for simplicity that all manifolds are oriented and that the connected sum is so defined as to be compatible with these orientations.

For $N = 1$, one sees easily that the connected sum of a real line and a circle is a real line:

$$\mathbf{R}^1 \# S^1 = \mathbf{R}^1 \quad . \tag{21.2}$$

Furthermore it is obvious for $N = 1$ that the connected sum is commutative and associative:

$$M_1 \# M_2 = M_2 \# M_1 \quad ,$$
$$(M_1 \# M_2) \# M_3 = M_1 \# (M_2 \# M_3) \quad . \tag{21.3}$$

We may therefore write either side of the last equation as $M_1 \# M_2 \# M_3$ without brackets.

Let us now consider $N = 2$. The connected sums involving the sphere S^2, the torus T^2 and the plane $\mathbf{R}^2$ are easily seen to fulfill the following:

$$S^2 \# S^2 = S^2 \quad ,$$
$$S^2 \# T^2 = T^2 \# S^2 = T^2 \quad ,$$
$$T^2 \# T^2 = \text{Sphere with two handles,}$$
$$(T^2 \# S^2) \# T^2 = T^2 \# (S^2 \# T^2) = T^2 \# T^2 \quad ,$$
$$\mathbf{R}^2 \# S^2 = S^2 \# \mathbf{R}^2 = \mathbf{R}^2$$
$$\quad = \text{Sphere with one point removed,}$$
$$\mathbf{R}^2 \# T^2 = T^2 \# \mathbf{R}^2 = \text{Torus with one point removed.}$$
$$\tag{21.4}$$

Such considerations are readily generalized when spheres with n handles are involved. They show the following basic facts about connected sums in two dimensions:

1. The two sphere acts as the identity with regard to connected sums:

$$M \# S^2 = S^2 \# M = M \quad . \tag{21.5}$$

2. The connected sum of $\mathbf{R}^2$ with any manifold M is a manifold obtained by removing a point from M.

3. The process of taking connected sums is commutative and associative:

$$M_1 \# M_2 = M_2 \# M_1 \quad , \tag{21.6}$$

$$(M_1 \# M_2) \# M_3 = M_1 \# (M_2 \# M_3). \tag{21.7}$$

We may thus define the connected sum of K distinct manifolds unambiguously as $M_1 \# M_2 \# \ldots \# M_K$.

These properties generalize to all dimensions N where of course the analogues of S^2 and $\mathbf{R}^2$ are S^N and $\mathbf{R}^N$.

Hereafter by a manifold, we shall mean a manifold without boundary. Further, as mentioned earlier, we restrict ourselves to orientable (and oriented) manifolds and assume that connected sums are taken so as to be compatible with these orientations.

We can now illustrate the prime decomposition theorem for $N = 2$. [The corresponding result for $N = 1$ is trivial, the only compact manifold there being a circle.] The oriented prime manifolds are S^2 and T^2. If M is compact, oriented and not a two sphere, it is a sphere with n handles (n being finite) and hence has the decomposition

$$M = T^2 \# T^2 \# \ldots \# T^2 \ (n \ \text{terms}) \tag{21.8}$$

into a finite number of prime summands. [If M is a two sphere, then its decomposition into prime summands is just $M = S^2$.] If M is oriented and has one asymptotic region, it can be obtained by removing a point from a compact oriented manifold $\overline{M}$ and hence

$$M = \mathbf{R}^2 \# \overline{M} \quad . \tag{21.9}$$

Now $\overline{M}$ can be decomposed into prime summands as above. Note that if $\overline{M} = S^2$, then $M = \mathbf{R}^2$.

The decompositions (21.8) and (21.9) into prime summands are not unique since we have the freedom to attach any number of S^2's to their right hand sides using $\#$ in view of (21.5). This ambiguity is clearly trivial and will not bother us in any of our considerations.

In $2+1$ quantum gravity, the analogue of the three dimensional geons of Friedman and Sorkin can be constructed from T^2. We consider $\mathbf{R}^2 \# T^2$ and define a metric $q = (q_{ab})$ on it in such a way that the handle is contained within a circle S^1 of small radius. Such a spatial slice along with its metric q and conjugate momentum $\pi = (\pi^{ab})$ may be regarded as defining a topological excitation which we identify with an elementary geon. [Here we are assuming that the constraints of the gravity theory in question are compatible with the existence of such q_{ab} and π^{ab}. This theory need not be Einstein's theory. We have also not discussed the role of the diffeomorphism group in these remarks. Ref. [154] may be consulted for clarification on this point.]

Let us now pass on to $3+1$ space time. Here as well, as we already mentioned, the obvious generalizations of properties 1, 2 and 3 are valid. The prime decomposition theorem is also valid so that any ^3M with one asymptotic region has the decomposition

$$^3M = \mathbf{R}^3 \# M_1 \# M_2 \# \ldots \# M_n \tag{21.10}$$

into prime summands M_i where the number of terms in (21.10) is finite. As explained earlier, elementary geons are constructed using prime manifolds.

In the case we are considering where all operations are compatible with orientation, this decomposition is known to be unique up to permutation of terms except for the trivial ambiguities due to the possibility of attaching S^3's.

The orientable prime three manifolds have been partially classified. We give below an incomplete list of such manifolds. Unlike for two manifolds, the number of distinct prime three manifolds is infinite.

The following are orientable prime three manifolds besides S^3:

1. $T^3 = S^1 \times S^1 \times S^1$.

2. $S^2 \times S^1$.

3. Spherical Space Forms:

Recall that the three sphere $S^3 = \{n \in \mathbf{R}^4 | n_0^2 + \vec{n}^2 = 1\}$ can be identified with the manifold of the group $SU(2)$ via the map $n \to n_0 \mathbf{1} + i\vec{\tau} \cdot \vec{n}$. Here $\mathbf{1}$ is the unit 2×2 matrix and τ_i are Pauli matrices. Now there is an action of $SU(2) \times SU(2) = \{(u, v)\}$ on $SU(2)$ [and hence on S^3] given by

$$SU(2) \ni g \to ugv^{-1} \in SU(2) \quad . \tag{21.11}$$

Since $(-\mathbf{1}, -\mathbf{1}) \in SU(2) \times SU(2)$ acts trivially, (20.11) can be regarded as defining the action of $SO(4) = [SU(2) \times SU(2)]/\mathbf{Z}_2$ on $SU(2)$, the elements of $\mathbf{Z}_2$ being $(\mathbf{1}, \mathbf{1})$ and $(-\mathbf{1}, -\mathbf{1})$. Let Γ be any discrete subgroup of $SO(4)$ which acts freely on $SU(2)$. Thus the only element of Γ which leaves any point of $SU(2)$ fixed is the identity. Such subgroups Γ can be obtained by considering the discrete subgroups $\overline{\Gamma}$ of $SU(2) \times SU(2)$ containing $(-u, -v)$, whenever they contain (u, v), and with the property

$$u_1 g v_1^{-1} = u_2 g v_2^{-1} \quad \Rightarrow$$
$$(u_1, v_1) = (u_2, v_2) \text{ or } (-u_2, -v_2) \tag{21.12}$$

for any g. The group Γ is then $\overline{\Gamma}/\mathbf{Z}_2$. The spherical space forms are defined to be $SU(2)/\Gamma$ or S^3/Γ.

The discrete subgroups of Γ which act freely on S^3 have been classified and are infinite in number. Thus there are an infinite number of spherical space forms.

In Chapter 8, we have discussed the spherical space form $SU(2)/H$, where H is the "quaternion" group. We have seen there that this space has a simple interpretation as the configuration space of a nucleus or a molecule with three distinct but fixed moments of inertia.

4. Hyperbolic Spaces:

Consider the upper sheet H^+ of the unit hyperboloid in four dimensional Minkowski space M^4:

$$H^+ = \{x \in M^4 | x_0^2 - \vec{x}^2 = 1, x_0 > 0\} \ . \tag{21.13}$$

The connected component $L_+^\uparrow$ of the Lorentz group [54] acts on H^+ transitively. Thus there are elements of $L_+^\uparrow$ which map any point of H^+ to any other point of H^+. Let Γ be any discrete subgroup of $L_+^\uparrow$ which acts freely on H^+. Then the hyperbolic spaces are H^+/Γ. They are also infinite in number.

We shall now briefly address the question as to which of these prime manifolds can occur classically. More precisely we comment on the following issue: If M is a prime manifold, then are there initial data q_{ab}, π^{ab} on $\mathbf{R}^3 \# M$ compatible with the boundary conditions (20.6) and the constraints (20.7)? If there are no such initial data already classically on a certain $\mathbf{R}^3 \# M$, then it is not probable that a quantum geon associated with $\mathbf{R}^3 \# M$ exists. It is known that $\mathbf{R}^3 \# M$ admits such initial data when M is $S^2 \times S^1$ or any one of the spherical space forms. The situations as regards the other manifolds still seems cloudy.

21.3 SPINORIAL STATES FROM PURE GRAVITY

According to the discussion in Chapter 20, the group $G = D^\infty / D_0^\infty$ can act nontrivially on the quantum states. The demonstration of Friedman and Sorkin that spinorial states can be constructed in quantum gravity involves the interrelation of this group with the 2π rotation of states.

For this purpose, we must first discuss the definition of spatial rotation of states. It is only asymptotically that the metric becomes flat [Cf. Eq. (20.6)] and it is only there that the rotation of points about an axis $\hat{n}$ by an angle θ can be so defined as to have a conventional meaning. Nonetheless it is possible to define spatial rotation of states in a meaningful way. Let $R_\theta(\hat{n})$ be any diffeomorphism of ^3M which asymptotically acts like a rotation by an angle θ about an axis $\hat{n}$. There are in general an infinity of such diffeomorphisms for a given θ and $\hat{n}$. Still the group of all such diffeomorphisms when acting on states subject to (20.8) generate either the group $SO(3)$ or the group $SU(2)$ as we show below. The restriction of the group of such diffeomorphisms to the states subject to (20.8) may thus be thought of as the realization of spatial rotations in quantum theory.

To show the stated property of the preceding diffeomorphisms, consider their infinitesimal forms corresponding to small rotations. They are associated with vector fields X_α which asymptotically generate spatial rotations. For each α, there are of course many such X_α, say X_α, X'_α, ... Now if X_α, X_β and X_γ are any three such vector fields which asymptotically are generators of rotations around α, β, and γ axes, then $[X_\alpha, X_\beta]$ and $\varepsilon_{\alpha\beta\gamma}X_\gamma$ act in the same way asymptotically, $\varepsilon_{\alpha\beta\gamma}$ being the Levi-Civita symbol. Therefore

$$[X_\alpha, X_\beta] = \varepsilon_{\alpha\beta\gamma}X_\gamma + Y \quad , \tag{21.14}$$

where Y is a vector field which asymptotically vanishes. Thus by (20.8) the operator which represents Y in quantum theory annihilates the physical states. A similar reasoning shows that if X_α and X'_α are two distinct vector fields both of which asymptotically represent the generator of rotations around α^{th} axis, then $X_\alpha - X'_\alpha$ annihilates the physical states. It follows that the Lie algebra of the group of diffeomorphisms of the form $R_\theta(\hat{n})$ when acting on physical states is isomorphic to the Lie algebra of $SU(2)$. This is the result we wanted to prove.

We may remark here that (21.14) defines an extension of the $SU(2)$ Lie algebra by the Lie algebra of D_0^∞.

We must of course make sure that for each θ and $\hat{n}$, there is at least one diffeomorphism $R_\theta(\hat{n})$ which asymptotically acts like rotation by angle θ

around $\hat{n}$. This is easy to show. Recall that by the assumption stated in Chapter 20, there is a compact set $\mathcal{C}$ such that its complement $^3M\backslash\mathcal{C}$ is diffeomorphic to the complement of a ball in $\mathbf{R}^3$. Thus all topological complications are contained in the set $\mathcal{C}$. Since $^3M\backslash\mathcal{C}$ is diffeomorphic to the complement of a ball $\mathbf{R}^3$, it admits a flat metric. We can thus choose an orthonormal frame with respect to this flat metric so that the metric components (suitable referred to this frame) are δ_{ab}. This frame at spatial infinity is the same frame for which the components q_{ab} of q are δ_{ab} [Cf. Eq. (20.6)]. Now fix a large enough two sphere S^2 [with respect to this flat metric] of radius say r_0 so that $\mathcal{C}$ is isolated in the interior of S^2. With these constructions, it is possible to explicitly define a diffeomorphism $R_\theta(\hat{n})$ of ^3M which can be identified with the "rotation by angle θ around axis $\hat{n}$" when acting on physical states. It acts as identity in the interior of S^2 as well as on S^2. On each sphere of radius $r > r_0$ (with respect to the flat metric), it acts as the rotation by an angle $\phi(r)$ about the axis $\hat{n}$ where the orthonormal frame we have chosen defines the axis $\hat{n}$. The angle $\phi(r)$ is so chosen as to change from 0 to θ as r increases from r_0 to ∞. We have now completed the definition of one diffeomorphism $R_\theta(\hat{n})$ with all the required properties. We shall use this $R_\theta(\hat{n})$ in the discussion that follows in order to be specific.

Consider now a 2π rotation $\hat{n}$. It is represented by the diffeomorphism $R_{2\pi}(\hat{n})$. It acts as identity at spatial infinity and is therefore an element of D^∞. If it is also an element of D_0^∞, it will act as identity on physical states. We shall see below however that this is not the case for certain ^3M and that this fact will lead to the possibility of spinorial states. On the other hand the 4π rotation $R_{4\pi}(\hat{n})$ is always an element of D_0^∞ and acts trivially on all physical states. This is because as r increases from r_0 to ∞, the rotations associated with $R_\theta(\hat{n})$ are of the form $e^{i\phi(r)\vec{J}\cdot\hat{n}}$ in terms of 3×3 matrices, where $\vec{J}$ are spin 1 angular momentum matrices, $\phi(r_0) = 0$ and $\phi(\infty) = 4\pi$. This closed curve in $SO(3)$ (traced out when r increases from r_0 to ∞) is well known to be smoothly deformable (homotopic) to the identity matrix. Consequently the element $R_{4\pi}(\hat{n}) \in D^\infty$ can be smoothly deformed to a diffeomorphism

$R'_{4\pi}(\hat{n})$ which acts as identity for $r \geq r_0$. Further during the process of this deformation, we can always stay within D^∞ [that is, maintain the condition $\phi(\infty) = 4\pi$] and we need not also alter the action within S^2. Thus $R_{4\pi}(\hat{n})$ is homotopic to the identity diffeomorphism and hence acts trivially on physical states.

Let us assume for a moment that there are ^3M for which $R_{2\pi}(\hat{n}) \notin D_0^\infty$. Then there are surely physical states ψ which are not invariant under $R_{2\pi}(\hat{n})$. For such states,

$$\hat{R}_{2\pi}(\hat{n})\psi \neq \psi \quad ,$$
$$\hat{R}_{4\pi}(\hat{n})\psi = \psi \quad . \tag{21.15}$$

Thus

$$[1 - \hat{R}_{2\pi}(\hat{n})]\psi \neq 0 \quad ,$$
$$\hat{R}_{2\pi}[1 - \hat{R}_{2\pi}(\hat{n})]\psi = -[1 - \hat{R}_{2\pi}(\hat{n})]\psi \tag{21.16}$$

or the state $[1 - \hat{R}_{2\pi}(\hat{n})]\psi$ is spinorial.

We may note here that the action of $R_{2\pi}(\hat{n})$ on physical states is in fact independent of $\hat{n}$. The proof of this independence is modelled on the proof that $R_{4\pi}(\hat{n})$ acts as identity on physical states. Thus if $R_{2\pi}(\hat{n})$ and $R_{2\pi}(\hat{m})$ are two such diffeomorphisms, the associated curves in $SO(3)$ as r increases from r_0 to ∞ are homotopic to each other. Therefore $R_{2\pi}(\hat{n}) = R_{2\pi}(\hat{m})\chi$, where $\chi \in D_0^\infty$ and $\hat{R}_{2\pi}(\hat{n})\psi = \hat{R}_{2\pi}(\hat{m})\psi$ on physical states ψ.

We next discuss the conditions under which $R_{2\pi}(\hat{n}) \notin D_0^\infty$. For this purpose, we first recall the definition of a two-Sylow subgroup of a finite group G. It is a subgroup of G whose order is a power of 2 and which is not properly contained in any larger such subgroup. All two-Sylow subgroups of a given group are known to be isomorphic.

The following result about $R_{2\pi}(\hat{n})$ is known for orientable manifolds ^3M with one asymptotic region. Let

$$^3M = \mathbf{R}^3 \# M_1 \# M_2 \# \cdots \quad , \tag{21.17}$$

where M_i are orientable prime manifolds. Then $R_{2\pi}(\hat{n}) \in D_0^\infty$ if and only if each of the prime summands is either $S^2 \times S^1$ or has a fundamental group with a cyclic two-Sylow subgroup.

From this result and from the partial list of prime manifolds earlier in this Chapter, we see that there are plenty of ^3M for which $R_{2\pi}(\hat{n}) \notin D_0^\infty$. For instance, let H be the quaternion group $\{\pm 1, \pm i\tau_i\}$, τ_i being Pauli matrices. It is of order 2^3 and hence is its own two-Sylow subgroup. It is nonabelian. It acts on $SU(2)$ by right multiplication and the space of cosets $SU(2)/H$ is a prime manifold with a noncyclic fundamental group H. Hence for the manifold

$$^3M = \mathbf{R}^3 \# [SU(2)/H] \quad , \tag{21.18}$$

$R_{2\pi}(\hat{n}) \notin D_0^\infty$ and quantum gravity based on this manifold admits spinorial states.

We draw attention to two results concerning geons in conclusion: 1) There is an analogue in $2+1$ dimensions of the possibility of spinorial states in $3+1$ dimensional pure gravity. Suppose for example that the two manifold at constant time in $2+1$ dimensions is $\mathbf{R}^2 \# T^2$. Then a pure gravity Lagrangian based on this manifold can be quantized so that the quantum states have any fractional spin, a particular quantum gravity theory being associated with a particular fractional spin. For a detailed discussion of this result, see Ref. [154]. 2) There are identical quantum geons in two and three spatial dimensions with remarkable statistical properties. For instance, it is possible to quantize certain of these geons in three spatial dimensions in such a way that they violate the familiar spin-statistics theorem. It is also possible to quantize certain of these geons so that they are neither bosons, nor fermions nor paraparticles. It is striking that strings in three spatial dimensions also admit quantization with somewhat similar unusual statistical features. The next Chapter is devoted to a brief discussion of strings and geons from the point of view of statistics.

Chapter 22

STATISTICS, STRINGS AND GRAVITY

22.1 INTRODUCTION

In this Chapter, we will briefly describe the results of Refs. [62,154-155] on the nature of statistics in string models and in generally covariant theories in $3 + 1$ dimensions. (The results on strings use certain mathematical results proved by Goldsmith [157]. See also Ref. [156].) We shall see in both cases that their statistical properties are not as a rule governed by the permutation group S_N as is the case for N identical particles in $3 + 1$ dimensions. The group which assumes the role of S_N is instead often an infinite nonabelian group [which depends on the details of the string or of the gravitational model and on the number N of identical constituents being considered] which does not even contain S_N as a subgroup. As a result, there are quantum theories associated with such models with unusual statistical features, such as the failure of the normal spin-statistics relation and the existence of systems which are not bosons, fermions or paraparticles. The implications of these novel possibilities are not well understood. Similar possibilities are known to occur for a system of identical particles in $2 + 1$ dimensions, where the braid group which governs statistics also does not contain S_N as a subgroup. They can also occur for identical particles in any number of space-time dimensions for suitable spatial topologies. [See Imbo et al. in Ref. [67].]

The strings we consider are not those which are considered in dual string models. Our strings are associated with unparametrized maps of circles to $\mathbf{R}^3$ at a given time, where we insist neither on the invariance of the action under transformations affecting the evolution coordinate nor on Poincaré invariance. Thus in our case there is no need to increase space-time dimension to a value beyond our current experience for reasons of consistency. Strings of this sort are of course physically important occurring as they do as defects in condensed matter systems as well as being predicted to occur in GUT models [see e.g. Refs. [22,136]].

There are many points of resemblance between the $3 + 1$ dimensional quantum strings and $3 + 1$ dimensional quantum gravities with interesting spatial topologies. We have alluded to a few of these features already. These similarities suggest that the study of these strings [which appear to be simpler dynamical systems than gravity] may also lead to insights into the nature of quantum gravity.

22.2 STATISTICS IN A PLANE

In this Section, we will briefly show that the principles of quantum mechanics allow for the existence of particle species which obey no definite statistics in the usual sense by examining the possible quantum mechanics of identical particles in a plane $\mathbf{R}^2$. For this purpose, let us recall a well known approach to the statistics of particles [55,61,96-99]. [See also Chapter 8.] The configuration space Q for N identical particles in $\mathbf{R}^2$ has coordinates $[x^{(1)}, x^{(2)}, \ldots, x^{(N)}]$ where $x^{(i)} \in \mathbf{R}^K$ and where we identify $[x^{(1)}, x^{(2)}, \ldots, x^{(i)}, \ldots, x^{(j)}, \ldots, x^{(N)}]$

with $[x^{(1)}, x^{(2)}, \ldots, x^{(j)}, \ldots, x^{(i)}, \ldots, x^{(N)}]$ in view of the identity of the particles. For technical convenience, the particles are also forbidden to occupy the same position so that $x^{(i)} \neq x^{(j)}$ if $i \neq j$. Such a Q is in general multiply connected. Now it is known [Cf. Section 8.5] that there are at least as many ways of quantizing the system as there are UIR's of $\pi_1(Q)$. In three or more dimensions, $\pi_1(Q) = S_N$ so that it is in fact the group which defines statistics in a conventional sense and its UIR's lead to quantum theories of Bose, Fermi or paraparticles. The situation however is quite different when $K = 2$. When $K = 2$, $\pi_1(Q)$ for two or more particles is an infinite discrete group called the braid group B_N. The group B_N does not contain a subgroup isomorphic to S_N so that the states of a particle species for a general UIR of B_N are not associated with any UIR of S_N. In the literature, it is customary to consider abelian UIR's of B_N. Even here we have the possibilities of "fractional statistics" which have no analogue in other dimensions [and which, barring exceptions, are not associated with UIR's of S_N]. As already mentioned in Chapter 11, such novel "statistics" have been found useful in the study of the quantum Hall effect [99] and high temperature superconductors [100,96]. [We remark that there is, of course, a way to get conventional "statistics" in two dimensions since S_N is a factor group of B_N. Thus any UIR of S_N can be found among those of B_N.]

Thus as these remarks indicate, and as it has been appreciated for some time, there exist novel statistical possibilities for point particles confined to a plane. [In this connection, note that a detailed discussion of the influence of the nonabelian Chern-Simons term on the statistics of identical particles (with internal symmetry) in a plane is contained Ref. [55]]. In the Sections which follow, we shall show that such novelties are to be found in three dimensions as well using the examples of strings [155,156] and of topological geons in gravity [154].

22.3 UNORIENTED STRINGS

We shall first consider the case of unoriented strings in three space dimensions.

The configuration space Q of a *single* unoriented string is constructed from maps $\mathcal{F}$ of the circle S^1 to $\mathbf{R}^3$ where we shall assume, for reasons of simplicity, that the image knots have no self linking, that is that they are "unknots". A point of Q may then be described as follows. Let σ be the coordinate on S^1 with $\sigma = 0$ and 2π being identified. Consider the group D of all diffeomorphisms (diffeos) $\sigma \to \varphi(\sigma)$ of S^1. These diffeos act on an element y in $\mathcal{F}$ according to $y \to \varphi y$ where $(\varphi y)(\sigma) = y[\varphi^{-1}(\sigma)]$. A point of Q is the equivalence class $< y >$ of all maps related to y by such a diffeomorphism.

As the group D contains diffeomorphisms such as $\sigma \to -\sigma$ reversing the orientation of S^1, the strings we consider at present are not oriented. Oriented strings will be briefly discussed on Section 4.

As we can infer from previous Chapters in this book, a principle topological property of interest for quantization is the fundamental group $\pi_1(Q)$. In the present case, it happens to be $\mathbf{Z}_2$. Let us describe the closed path or loop C in Q associated with the generator of this $\mathbf{Z}_2$ [157]. It is convenient to choose the base point which enters the definition of $\pi_1(Q)$ to be $< z >$ where

$$z(\sigma) = (\cos\sigma ,\ \sin\sigma ,\ 0) \ . \tag{22.1}$$

Thus the image of S^1 under z is the circle of unit radius in the 1-2 plane with origin as center. Let $R_1(\varphi)$ be the rotation by angle φ about 1-axis and consider $R_1(\varphi)z(\sigma)$. Since

$$R_1(\pi)z(\sigma) = (\cos\sigma ,\ -\sin\sigma ,\ 0) \ , \tag{22.2}$$

we have

$$< z >=< R_1(\pi)z > \ , \tag{22.3}$$

where $(R_1(\varphi)z)(\sigma) = R_1(\varphi)z(\sigma)$. Thus $< R_1(\varphi)z >$ describes a loop C in Q as φ changes from 0 to π and it is this loop which is associated with the generator of $\mathbf{Z}_2$. The following remarks may help to clarify why this is so. It can not be smoothly shrunk to a point in Q as will become explicit when we construct the universal covering space $\overline{Q}$ of Q below. On the other hand, the curve C^2 obtained by traversing C twice is $< R_1(\varphi)z >$ with φ changing from 0 to 2π and this curve *is* deformable to a point. To see this, let $J_\alpha(\alpha = 1, 2, 3)$ be the usual spin 1 angular momentum matrices $[(J_\alpha)_{ij} = -i\varepsilon_{\alpha ij}, \varepsilon_{\alpha ij}$ being the Levi-Civita symbol]. Then $R_1(\varphi) = e^{i\varphi J_1}$. Consider $e^{i\varphi \vec{n}(t)\cdot\vec{J}}$, where $\vec{n}(t)$ is a unit vector smoothly dependent on t such that $\vec{n}(0) = (1, 0, 0)$ and $\vec{n}(1) = (0, 0, 1)$. Since $< e^{i\varphi \vec{n}(1)\cdot\vec{J}}z >=< e^{i\varphi J_3}z >=< z >$ for all φ, the closed curves

$$C_t^2 = \{< e^{i\varphi \vec{n}(t)\cdot\vec{J}}z > ; \ 0 \leq \varphi \leq 2\pi\} \ , \tag{22.4}$$

provide a homotopy of C^2 to a point as t varies from 0 to 1. Thus the group generated by the homotopy class τ of C is $\mathbf{Z}_2$.

The universal covering space $\overline{Q}$ of Q is just the space of single oriented strings. An element of $\overline{Q}$ is the equivalence class $[y]$ of all strings related to y by the identity component D_0 of D. The projection map $\pi : \overline{Q} \rightarrow Q$ consists in identifying two strings with the same location in $\mathbf{R}^3$ but with distinct orientations. The generator τ of $\pi_1(Q) = \mathbf{Z}_2$ $[\tau^2 = $ identity $e]$ acts by permuting the fibres over $< y >$ so that the inverse image of $< y >$ is the set $\{[y] , \tau[y]\}$.

Note that since z and $R_1(\pi)z$ are oppositely oriented, the lift of the loop C from Q to $\overline{Q}$ starting at $[y]$ is an open curve from $[y]$ to $\tau[y]$. From this fact, and an elementary and familiar argument, it follows that C is not homotopic to a trivial curve.

The group $\mathbf{Z}_2$ has two distinct UIR's. The fact that $\pi_1(Q) = \mathbf{Z}_2$ for single unoriented strings means therefore that there are two ways of quantizing such a string, one for each UIR. The image of τ in one of these ways is $+1$ while in the other, it is -1. [The analogous distinct quantum theories in QCD are associated with the distinct UIR's $\mathbf{Z} \ni n \rightarrow e^{in\theta}$ $(0 \leq \theta < 2\pi)$ of $\mathbf{Z}$ and characterized by the familiar vacuum angle θ.] We will not further develop these distinct theories of unoriented strings here as our principal interest is in illustrating certain novel possibilities involving statistics.

Let us next consider N of these strings which we regard as identical. As for point particles, they are not allowed to overlap. This means that if $y^{(i)}$ is the map associated with the ith string, the intersection of the images of S^1 under $y^{(i)}$ and $y^{(j)}$ is null if $i \neq j$. A point of the configuration space Q for N identical strings is thus the unordered set

$$[< y^{(1)} >, \ldots, < y^{(i)} >, \ldots, < y^{(j)} >, \ldots, < y^{(N)} >] =$$
$$[< y^{(1)} >, \ldots, < y^{(j)} >, \ldots, < y^{(i)} >, \ldots, < y^{(N)} >] \qquad (22.5)$$

subject to the condition $\{y^{(i)}(\sigma)\} \cap \{y^{(j)}(\sigma)\} = \phi$ if $i \neq j$. This space Q is in general quite involved, so we shall make a further simplification by requiring that the strings are unlinked.

Let us now specialize to the case of two strings. The "motion" group [7] $\pi_1(Q)$ for this case is known to be generated by three elements T, E and S and to have the presentation

$$< T, E, S | T^2 = E^2 = (TE)^4 = (TESE)^2 = e > \quad , \qquad (22.6)$$

e being the identity. Let us now describe these elements.

The element T in (22.6) is associated with a nontrivial closed curve for an individual string described earlier. If z is the string (22.1), and if we set $y^{(1)} = z$, this curve is traced by

$$[< R_1(\varphi)y^{(1)} > , < y^{(2)} >] \qquad (22.7)$$

as φ increases from 0 to π. There is a similar closed curve which comes from $y^{(2)}$, but as the exchange operation E described below generates it from (22.7), there is no need to list it separately. The relation $T^2 = e$ can be proved following our earlier discussion.

The element E in (22.6) describes the exchange of the two strings. Let us take the base point for homotopy to be $[< z^{(1)} > , < z^{(2)} >]$ where both strings $z^{(i)}$ are circular and of unit radius, are in the 1-2 plane and have centers at

$(\pm 3, 0, 0)$ say. The closed path in Q which describes E corresponds to gradually moving one string to the position of the second string while at the same time moving the second string to the position of the first string, see Fig.1.

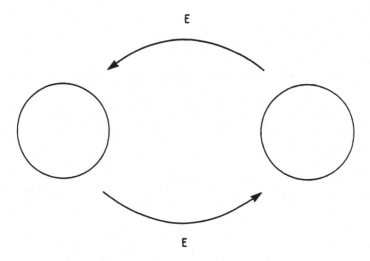

E

E

Figure 1

Now if the exchange is repeated twice, the resultant closed path or loop in Q can be deformed to a point. This is so because under E^2, each string would have travelled in a closed loop. Each of these loops can be deformed to a point, and it is trivial to arrange matters so that the strings do not overlap during this process. [For example, choose one of the loops to be in the 1-2 plane and choose the second loop in a plane which is away from this plane.] Thus $E^2 = e$.

It is easy to see now that ETE^{-1} is associated with the nontrivial loop for $y^{(2)}$ as was remarked earlier. It is also not difficult to see that T and

$ETE^{-1} = ETE$, which are associated with the nontrivial loops for $y^{(1)}$ and $y^{(2)}$, commute. This leads to the relation $(TE)^4 = e$ in Eq. (22.6).

The element S in (22.6) may be called as a slide. It is associated with the loop in Q got by transporting one string $y^{(1)}$ through a loop L which goes through the middle of the second string $y^{(2)}$, see Figure 2.

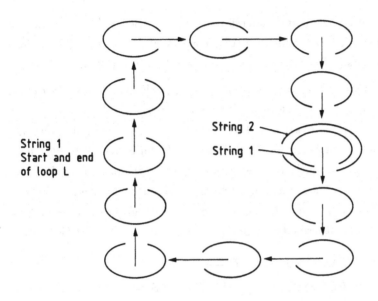

String 1
Start and end
of loop L

String 2
String 1

Figure 2

The inverse S^{-1} of S is obtained by traversing L in the opposite sense while the slides of $y^{(2)}$ through $y^{(1)}$ are ESE^{-1} and $ES^{-1}E^{-1}$.

We refer to Ref. [155] for a proof of the relation $(TESE)^2 = e$.

We may remark here that the definition of the exchange E, for instance, is not unique because we can identify ET or SES^{-1} equally well with the physical operation of the exchange of strings. The statements we shall make about statistics below will be insensitive to such redefinitions.

The possible quantum theories are associated with the UIR's of $\pi_1(Q)$. Here we shall not find all UIR's of $\pi_1(Q)$ for two strings, but rather find those where the additional relations

$$T = S^2 = e \qquad (22.8)$$

are fulfilled. These are enough to bring out the physical points. Further, the group

$$G = < T, E, S | T = E^2 = S^2 = e > \qquad (22.9)$$

that results with these additional relations can be used to discuss oriented strings and geons as well, which is rather convenient. Note that the relations for the generators in Eq. (22.9) imply the relations in Eq. (22.6).

The UIR's of G are easy to find [154]. The element $SE + ES$ can be verified to belong to the centre of the group algebra $\mathbf{C}(G)$ so that it is a multiple of e in a UIR, say $SE + ES = \beta e$, β being a complex number. Using this relation, any element of $\mathbf{C}(G)$ in a UIR can be reduced to the form $ae + bE + cS + dSE$ which involves at most four linear operators e, E, S and SE. Hence, the UIR's are at most two-dimensional by Burnside's theorem [158]. There are four obvious one-dimensional UIR's given the choices $E \to \pm 1, S \to \pm 1$. They describe two types of bosons and two types of fermions:

$$\textbf{BOSONS}: \quad E \longrightarrow 1 \,, \ S \longrightarrow \pm 1 \ . \qquad (22.10)$$

$$\textbf{FERMIONS}: \quad E \longrightarrow -1 \,, \ S \longrightarrow \pm 1 \ . \qquad (22.11)$$

Next consider the two-dimensional UIR's. Here $E, S \neq \pm 1_{2\times 2}$ as otherwise the representation will be reducible. Since $E^2 = 1_{2\times 2}$, we can therefore set $E = \sigma_3$ after a choice of basis, σ_i being the Pauli matrices. As regards S, since it is unitary, we have $S = e^{i\alpha}[a_0 1_{2\times 2} + i\vec{\sigma} \cdot \vec{a}]$, $a_0^2 + \vec{a} \cdot \vec{a} = 1$, α and a_μ, $\mu = 0, 1, 2, 3$, being real. Using $S^2 = 1_{2\times 2}$ and $S \neq \pm 1_{2\times 2}$, we thus find that $S = \vec{\sigma} \cdot \vec{n}$ where $\vec{n}$ is a real unit vector. By a rotation around the third axis which leaves E unaffected, we can reduce S to $S = \cos\theta\sigma_3 + \sin\theta\sigma_1$, $\sin\theta > 0$. [Sin $\theta = 0$ is forbidden for then the representation becomes reducible.] Thus we have the one parameter family of two-dimensional UIR's

$$E = \tau_3 \,, \ S = \cos\theta\sigma_3 + \sin\theta\sigma_1 \,, \ 0 < \theta < \pi \quad . \qquad (22.12)$$

The states of a single string are certainly always tensorial since as we saw, the loop C^2 associated with 2π rotation is homotopically trivial. Nevertheless, (22.10) and (22.11) show that quantum theories exist where it is either a boson or a fermion. The second possibility suggests a failure of the spin-statistics connection. Even more remarkable are the quantum theories associated with (22.12). In these theories, E has both the eigenvalues ± 1. Thus in such theories, the strings are *neither* bosons *nor* fermions. If initially there is a symmetric state with $E = 1$, interactions associated with slides can mix it with states with $E = -1$. Furthermore, a generic state is not even an eigenstate of exchange. [Note that the states of *two* identical particles in $\mathbf{R}^K (K \geq 3)$ are always *either* symmetric or antisymmetric, but not both. This is so even if the particles obey parastatistics. Likewise the states of two identical particles in $\mathbf{R}^2$ are always associated with a definite eigenvalue for exchange.]

It may be that second quantization will eliminate some or all of the exotic possibilities for string statistics indicated above (and in the following Section). This is made plausible by the fact that certain "topological" spin-statistics theorems have already been proved in Ref. [161] which establish the normal spin-statistics connection for strings (and particles as well) and eliminate all the exotic possibilites found above. The proof for strings requires the assumption that there is an "antistring" for every string with which it can annihilate (or pairproduced). The work in these papers [161] is based on Ref. [112] and the work by Sorkin [24] and Tscheuschner [96].

We mention in this connection that the analogue of the Chern-Simons terms [see Chapter 11] for strings which produce some of these novel statistical features have been constructed [155] [see also Harvey and Liu in Ref. [156]]. Also unusual statistical possibilities exist for identical particles in one-dimensional "mesoscopic " networks and topological spin-statistics theorems can be proved for these systems as well [162].

22.4 ORIENTED STRINGS

The configuration space for our single oriented strings is the universal cover $\overline{Q}$ of the unoriented single string configuration space Q. For a given geographical location in $\mathbf{R}^3$, we can regard the oriented string as having two internal states corresponding to the two orientations. Further, it is possible to continuously transform one internal state to the other by using $R_1(\varphi)$ or its suitable generalizations. The space $\overline{Q}$ is the quotient of $\mathcal{F}$ by the identity component D_0 of D. [D_0 does not contain orientation reversing diffeos of S^1.] It is simply connected so that there is no ambiguity in the construction of its quantum theory caused by a nontrivial $\pi_1(\overline{Q})$.

It may be emphasized that just as for the unoriented strings, the single oriented strings we consider are unknots. We shall also assume the absence of links for N identical oriented strings just as we did for the unoriented case.

Consider next two identical oriented strings. The points of the appropriate configuration space Q are then the unordered pairs $[[y^{(1)}], [y^{(2)}]] = [[y^{(2)}], [y^{(1)}]]$ where $\{y^{(1)}(\sigma)\} \cap \{y^{(2)}(\sigma)\} = \phi$. It can be shown that $\pi_1(Q)$ has the presentation

$$< E, S | E^2 = e > \quad , \tag{22.13}$$

where the relation $E^2 = e$ can be proved as before. There is only one point which merits emphasis here as regards the definitions of E and S. The orientation of the strings must appropriately match at the end of the processes associated with E and S. For instance, in the case of E, when string i is moved to the position of string j, the transformed $\{y^{(i)}(\sigma)\}$ must be identical to $\{y^{(j)}(\sigma)\}$ both in geographical location and in orientation. This can always be achieved, if necessary by using an operation such as the one associated with $R_1(\varphi)$.

The group G is obtained from (22.13) by imposing the additional relation $S^2 = e$. We can thus use the UIR's of G found earlier to reach similar conclusions about the spin and statistical properties of these strings as well.

22.5 QUANTUM GRAVITY

We shall now discuss the question of statistics in generally covariant theories in $K + 1$ dimensions built on asymptotically flat spatial slices M. For purposes of presentation, we will assume that there are no matter fields present. The configuration space Q is then $\tilde{Q}/D^\infty$ where $\tilde{Q}$ is the collection of asymptotically flat Riemannian metrics on the K dimensional manifold M and D^∞ is the group of diffeos of M leaving infinity and a frame at infinity invariant. [See Chapter 20 and also Refs. [62,69,151,154,156].] Using a specific choice of M for $K = 3$, we shall show that it is not in general possible to associate any definite statistics at all in the usual sense to quantum theories on Q. It has been shown elsewhere [154] that similar results are valid in $2 + 1$ dimensions as well.

The novel statistical possibilities available in quantum gravity will be shown below using the topological geon excitations of Chapter 21. As a prelude to their discussion, we briefly recall the definition and properties of the connected sum $M\#M'$ of two boundaryless manifolds of dimension K from that Chapter. It is defined by first removing two K dimensional balls B and B' from M and M'. The resultant sets $M\backslash B$ and $M'\backslash B'$ have $(K-1)$ dimensional spheres as boundaries. On identifying these boundaries, we get $M\#M'$. A space with one asymptotic region can always be thought of as the connected sum of a compact and boundaryless (that is to say, closed) manifold with $\mathbf{R}^K$. In dimensions K up to three, there exists a class of closed manifolds $\mathcal{P}_\alpha$ called prime manifolds so that any manifold with one asymptotic region is a connected sum of $\mathcal{P}_\alpha$'s and $\mathbf{R}^K$. For $K = 1$, the only prime manifold is the circle S^1 while for $K = 2$ there are three prime manifolds, namely S^2, T^2 and $\mathbf{R}P^2$. [Note that $\mathbf{R}P^2$ is not orientable. It is relevant only when considering the prime decomposition of nonorientable manifolds.] For $K = 3$, there are infinitely many prime manifolds $\mathcal{P}_\alpha$, and they have been partially classified. Since $M\#S^K = M$ for any M, spheres are usually omitted in the prime decomposition of any manifold other than a sphere itself. With this convention, a prime manifold can not be written as the connected sum of other manifolds and is in this sense elementary.

As discussed earlier, prime manifolds are elementary building blocks for constructing the topological excitations called geons with particle like characteristics. For instance, in $3 + 1$ dimensions, $M = \mathbf{R}^3 \# \mathcal{P}$ is a possible spatial slice for a generally covariant theory where we take $\mathcal{P}$ to be prime. In this slice, we can enclose or isolate $\mathcal{P}$ in a region B bounded by a sphere S^2 in the sense that $M \backslash B$ is homeomorphic to the complement of a ball in $\mathbf{R}^3$. We can also choose the metric in such a way that S^2 is of arbitrarily small size so that $\mathcal{P}$ is associated with a particle like excitation. [The definition of size here is perhaps a little too crude, as it is not D^∞ invariant. For a D^∞ invariant definition of size, see the second paper in Ref. [154].] This excitation can be regarded as elementary as $\mathcal{P}$ can not be decomposed as the connected sum of other manifolds. If $\mathcal{P}_1$ and $\mathcal{P}_2$ are both prime, we can similarly consider $\mathbf{R}^3 \# \mathcal{P}_1 \# \mathcal{P}_2$, isolate $\mathcal{P}_1$ and $\mathcal{P}_2$ in two small regions and interpret the situation as describing two well separated elementary geons. It will describe two classically identical geons when $\mathcal{P}_1 = \mathcal{P}_2$. [Two classically identical geons may lose their identity in quantum theory. See Ref. [62]. This does not happen in the examples in this Chapter.]

Now, how do we discuss the statistics of these particles? For this purpose, recall (as discussed above) that in particle and string mechanics, statistics can be understood using the UIR's of $\pi_1(Q)$. Likewise, here as well, the relevant group for discussing statistics is $\pi_1(\tilde{Q}/D^\infty)$ which we have seen is equal to $\pi_0(D^\infty) = D^\infty/D_0^\infty$, D_0^∞ being the identity component of D^∞ [see Chapter 20 and Refs. 62,69,151,154,156].] Exchanges of identical geons (or the appropriate braid operations in two dimensions) have thus to be identified with elements of this group. A quantum theory is associated with a definite UIR of this group and the statistics of geons are to be understood from the properties of these elements in this UIR.

Let us consider an example in $3+1$ dimensions where the three manifold is $M = \mathbf{R}^3 \# \mathbf{R}P^3 \# \mathbf{R}P^3$. The elementary geon is then associated with the prime manifold $\mathbf{R}P^3 = S^3/\mathbf{Z}_2$. It has been shown elsewhere [154] that D^∞/D_0^∞ for this M is generated by two elements E and S which are subject to the relations $E^2 = S^2 = e$, e being the identity of D^∞/D_0^∞. It is thus precisely the group G. E is to be thought of as the exchange of the geons and is defined as follows: Let C_1 and C_2 be two concentric tori where C_1 encloses both prime summands and C_2 encloses C_1. The diffeomorphism E "rotates" C_1 and its interior by π so that the prime summands exchange places. The rotation angle decreases from C_1 to C_2 and outside C_2, E acts as the identity. The slide S is similar except that the two tori "thread through" the second summand instead of enclosing it, and the inner torus "rotates" by 2π instead of by π. It can be interpreted as the result of sliding the first prime summand through a homotopically nontrivial loop associated with the second prime summand.

The following points are noteworthy in connection with the definition of E and S above. Let G_α denote the group D^∞/D_0^∞ for the connected sum $\mathbf{R}^K\#\mathcal{P}_\alpha$, $\mathcal{P}_\alpha$ being prime. Then D^∞/D_0^∞ for the connected sum $\mathbf{R}^K\#\mathcal{P}_\alpha\#\mathcal{P}_\beta$ contains G_α and G_β as subgroups. These subgroups may be regarded as internal diffeomorphisms associated with the individual geons. The group G_α, for instance, may be thought of as acting on one of the prime summands in the following way. Let B be a region with boundary $\partial B = S^{K-1}$ which isolates $\mathcal{P}_\alpha$ in its interior as before. Then G_α acts as identity on the exterior and boundary of B. The interior of B is diffeomorphic to $\mathbf{R}^K\#\mathcal{P}_\alpha$ and the action of G_α it carries completes the definition of the global action of G_α on $\mathbf{R}^K\#\mathcal{P}_\alpha\#\mathcal{P}_\beta$. [Here as elsewhere we are trying to realize D^∞/D_0^∞ by elements of D^∞. This will involve uncertainties by elements of D_0^∞, but such uncertainties will not be a cause for concern in what follows.] Now in the general case of a manifold $\mathbf{R}^K\#\mathcal{P}\#\mathcal{P}$ with a prime $\mathcal{P}$, the definition of E or S along the lines outlined above is not unique, as any such definition can be altered by conjugating it with nontrivial internal diffeomorphisms $(\notin D_0^\infty)$. For the manifold $\mathbf{R}^3\#\mathbf{R}P^3$, however, the group D^∞/D_0^∞ consists of the identity alone [10] so that this ambiguity is absent. There still remain ambiguities in the definition of E or S since as remarked earlier, we can, for instance, replace E

by SES^{-1}. But they will not affect our conclusions as explained in Ref. [62] and more comprehensively in Ref. [159].

Since the group $\pi_1(Q)$ for the two geon system under consideration is G, we can reach conclusions about the statistical properties of this system similar to those in the last two Sections. Note the following in this connection. We have seen in Chapter 21 that the spin-type of a geon can be defined by considerating $\mathbf{R}^K \# \mathcal{P}$, $\mathcal{P}$ being prime, and considering the diffeomorphism describing the 2π rotation of the geon. We have also seen that it can be identified with an element (call it $R_{2\pi}$) of D^∞/D_0^∞ in so far as its action on quantum states is concerned. The chosen UIR of D^∞/D_0^∞ may be such that $R_{2\pi}$ does not act as identity on the states. Then we will have intrinsic spin for geons and the possibility of constructing spinorial states in pure gravity [69]. However, the one geon D^∞/D_0^∞ is trivial in the preceding example [151] so that the states of these geons are tensorial. Still, as we have seen, this fact does not fix their statistical properties. Therefore, as Sorkin has emphasized elsewhere [62,156,159], normal correlations between spin and statistics do not seem valid in generally covariant theories [without topology change] just as we have seen it to be the case for strings [without antistrings with which it can annihilate or be pair-produced [161]].

Chapter 23

CONCLUDING REMARKS

The quantum state vector of a physical system under many circumstances is best represented as a wave function not on the classical configuration space Q, but on a suitable $U(1)$ bundle $\hat{Q}$ over Q. The physical system also exhibits many novel features when the bundle $\hat{Q}$ is twisted. In particular, if a group G acts on Q and is a symmetry of the classical dynamics, there is no guarantee that G will act on $\hat{Q}$ and be a symmetry of the quantum dynamics. Rather, from the action of G on Q, it is necessary to construct the action of a new group $\hat{G}$ on $\hat{Q}$ which fulfills the physical principles outlined in Chapter 7. The group $\hat{G}$ may or may not exist. If it exists, the group G_{QM} which assumes the role of G in quantum theory is either the group $\hat{G}$ or a subgroup of $\hat{G}$.

One of the primary purposes of this book is to explain the basic mathematical features of $\hat{Q}$ and $\hat{G}$ and their importance for physically interesting quantum theories. After the introductory material of the first two Chapters, we have devoted Chapters 2 to 7 to an elementary discussion of $\hat{Q}$ and $\hat{G}$. This discussion has also been illustrated by simple physical systems such as particles of fixed spin and the charge-monopole system.

Another problem we have treated in this book is the quantization of classical systems with multiply connected configuration spaces. Nonabelian gauge theories are examples with such configuration spaces. In these examples, the fundamental group of the configuration space is abelian. It is known from work in nonabelian gauge theories for example that when the configuration space Q is multiply connected with an abelian fundamental group $\pi_1(Q)$, the state vector is best regarded as a function on the universal covering space $\overline{Q}$ of Q. Since $\overline{Q}$ is a $\pi_1(Q)$ bundle over Q, these cases are seen to involve principal fibre bundles with abelian structure groups.

Quantum theory however in principle allows the possibility of wave functions which are functions on any G bundle over Q where there is no obvious restriction on the nature of the group G. Such more general G bundles become relevant when the fundamental group $\pi_1(Q)$ of the configuration space is nonabelian. In Chapters 8, 13, 17 and 20 to 22, we have discussed systems for which $\pi_1(Q)$ is abelian or nonabelian, [the treatment of the latter following the work of Sorkin] and explained certain remarkable features in the quantum theories of systems with nonabelian $\pi_1(Q)$. These features are sufficiently novel and interesting that they merit further study.

Many of the topological ideas related to quantization find applications in soliton physics. Chapters 9 to 19 of this book have been devoted to soliton physics and in particular to Skyrmion physics. Our discussion of Skyrmions was not confined to its purely formal aspects although it is true that our treatment of Skyrmion phenomenology is not adequate. The reader is referred to the many excellent reviews [58,122,123] of the Skyrme model for a better appreciation of its successes and failures. In Chapter 19, we described electroweak Skyrmions. In a sense, this Chapter is an anomaly since these are not topological solitons. Nonetheless, as they bear a strong resemblance to Skyrmions, it seemed to us reasonable to discuss them as well.

Important and related subjects we have discussed in Chapters 20 to 22 concern certain topological aspects of quantum gravity and quantum strings in $3+1$ dimensional space time. We have shown that in $3+1$ dimensional gravity, the configuration space Q can be multiply connected with $\pi_1(Q)$ abelian or nonabelian according to our choice of the three manifold ${}^3M$. Thus, $3+1$

gravity provides us with examples where the group $\pi_1(Q)$ is nonabelian. We have also shown that the same is the case for a system of two or more identical strings in $3 + 1$ dimensions.

Nontrivial fundamental groups in $3 + 1$ gravity are typically associated with certain remarkable topological excitations called "topological geons" which have particle like characteristics and which may indeed admit interpretation as particles. In Chapter 21, we have described how these geons arises in $3 + 1$ gravity and have also indicated a proof of the striking result that the quantum states of certain of this geons can be spinorial in nature. The possibility of constructing such spinorial states in pure gravity is decisively influenced by the nontrivial nature of $\pi_1(Q)$. Indeed the existence of these states is best understood within the framework of the general approach to quantization on multiply connected spaces. In Chapter 22, we also discussed the statistics of geons and strings and shown by specific examples that they exhibit novel features not to be found in systems of identical particles.

There are many interesting problems related to geons and space-time topology which have not found a place in this book. It is known for instance that gauge theories based on space-time manifolds with nontrivial topology have rich topological features. We refer the reader to the literature to the discussion of such results [148,149]. Unusual statistical possibilities arise not only for strings and geons, but also for instance for identical particles on one-dimensional "mesoscopic" networks [162] and we have not discussed them in this book. We have alluded to topological spin-statistics theorems [161,112,24,96,162] which rule out several of these possibilities without explaining the theorems themselves. There are also many other problems suggested by geons which have not been adequately treated here or elsewhere. The phenomenology of the geons has hardly been explored and there is very little known of even the qualitative properties of geons [see, however, Ref. [160]]. For example, we do not even know if there is an antigeon for every geon with which it can annihilate. [This is perhaps because of our poor understanding of the role of topological change in quantum gravity.] This situation may be contrasted with the situation in the physics of Skyrmions which may be thought of as the analogues of geons in the chiral model. A good deal of further work is clearly necessary before we can claim an adequate understanding of the properties of geons.

REFERENCES

1. J. Wess and B. Zumino, *"Consequences of Anomalous Ward Identities "*, Phys. Lett. **37B** (1971) 95-97.

2. D. J. Simms and N. Woodhouse, *"Lectures on Geometric Quantization "*, Lecture Notes in Physics, **53** (Springer Verlag, Berlin, 1977);
   N. Woodhouse, *"Geometric Quantization "* (Oxford Mathematical Monographs, Clarendon, Oxford, 1980);
   J. Sniatycki, *"Geometric Quantization and Quantum Mechanics "* (Springer Verlag, Berlin, 1980).

3. A. P. Balachandran, G. Marmo, B.-S. Skagerstam and A. Stern, *"Spinning Particles in General Relativity "*, Phys. Lett. **89B** (1980) 199-202.

4. R. Soldati and S. Zerbini, *"Spinning Test Particles in the Manifold of Reference Frames "*, Nuovo Cim. Lett. **27** (1980) 575-580.

5. B.-S. Skagerstam and A. Stern, *"Lagrangian Descriptions of Classical Charged Particles with Spin "*, Physica Scripta **24** (1981) 493-497.

6. A. P. Balachandran, G. Marmo, B.-S. Skagerstam and A. Stern, *"Gauge Symmetries and Fibre Bundles - Applications to Particle Dynamics "*, Lecture Notes in Physics, **188** (Springer Verlag, Berlin, 1983).

7. C. Cognola, R. Soldati, L. Vanzo and S. Zerbini, *"Lagrangian Dynamics of a Classical Charged Spinning Particle with Dipole Moments "*, Nuovo Cim. **76B** (1983) 109-129.

8. M. V. Atre, A. P. Balachandran and T. R. Govindarajan, *"Massless Spinning Particles in All Dimensions and Novel Magnetic Monopoles "*, Int. J. Mod. Phys. **A2** (1987) 453-483.

9. B.-S. Skagerstam and A. Stern, *"Light-Cone Versus Proper-Time Gauge for Massless Spinning Particles "*, Nucl. Phys. **B294** (1987) 636-670.

10. H. Arodź, *"Lagrangian and Hamiltonian Formulations of Dynamics of Classical Particles With Spin and Colour "*, Acta Phys. Polon. **B19** (1988) 697-708.

11. J. R. Klauder, *"Quantization Is Geometry, after All "*, Ann. Phys. (N.Y.) **188** (1988) 120-141.

12. H. B. Nielsen, D. Rohrlich, *"A Path Integral to Quantize Spin "*, Nucl. Phys. **B299** (1988) 471-483.

13. M. V. Berry, *"Quantal Phase Factors Accompanying Adiabatic Changes "*, Proc. R. Soc. **A 392** (1984) 45-57;
    B. Simon, *"Holonomy, the Quantum Adiabatic Theorem, and Berry's Phase "*, Phys. Rev. Lett. **51** (1983) 2167-2170;
    I. J. R. Aitchison, *"Berry Phases, Magnetic Monopoles, and Wess-Zumino Terms or How the Skyrmion Got Its Spin "*, Acta. Phys. Polon. **B18** (1987) 207-235; *"Berry's Topological Phase in Quantum Mechanics and Quantum Field Theory "*, Physica Scripta **T23** (1988) 12-20.

14. M. Stone, *"Supersymmetry and the Quantum Mechanics of Spin "*, Nucl. Phys. **B314** (1989) 557-586.

15. B.-S. Skagerstam and A. Stern, *"Topological Quantum Mechanics in 2+1 Dimensions "*, Int. J. Mod. Phys. **A5** (1990) 1575-1595.

16. P. Orland, *"Bosonic Path Integrals for Four-Dimensional Dirac Particles "*, Int. J. Mod. Phys. **A4** (1989) 3615-3628.

17. P. Orland and D. Rohrlich, *"Lattice Magnets: Local Spin from Isospin "*, Nucl. Phys. **B338** (1990) 647-672.

18. G. Marmo, E. J. Saletan, G. Simoni and B. Vitale, *"Dynamical Systems - A Differential Geometric Approach to Symmetry and Reduction "* (John Wiley, Chichester, 1985);
    G. Marmo and C. Rubano, *"Particle Dynamics on Fibre Bundles "* (Bibliopolis, Naples, 1988).

19. P. A. M. Dirac, "*Quantized Singularities in the Electromagnetic Field*", Proc. Roy. Soc. (London) **A133** (1931) 60-72; "*The Theory of Magnetic Poles*", Phys. Rev. **74** (1948) 817-830.

20. T. T. Wu and C. N. Yang, "*Concept of Nonintegrable Phase Factors and Global Formulation of Gauge Fields*", Phys. Rev. **D12** (1975) 3845-3857; "*Dirac's Monopole Without Strings: Classical Lagrangian Theory*", Phys. Rev. **D14** (1976) 437-445; "*Dirac Monopole Without Strings: Monopole Harmonics*", Nucl. Phys. **B107** (1976) 365-380.

21. A. Trautman, "*Solutions of the Maxwell and Yang-Mills Equations Associated With Hopf Fibrings*", Int. J. Theor. Phys. **16** (1077) 561-565; J. Nowakowski and A. Trautman, "*Natural Connections on Stiefel Bundles are Sourceless Gauge Fields*", J. Math. Phys. **19** (1978) 1100-1103.

22. P. Goddard and D. I. Olive, "*Magnetic Monopoles in Gauge Fields Theories*", Rep. Prog. Phys. **41** (1978) 1357-1437.

23. A. P. Balachandran, G. Marmo, B.-S. Skagerstam and A. Stern, "*Magnetic Monopoles With No Strings*", Nucl. Phys. **B162** (1980) 385-396; "*Supersymmetric Point Particles and Monopoles With No Strings*", Nucl. Phys. **B164** (1980) 427-444.

24. J. L. Friedman and R. D. Sorkin, "*Dyon Spin and Statistics: A Fibre-Bundle Theory of Interacting Magnetic and Electric Charges*", Phys. Rev. **D20** (1979) 2511-2525; "*A Spin-Statistics Theorem for Composites Containing both Electric and Magnetic Charges*", Commun. Math. Phys. **73** (1980) 161-196; "*Kinematics of Yang-Mills Solitons*", Commun. Math. Phys. **89** (1983) 483-499; "*Statistics of Yang-Mills Solitons*", Commun. Math. Phys. **89** (1983) 501-521;
R. D. Sorkin, "*A General Relation Between Kink-Exchange and Kink Rotation*", Commun. Math. Phys. **115** (1988) 412-434.

25. A. P. Balachandran, F. Lizzi and G. Sparano, "*A New Approach to Strings and Superstrings*", Nucl. Phys. **B277** (1987) 359-387;

A.P.Balachandran, F.Lizzi, G.Sparano and R. D. Sorkin, "*Topological Aspects of String Theories* ", ibid. **B287** (1987) 508-550.

26. P. Salomonson, B.-S. Skagerstam and A. Stern, "*General Coupling of Strings to the Low-Energy Effective Theory* ", Phys. Rev. **D37** (1988) 3628-3634.

27. E. Witten, "*Current Algebra, Baryons, and Quark Confinement* ", Nucl. Phys. **B223** (1983) 433-444.

28. "*Symposium on Anomalies, Geometry, Topology* ", Eds. W. A. Bardeen and A. R. White (World Scientific, Singapore, 1985).

29. L. Alvarez-Gaumé, "*An Introduction to Anomalies* ", in the Proceedings of "*The International School on Mathematical Physics* ", Erice, 1985.

30. P. Ginsparg, "*Applications of Topological and Differential Geometric Methods to Anomalies in Quantum Field Theory* ", in "*Proceedings of the XVI GIFT Seminar* ", Jaca, 1985.

31. L. Bonora, "*Anomalies and Cohomolgy* " in the Proceedings of the Second Ferrara School in Theoretical Physics, "*Anomalies, Phases and Defects* ", Ferrara 1989, Eds. M. Bregola, G. Marmo and G. Morandi (Bibliopolis, Naples, 1990).

32. A. Yu Morozov, "*Anomalies in Gauge Theories* ", Sov. Phys. Usp., **29** (1986) 993-1039;
M. A. Shifman, "*Anomalies and Low-Energy Theorems of Quantum Chromodynamics* ", Sov. Phys. Usp. **32** (1989) 289-309.

33. P. A. M. Dirac, "*Generalized Hamiltonian Dynamics* ", Can. J. Math. **2** (1950) 129-148; "*Generalized Hamiltonian Dynamics* ", Proc. Roy. Soc. **A246** (1958) 326-332; "*Lectures on Quantum Mechanics* ", Yeshiva University (Academic Press, New York, 1967).

34. A. J. Hanson, T. Regge and C. Teitelboim, "*Constrained Hamiltonian Systems* " (Accademia Nazionale dei Lincei, Roma 1976).

35. K. Sundermeyer, "*Constrained Dynamics* ", Lecture Notes in Physics **169** (Springer Verlag, Berlin 1982).

36. P. G. Bergmann, "*Observables in General Relativity* ", Rev. Mod. Phys. **33** (1961) 510-514.

37. G. Marmo, N. Mukunda and J. Samuel, "*Dynamics and Symmetry for Constrained Systems: A Geometrical Analysis* ", Rivista Nuovo Cimento **6** (no.2) (1983) 1-62.

38. C. Becchi, A. Rouet and R. Stora, "*The Abelian Higgs-Kibble Model, Unitary of the S-operator* ", Phys. Lett. **52B** (1974) 344-346; "*Renormalization of the Abelian Higgs-Kibble Model* ", Commun. Math. Phys. **42** (1975) 127-162; "*Renormalization of Gauge Theories* ", Ann. Phys. **98** (1976) 287-321.

39. I. V. Tyutin, "*Gauge Invariance in Field Theory and in Statistical Mechanics in the Operator Formalism* ", Lebedev preprint FIAN n.**39** (1975) (unpublished).

40. M. Henneaux, "*Classical Foundations of BRST Symmetry* " (Bibliopolis, Naples, 1988); "*Hamiltonian Form of the Path Integral for Theories With a Gauge Freedom* ", Phys. Rep. **126** (1985) 1-66.

41. E. S. Fradkin and G. A. Vilkovisky, "*Quantization of Relativistic Systems with Constraints* ", Phys. Lett. **55B** (1975) 224-226;
I. A. Batalin and G. A. Vilkovisky, "*Relativistic S-Matrix of Dynamical Systems with Boson and Fermion Constraints* ", Phys. Lett. **69B** (1977) 224-226;
E. S. Fradkin and T. E. Fradkina, "*Quantization of Relativistic Systems With Boson and Fermion First- And Second-Class Constraints* ", Phys. Lett. **72B** (1978) 343-348;
I. A. Batalin, E. S. Fradkin, "*Operator Quantization And Abelization of of Dynamical Systems Subject to First-Class Constraints* ", Riv. Nuovo Cim. **9** (1986) 1-48;

I. A. Batalin, E. S. Fradkin and T. E. Fradkina, "*Another Version for Operatorial Quantization of Dynamical Systems With Irreducible Constraints* ", Nucl. Phys. **B314** (1989) 158-174.

42. N. Steenrood, "*Topology of Fibre Bundles* " (Princeton University Press, Princeton, New Jersey, 1951).

43. D. Husemoller, "*Fibre Bundles* ", 2nd ed. (Springer Verlag, Berlin, 1975).

44. T. Eguchi, P. G. Gilkey and A. J. Hanson, "*Gravitation, Gauge Theories and Differential Geometry* ", Phys. Rep. **66C** (1980) 213-393.

45. Y. Choquet-Bruhat, C. DeWitt-Morette and M. Dillard-Bleick, "*Analysis, Manifolds and Physics* " (North-Holland, Amsterdam, 1977);
C. Nash and S. Sen, "*Topology and Geometry for Physicists* " (Academic Press, London, 1983);
C. J. Isham, "*Modern Differential Geometry for Physicists* " (World Scientific, Singapore, 1989).

46. I. M. Singer, "*Some Remarks on the Gribov Ambiguity* ", Commun. Math. Phys. **60** (1978) 7-12.

47. A. Stern, "*General Action Principle for Supersymmetric Particles* ", Phys. Rev. Lett. **55** (1985) 1351-1354.

48. F. A. Berezin and M. S. Marinov, "*Classical Spin and Grassmann Algebra* ", JETP Lett. **21** (1975) 320-321; "*Particle Spin Dynamics as the Grassmann Variant of Classical Mechanics* ", Ann. Phys. (N.Y.) **104** (1977) 336-362.

49. R. Casalbuoni, "*Relativity and Supersymmetries* ", Phys. Lett. **62B** (1976) 49-50; "*The Classical Mechanics for Bose-Fermi Systems* ", Nuovo Cimento **33A** (1976) 389-431.

50. A. Barducci, R. Casalbuoni and L. Lusanna, "*Supersymmetries and the Pseudoclassical Relativistic Electron* ", Nuovo Cimento **35A** (1976) 377-399.

51. L. Brink, S. Deser, B. Zumino, P. Di Vecchia and P. S. Howe, *"Local Supersymmetry for Spinning Particles "*, Phys. Lett. **64B** (1976) 435-438.

52. L. Brink, P. Di Vecchia and P. S. Howe, *"A Lagrangian Formulation of the Classical and Quantum Dynamics of Spinning Particles "*, Nucl. Phys. **B118** (1977) 76-94.

53. A. P. Balachandran, S. Borchardt and A. Stern, *"Lagrangian and Hamiltonian Descriptions of Yang-Mills Particles "*, Phys. Rev. **D17** (1978) 3247-3256.

54. See e.g. J. D. Talman, *"Special Functions - A Group Theoretic Approach "* (W. A. Benjamin, New York, 1968);
A. O. Barut and R. Raczka, *"Theory of Group Representation and Applications "* (PWN-Polish Scientific Publishers, 1980);
A. P. Balachandran and C. G. Trahern, *"Lectures on Group Theory for Physicists "* (Bibliopolis, Naples, 1984).

55. A. P. Balachandran, M. Bourdeau and S. Jo, *"The Non-Abelian Chern-Simons Term With Sources and Exotic Source Statistics "*, Mod. Phys. Lett. **4** (1989) 1923-1935; *"Braid Statistics for the Sources of the Non-Abelian Chern-Simons Term "*, Int. J. Mod. Phys. **A5** (1990) 2423-2470 and ibid **A5** (1990) 3461(E).

56. See e.g. A. A. Kirillov, *"Elements of the Theory of Representations "*, A Series of Comprehensive Studies in Mathematics **220** (Springer Verlag, Berlin 1976). See especially Chapt. 15. Also see O. Alvarez, I. M. Singer and P. Windey, *"Quantum Mechanics and the Geometry of the Weyl Character Formula "*, Nucl. Phys. **B337** (1990) 467-486.

57. F. Zaccaria, E. C. G. Sudarshan, J. S. Nilsson, N. Mukunda, G. Marmo and A. P. Balachandran, *"Universal Unfolding of Hamiltonian Systems: From Symplectic Structure to Fibre Bundles "*, Phys. Rev. **D27** (1983) 2327-2340.

58. For reviews see e.g. *"Solitons in Nuclear and Elementary Particle Physics"*, Eds. A. Chodos, E. Hadjimichael and C. Tze (World Scientific, Singapore, 1984);

A. P. Balachandran, *"The Skyrme Soliton"* in *"Recent Advances in Theoretical Physics"*, Proceedings of the *"Silver Jubilee Workshop"* held at IIT, Kanpur, 1984, Ed. R. Ramachandran (World Scientific, Singapore, 1985); *"Skyrmions"*, in *"High Energy Physics 1985"*, Eds M. J. Bowick and F. Gürsey, Vol.1 (World Scientific, Singapore, 1986);

A. Dhar, R. Shankar and S. R. Wadia, *"Nambu-Jona-Lasinio-Type Effective Lagrangians: Anomalies and Nonlinear Lagrangian of Low-Energy Large-N QCD"*, Phys. Rev. **D31** (1985) 3256-3267;

M. E. Peskin, *"Pion-Skyrmion Scattering: Collective Coordinates at Work"* in *"Recent Developments in Quantum Field Theory"*, Eds. J. Ambjørn, B. J. Durhus and J. L. Petersen (North-Holland, Amsterdam, 1985);

M. Abud, G. Maiella, F. Nicodemi, R. Pettorino and K. Yoshida, *"Hidden Symmetry and Skyrmions in the Effective Lagrangian Approach to QCD"*, Phys. Lett. **159B** (1985) 155-160; *"Hidden Symmetry and Skyrmions"* in *"Quantum Field Theory"*, Proceedings of the International Workshop in Honour of H. Umezawa, Ed. F. Mancini (North-Holland, Amsterdam, 1986);

I. Zahed and G. E. Brown, *"The Skyrme Model"*, Phys. Rep. **142** (1986) 1-102;

G. Holzwarth and B. Schwesinger, *"Baryons in the Skyrme model"*, Rep. Prog. Phys. **49** (1986) 825-871;

"Chiral Solitons", Ed. K.-F. Liu (World Scientfic, Singapore, 1987);

Y. Dothan and L. C. Biedenharn, *"Old Models Never Die: The Revival of the Skyrme Model"*, Comments Nucl. Part. Phys. **2** (1987) 63-91;

S. Gupta and R. Shankar, *"Large-N Baryons: From Quarks to Solitons"*, Phys. Rev. Lett. **58** (1987) 2178-2181;

"Workshop on Skyrmions and Anomalies", Eds. M. Jezabek and M. Prasza\owicz (World Scientific, Singapore, 1987);

Proceedings of the "*1987 International Workshop on Low Energy Effective Theory of QCD* ", Eds. S. Saito and K. Yamawaki (Nagoya University, 1987);

E. Sorace and M. Tarlini, "*Energy of Skyrmionic Configurations* " in "*Symmetries in Science III* ", Eds. B. Gruber and I. Iachello (Plenum Press, New York, 1989);

B. Schwesinger, H. Wiegel, G. Holzwart and A. Hayashi, "*The Skyrme Soliton in Pion, Vector- and Scalar-Meson Fields:* πN-*Scattering and Photoproduction* ", Phys. Rep. **173C** (1989) 173-255;

M. Rho, "*Skyrmions Revisited* ", Mod. Phys. Lett: **4** (1989) 2571-2587;

M. C. Birse, "*Soliton Models for Nuclear Physics* ", Progress in Particle and Nuclear Physics (in press).

59. W. Marciano and H. Pagels, "*Quantum Chromodynamics* ", Phys. Rep. **36C** (1978) 137-276;

F. J. Ynduráin, "*Quantum Chromdynamics* " (Springer-Verlag, Berlin, 1983).

60. A. P. Balachandran, G. Marmo, A. Simoni and G. Sparano (under preparation).

61. See e.g. A. P. Balachandran, "*Wess-Zumino Terms and Quantum Symmetries, A Review* ", in "*Conformal Field Theory, Anomalies and Superstrings* ", Eds. B. E. Baaquie, C. K. Chew, C. H. Oh and K. K. Phua (World Scientific, Singapore, 1988); "*Classical Topology and Quantum Phases: Quantum Mechanics* ", in "*Geometric and Algebraic Aspects of Nonlinear Field Theories* ", Eds. S. De Filippo, M. Marinaro, G. Marmo and G. Vilasi (North-Holland, Amsterdam, 1989); "*Classical Topology and Quantum Phases* " in the Proceedings of the Second Ferrara School in Theoretical Physics, "*Anomalies, Phases and Defects* ", Ferrara 1989, Eds. M. Bregola, G. Marmo and G. Morandi (Bibliopolis, Naples, 1990) and references cited therein.

62. R. D. Sorkin, "*Introduction to Topological Geons* ", in "*Topological Prop-

erties and Global Structure of Spacetime ", Eds P. G. Bergmann and V. de Sabata (Plenum, 1986); "*Classical Topology and Quantum Phases: Quantum Geons* " in "*Geometric and Algebraic Aspects of Nonlinear Field Theories* ", Eds. S. De Filippo, M. Marinaro, G. Marmo and G. Vilasi (North-Holland, Amsterdam, 1989).

63. J. B. Hartle and J. R. Taylor, "*Quantum Mechanics of Paraparticles* ", Phys. Rev. **178** (1969) 2043-2051;
R. H. Stolt and J. R. Taylor, "*Classification of Paraparticles* ", Phys. Rev. **D1** (1970) 2226-2228.

64. Y. Aharonov and D. Bohm, "*Significance of Electomagnetic Potentials in the Quantum Theory* ", Phys. Rev. **115** (1959) 485-491; "*Further Considerations of Electromagnetic Potentials in the Quantum Theory* ", Phys. Rev. **123** (1961) 1511-1524; "*Remarks on the Possibility of Quantum Electrodynamics without Potentials* ", Phys. Rev.**125** (1962) 2192-2193; "*Further Discussions on the Role of Electromagnetic Potentials in the Quantum Theory* ", Phys. Rev. **130** (1963) 1625-1632.

65. A. Bohr and B. R. Mottleson, "*Nuclear Structure, Volume II: Nuclear Deformations* " (W. A. Benjamin, Inc., Advanced Book Program, Reading, Massachusetts, 1975);
P. Ring and P. Schuck, "*The Nuclear Many-Body Problem* " (Springer-Verlag New York Inc., 1980);
L. D. Landau and E. W. Lifshitz, "*Quantum Mechanics, Non-relativistic Theory* " (Addison-Wesley Publishing Company, Inc., Reading, Massachusetts, 1958).

66. J. Birman, "*Braids, Links and Mapping Class Groups* ", Annals of Math. Studies #82 (Princeton University Press, Princeton, 1973).

67. "*Braid Group, Knot Theory and Statistical Mechanics* ", Eds. C. N. Yang and M. L. Ge (World Scientific, Singapore, 1989);
T. D. Imbo, C. S. Imbo and E. C. Sudarshan, "*Identical Particles, Exotic*

Statistics and Braid Groups ", Phys. Lett. **B234** (1990) 103-107;

R. Cappuccio and E. Guadagnini, "*On Braid Statistics* ", preprint, CERN-TH.5772/90, Geneva, 1990.

68. A. P. Balachandran, G. Marmo, N. Mukunda, J. S. Nilsson, E. C. G. Sudarshan and F. Zaccaria, "*Monopole Topology and the Problem of Color*", Phys. Rev. Lett. **50** (1983) 1553-1555; "*Non-Abelian Monopoles Break Color. I. Classical Mechanics*", Phys. Rev. **D29** (1984) 2919-2935; "*Non-Abelian Monopoles Break Color. II. Field Theory and Quantum Mechanics* ", Phys. Rev. **D29** (1984) 2936-2943;

P. Nelson and A. Manohar, "*Global Color is Not Always Defined* ", Phys. Rev. Lett. **50** (1983) 943-945;

A. Abouelsaood, "*Chromodyons and Equivariant Gauge Transformations* ", Phys.Lett. **125** (1983) 467-469; "*Are There Chromodyons?* ", Nucl. Phys. **B226** (1984) 309-338;

P. Nelson and S. Coleman, "*What Becomes of Global Color* ", Nucl. Phys. **B237** (1984) 1-31;

A. P. Balachandran, F. Lizzi and V. G. J. Rodgers, "*Topological Symmetry Breaking in Nematics and* ^3He ", Phys. Rev. Lett. **52** (1984) 1818-1821;

A. Stern and U. Yajnik, "*SO(10) Vortices and the Electroweak Phase Transition* ", Nucl. Phys. **B267** (1986) 158-180;

L. M. Krauss and F. Wilczek, "*Discrete Gauge Symmetry in Continuum Theories* ", Phys. Rev. Lett. **62** (1989) 1221-1223;

M. G. Alford, K. Benson, S. Coleman, J. March-Russell and F. Wilczek, "*Interactions and Excitations of Non-Abelian Vortices* ", Phys. Rev. Lett. **64** (1990) 1632-1635; "*Zero Modes of Non-Abelian Vortices* ", preprint HUTP-89/A052 and IASSNS-HEP-89-93, Harvard, 1990;

J. Preskill and L. M. Krauss, "*Local Discrete Symmetry and Quantum-Mechanical Hair* ", Nucl. Phys. **B341** (1990) 50-100;

J. Preskill, "*Quantum Hair* ", in the Proceedings of the Nobel Symposium "*The Birth and Early Evolution of Our Universe* ", Gräftåvallen,

1990, Eds. J. S. Nilsson, B. Gustafsson and B.-S. Skagerstam (World Scientific, to appear).

69. J. L. Friedman and R. D. Sorkin, "*Spin $\frac{1}{2}$ From Gravity* ", Phys. Rev. Lett. **44** (1980) 1100-1103 and ibid. **45** (1980) 148(E); "*Half-Integral Spin From Quantum Gravity* ", General Relativity and Gravitation **14** (1982) 615-620.

70. R. Rajaraman, "*Solitons and Instantons - An Introduction to Solitons and Instantons in Quantum Field Theory* " (North-Holland Personal Library, Amsterdam, 1982).

71. C. Rebbi and G. Soliani, "*Solitons and Particles* " (World Scientific, Singapore, 1984).

72. S. Coleman, "*Aspects of Symmetry, Selected Erice Lectures* ", (Cambridge University Press, Cambridge, 1985).

73. E. B. Bogomol'nyi, "*The Stability of Classical Solutions* ", Sov. J. Nucl. Phys. **24** (1976) 449-454.

74. M. K. Prasad and C. M. Sommerfield, "*Exact Classical Solutions for the 't Hooft Monopole and the Julia-Zee Dyon* ", Phys. Rev. Lett. **35** (1975) 760-762.

75. A. A. Belavin and A. M. Polyakov, "*Metastable States of Two-Dimensional Isotropic Ferromagnets* ", JETP Lett. **22** (1975) 245-247.

76. G. H. Derrick, "*Comments on Nonlinear Wave Equations as Models of Elementary Particles* ", J. Math. Phys. **5** 1252-1254.

77. H. B. Nielsen and P. Olesen, "*Local Field Theory for the Dual String* ", Nucl. Phys. **B57** (1973) 367-380.

78. S. Coleman, J. Wess and B. Zumino, "*Structure of Phenomenological Lagrangians. I* ", Phys. Rev. **177** (1969) 2239-2247;

C. Callan, S. Coleman, J. Wess and B. Zumino, "*Structure of Phenomenological Lagrangians. II* ", Phys. Rev. **177** (1969) 2247-2250;

D. V. Volkov, "*Phenomenological Lagrangians* ", Sov. J. Particles and Nuclei **4** (1973) 1-17.

79. S. Gasiorowicz and D. Geffen, "*Effective Lagrangians and Field Algebras With Chiral Symmetry* ", Rev. Mod. Phys. **41** (1969) 531-573;

    B. W. Lee, "*Chiral Dynamics* " (Gordon and Breach, New York, 1971).

80. K. Pohlmeyer, "*Integrable Hamiltonian Systems and Interactions Through Quadractic Constraints* ", Commun. Math. Phys. **46** (1976) 207-221;

    M. Lüscher and K. Pohlmeyer, "*Scattering of Massless Lumps and Non-Local Charges in the Two-Dimensional Classical Non-Linear-Model* ", Nucl. Phys. **B137** (1978) 46-54;

    A. M. Polyakov, "*Hidden Symmetry of the Two-Dimensional Chiral Fields* ", Phys. Lett. **72B** (1977) 224-226;

    H. Eichenherr, "*SU(N) Invariant Non-Linear σ Models* ", Nucl. Phys. **B146** (1978) 215-223;

    P. B. Wiegmann, "*Exact Solution of the O(3) Nonlinear σ-Model*", Phys. Lett. **152B** (1985) 209-215. For a review see e.g. A. M. Perelomov, "*Instanton-Like Solutions in Chiral Models* ", Physica **4D** (1981) 1-25.

81. A. M. Polyakov, "*Interaction of Goldstone Particles in Two Dimensions. Applications to Ferromagnetics and Massive Yang-Mills Fields* ", Phys. Lett. **59B** (1975) 79-81. For a review see e.g. V. A. Novikov, M. A. Shifman, A. I. Vainshtein and V. I. Zakharov, "*Two-Dimensional Sigma Models: Modelling Non-Perturbative Effects in Quantum Chromodynamics* ", Phys. Rep. **116** (1984) 103-171.

82. A. D'Adda, M. Lüscher and P. Di Vecchia, "*A 1/N Expandable Series of Non-Linear σ Models With Instantons* ", Nucl. Phys. **B146** (1978) 63-76.

83. A. M. Polyakov, "*String Representations and Hidden Symmetries for Gauge Fields* ", Phys. Lett. **82B** (1979) 247-250.

328

84. C. Misner, "*Harmonic Maps as Models for Physical Theories*", Phys. Rev. **D18** (1978) 4510-4524;

    M. Hirayama, H. Chia Tze, J. Ishida and T. Kawabe, "*On the Chiral Connection Between the Ferromagnet, the Axisymmetric Gravitational Problem and the SU(2) Vacuum Gauge Field*", Phys. Lett. **66A** (1978) 352-356.

85. "*Conformal Invariance and Applications to Statistical Mechanics*", Eds. C. Itzykson, H. Saleur and J.-B. Zuber (World Scientific, Singapore, 1988);

    P. Ginsparg, "*Applied Conformal Field Theory*", in "*Fields, Strings and Critical Phenomena*", Eds. E. Brézin and J. Zinn-Justin (Elsevier Science Publications BV, 1989).

86. M. B. Green, J. H. Schwarz and E. Witten, "*Superstring Theory*", Vols. 1 and 2 (Cambridge University Press, Cambridge, 1987).

87. D. Nemeschansky and S. Yankielowicz, "*Critical Dimension of String Theories in Curved Space*", Phys. Rev. Lett. **54** 1985) 620-623;

    D. Gepner and E. Witten, "*String Theory on Group Manifolds*", Nucl. Phys. **B278** (1986) 493-549;

    J. Mickelsson, "*Strings on a Group Manifold, Kac-Moody Groups, and Anomaly Cancellation*", Phys. Rev. Lett. **57** (1986) 2493-2495.

88. S. P. Novikov, "*The Hamiltonian Formalism and Multivalued Analogue of Morse Theory*", Russian Math. Surveys **37** (1982) 1-56;

    E. Witten, "*Non-Abelian Bosonization in Two Dimensions*", Commun. Math. Phys. **92** (1984) 544-472;

    A. Polyakov and P. B. Wiegman, "*Theory of Nonabelian Goldstone Bosons in Two Dimensions*", Phys. Lett. **131B** (1983) 121-126;

    V. G. Knizhnik and A. B. Zamolodchikov, "*Current Algebra and the Wess-Zumino Model in Two Dimensions*", Nucl. Phys. **B247** (1984) 83-103.

89. A. Aleksev, L. Fadeev and S. Shatashvili, "*Quantization of the Symplectic Orbits of the Compact Lie Groups by Means of the Functional Integral*", J. Geom. Phys. **1** (in press);
A. Aleksev and S. Shatashvili, "*Path Integral Quantization of the Coadjoint Orbits of the Virasoro Group and 2D Gravity*", Nucl. Phys. **B323** (1989) 719-733;
B. Rai and V. G. J. Rodgers, "*From Coadjoint Orbits to Scale Invariant WZNW Type Actions and 2-D Quantum Gravity Action*", Nucl. Phys. **B341** (1990) 119-133;
G. W. Delius, P. van Nieuwenhuizen and V. G. J. Rodgers, "*The Method of Coadjoint Orbits: An Algorithm for the Construction of Invariant Actions*", Int. J. Mod. Phys. **A5** (1990) 3943-3983.

90. A. P. Balachandran, A. Stern and G. Trahern, "*Nonlinear Models as Gauge Theories*", Phys. Rev. **D19** (1978) 2416-2425.

91. M. A. Semenov-Tjan-Sănskǔ and L. D. Faddeev, "*On the Theory of Non-Linear Models*", Vestnik Leningrad Univ. Math. **10** (1982) 319-327.

92. A. D'Adda, M. Lüscher and P. Di Vecchia, "*Topology and Higher Symmetries of the Two-Dimensional Non-Linear σ Model*", Phys. Rep. **49** C (1979) 239-244;
Chan Kuang Chao, Tu Tung Sheng and Yean Tu Nam, "*The Pure Gauge Field on a Coset Space*", Sci. Sinica **22** (1979) 37-52;
F. Gürsey and H. Chia Tze, "*Complex and Quaternionic Analyticity in Chiral and Gauge Theories. I*", Ann. Phys. (N.Y.) **128** (1980) 29-130.

93. H. Eichenherr, Ref. [80];
V. L. Golo and A. M. Perelomov, "*Solution of the Duality Equation for the Two-Dimensional SU(N)-Invariant Chiral Model*", Phys. Rev. Lett. **79** (1978) 112-113.
For a review see W. J. Zakrewski, "*Low Dimensional Sigma Models*" (Adam Hilger, Bristol and Philadelphia, 1989).

94. A. D'Adda, M. Lüscher and P. Di Vecchia, "*Confinement and Chiral Symmetry Breaking in* $\mathbf{CP}^{n-1}$ *Models with Quarks*", Nucl. Phys. **B152** (1979) 125-144.

95. A. Schwarz, "*The Partition Function of Degenerate Quadratic Functional and Ray-Singer Invariants*", Lett. Math. Phys. **2** (1978) 247-252; "*The Partition Function of a Degenerate Functional*", Commun. Math. Phys. **67** (1979) 1-16;

   R. Jackiw and S. Templeton, "*How Super-Renormalizable Interactions Cure Their Infrared Divergences*", Phys. Rev. **D23** (1981) 2291-2304;

   J. F. Schonfeld, "*A Mass Term for Three-Dimensional Gauge Field*", Nucl. Phys. **B185** (1981) 157-171;

   S. Deser, R. Jackiw and S. Templeton, "*Three-Dimensional Massive Gauge Theories*", Phys. Rev. Lett. **48** (1982) 1975-978 and "*Topological Massive Gauge Theories*", Ann. Phys. (N.Y.) **140** (1982) 372-411;

   I. Affleck, J. Harvey and E. Witten, "*Instantons and (Super-) Symmetry Breaking in (2+1) Dimensions*", Nucl. Phys. **B206** (1982) 413-439;

   L. Alvarez-Gaumé and E. Witten, "*Gravitational Anomalies*", Nucl. Phys. **B234** (1983) 269-330;

   S. Deser and R. Jackiw, ""*Self-Duality*" of Topologically Massive Gauge Theories", Phys. Lett. **139B** (1984) 371-373;

   C. R. Hagen, "*A New Gauge Theory Without an Elementary Photon*", Ann. Phys. **157** (1984) 342-359;

   A. Niemi and G. W. Semenoff, "*Axial-Anomaly-Induced Fermion Fractionzation and Effective Gauge-Theory Actions in Odd-Dimensional Space-Times*", Phys. Rev. Lett. **51** (1983) 2077-2080:

   A. N. Redlich, "*Gauge Noninvariance and Parity Nonconservation of Three-Dimensional Fermions*" Phys. Rev. Lett. **52** (1984) 18-21; "*Parity Violation and Gauge Noninvariance of the Effective Gauge Field Action in Three Dimensions*" Phys. Rev. **D29** (1984) 2366-2374;

   R. D. Pisarski, "*Chiral-Symmetry Breaking in Three-Dimensional Electrodynamics*", Phys. Rev. **D29** (1984) 2423-2426;

R. J. Hughes, "*The Effective Action for Photons in (2 + 1) Dimensions*", Phys. Lett. **148B** (1984) 215-219;

O. Boyanovsky and R. Blankenbecler, "*Axial and Parity Anomalies and Vacuum Charge: A Direct Approach* ", Phys. Rev. **D31** (1985) 3234-3250;

A. N. Redlich and L. C. R. Wijewardhana, "*Induced Chern-Simons Terms at High Temperatures and Finite Densities* ", Phys. Rev. Lett. **54** (1985) 970-973;

A. J. Niemi and G. W. Semenoff, "*Comment on "Induced Chern-Simons Terms at High Temperatures and Finite Densities* "", Phys. Rev. Lett. **54** (1985) 2166;

K. Tsokos, "*Topological Mass Terms and the High Temperature Limit of Chiral Gauge Theories* ", Phys. Lett. **157B** (1985) 413-415;

Y.-C. Kao and M. Suzuki, "*Radiatively Induced Topological Mass Terms in (2+1)-Dimensional Gauge Theories* ", Phys. Rev. **D31** (1985) 2137-2138;

M. D. Bernstein and T. Lee, "*Radiative Corrections to the Topological Mass in (2+1)-Dimensional Electrodynamics* ", Phys. Rev. **D32** (1985) 1020;

S. Coleman and B. Hill, "*No More Corrections to the Topological Mass Term in QED_3* ", Phys. Lett. **159B** (1985) 184-188;

V. A. Rubakov and A. N. Tavkhelidze, "*Stable Anomalous States of Superdense Matter in Gauge Theories*'", Phys. Lett. **165B** (1985) 109-112;

S. Rao and R. Yahalom, "*Parity Anomalies in Gauge Theories in 2 + 1 Dimensions* ", Phys. Lett. **B172** (1986) 227-230;

P. K. Panigrahi, S. Roy and W. Scherer, "*Rotational Anomaly and Fractional Spin in the Gauged CP^1 Nonlinear σ Model With the Chern-Simons Term* ", Phys. Rev. Lett. **61** (1988) 2827-2830;

G. W. Semenoff, P. Sodano and Y.-S. Wu, "*Renormalization of the Statistics Parameter in Three-Dimensional Electrodynamics* ", Phys. Rev. Lett. **62** (1989) 715-718;

E. Guadagnini, M. Martellini and N. Mintchev, "*Perturbative Aspects of* the *Cern-Simons Field Theory* ", Phys. Lett. **B227** (1989) 111-117;

L. Alvarez-Gaumé, J. M. F. Labastida and A. V. Ramallo, "*A Note on Perturbative Chern-Simons Theory* ", Nucl. Phys. **B334** (1990) 103-124;

V. P. Spiridonov, "*Quantum Dynamics of the D=3 Abelian Higgs Model and Spontaneous Breaking of Parity* ", preprint, CALT-63-558, California Institute of Technology, 1990;

G. P. Korchemsky, "*Parity Anomaly in D=3 Chern-Simons Gauge Theory* ", preprint, JINR-E2-90-15, Dubna, 1990.

96. Cf. L. S. Schulman, "*A Path Integral For Spin* ", Phys. Rev. **176** (1968) 1558-1569, and "*Techniques and Applications of Path Integration* " (J. Wiley, New York, 1981);

M. G. G. Laidlaw and C. Morette De Witt, "*Feynman Functional Integrals for Systems of Indistinguishable Particles* " Phys. Rev. **D3** (1971) 1375-1378;

J. S. Dowker, "*Quantum Mechanics and Field Theory on Multiply Connected an on Homogeneous Spaces* ", J. Phys. **A**: Vol. **5** (1972) 936-943;

J. Leinaas and J. Myrheim, "*On the Theory of Identical Particles* ", Nuovo Cimento, **B37** (1977) 1-23;

G. A. Goldin, R. Menikoff and D. H. Sharp, "*Representations of Local Current Algebra in Nonsimply Connected Space and the Ahoronov-Bohm Effect* ", J. Math. Phys. **88** (1981) 1664-1668;

F. Wilczek, "*Magnetic Flux, Angular Momentum, and Statistics*", Phys. Rev. Lett. **48** (1982) 1144-1146; "*Remarks on Dyons* ", ibid. **48** (1982) 1146-1149; "*Quantum Mechanics of Fractional-Spin Particles* ", ibid. **49** (1982) 957-959;

G. A. Goldin and D. H. Sharp, "*Rotation Generators in Two-Dimensional Space and Particles Obeying Unusual Statistics* ", Phys. Rev. **D28** (1983) 830-832;

R. Jackiw and R. N. Redlich, "*Two-Dimensional Angular Momentum in the Presence of Long-Range Magnetic Flux* ", Phys. Rev. Lett. **50** (1983) 555-559;

F. Wilczek and A. Zee, "*Linking Numbers, Spin, and Statistics of Solitons* ", Phys. Rev. Lett. **51** (1983) 2250-2252;

R. D. Sorkin, "*Particle Statistics in Three Dimensions* ", Phys. Rev. **D27** (1983) 1787-1797;

Y.-S. Wu, "*General Theory for Quantum Statistics in Two Dimensions* ", Phys. Rev. Lett. **52** (1984) 2103-2106;

Y.-S. Wu and A. Zee, "*Comments on the Hopf Lagrangian and Fractional Statistics of Solitons* ", Phys. Lett. **147B** (1984) 325-329;

Y.-S. Wu, "*Multiparticle Quantum Mechanics Obeying Fractional Statistics* ", Phys. Rev. Lett. **53** (1984) 111-114;

D. C. Ravenel and A. Zee, "*Spin and Statistics of Solitons and J-Homomorphism* ", Commun. Math. Phys. **98** (1985) 239-243;

R. Chia, A. Hansen and A. Moulthrop, "*Fractional Statistics of the Vortex in Two-Dimensional Superfluids* ", Phys. Rev. Lett. **54** (1985) 1339-1342; "*N-Dependent Fractional Statistics of N Vortices* ", ibid. **55** (1985) 1431-1434;

M. B. Paranjape, "*Induced Angular Momentum in (2+1)-Dimensional QED* ", Phys. Rev. Lett. **55** (1985) 2390-2393;

A. M. Din and W. J. Zakrewski, "*The Point-Particle Limit of* **CP**^{1} *Skyrmions* ", Nucl. Phys. **B253** (1985) 77-92;

D. P. Arovas, R. Schrieffer, F. Wilczek and A. Zee, "*Statistical Mechanics of Anyons* ", Nucl. Phys. **B251 [FS13]** (1985) 117-126;

G. W. Semenoff and L. C. R. Wijewardhana, "*Induced Fractional Spin and Statistics in Three-Dimensional QED* ", Phys. Lett. **B184** (1987) 397-402;

S. A. Kivelson, D. S. Rokshar and J. P. Sethna, "*Topology of the Resonating Valence-Bond State: Solitons and High-T_c Superconductivity* ", Phys. Rev. **B35** (1987) 8865-8868;

A. M. Polyakov, "*Fermi-Bose Transmutations Induced By Gauge Fields* ", Mod. Phys. Lett. **A3** (1988) 325-328;

C. H. Tze, "*Manifold-Splitting Regularization, Self-Linking, Twisting, Writhing Numbers of Space-Time Bosons and Polyakov's Proof of Fermi-*

Bose Transmutations ", Int. J. Mod. Phys. **A3** (1988) 1959-1979;

J. Fröhlich and P. A. Marchetti, "*Quantum Field Theory of Anyons* ", Lett. Math. Phys. **16** (1988) 347-358;

G. W. Semenoff, "*Canonical Quantum Field Theory and Exotic Statistics* ", Phys. Rev. Lett. **61** (1988) 517-520;

R. Mackenzie and F. Wilczek, "*Peculiar Spin and Statistics in Two Dimensions* ", Int. J. Mod. Phys. **A3** (1988) 2827-2853;

A. P. Balachandran, M. J. Bowick, K. S. Gupta and A. M. Srivastava, "*Symmetries of the Hubbard Model and the θ-Term in the σ-Model* ", Mod. Phys. Lett. **A3** (1988) 1725-1732;

E. C. G. Sudarshan, T. D. Imbo and T. R. Govindarajan, "*Configuration Space Topology and Quantum Internal Symmetries* ", Phys. Lett. **213B** (1988) 471-476;

A. S. Goldhaber, R. MacKenzie and F. Wilczek, "*Field Corrections to Induced Statistics* ", preprint, HUTP-88/A044, Harvard, 1988;

T. H. Hansson, M. Roček, I. Zahed and S. C. Zhang, "*Spin and Statistics in Massive (2+1) Dimensional QED* ", Phys. Lett. **B214** (1988) 475-479;

T. Jacobson, "*Bosonic Path Integral for Spin-$\frac{1}{2}$ Particles* ", Phys. Lett. **B216** (1989) 150-154;

J. Grundberg, T. H. Hansson, A. Karlhede and U. Lindström, "*Spin, Statistics and Linked Loops* ", Phys. Lett. **B218** (1989) 321-325;

T. H. Hansson, A. Karlhede and M. Roček, "*On Wilson Loops in Abelian Chern-Simons Theories* ", Phys. Lett. **B225** (1989) 92-94;

A. Foerster and H. O. Girotti, "*Anyons and Singular Gauge Transformations* ", Phys. Lett. **B230** (1989) 83-87;

X. G. Wen and A. Zee, "*Quantum Disorder, Duality, and Fractional Statistics in 2+1 Dimensions* ", Phys. Rev. Lett. **62** (1989) 1937-1940;

E. Fradkin, "*Jordan-Wigner Transformation for Quantum-Spin Systems in Two Dimensions and Fractional Statistics* ", Phys. Rev. Lett. **63** (1989) 322-325;

C. Aneziris, A. P. Balachandran and A. M. Srivastava, "*The Half-Filled Hubbard Model and the 2+1 Dimensional Dirac Lagrangian* ", preprint,

SU-4228-422, Syracuse, 1989:

A. P. Balachandran, G. Landi and B. Rai, *"A Modified Chern-Simons Term: Chiral Spins Systems and Anyon Superconductivity* , preprint, SU-4228-423, Syracuse, 1989 and Int. J. Mod. Phys. **A** (in press);

M. Lüscher, *"Bosonization in 2+1 Dimensions* ", Nucl. Phys. **B236** (1989) 557-582;

R. D. Tscheuschner, *"Towards a Topological Spin-Statistics Theorem in Quantum Field Theory* ", Int. J. Th. Phys. **28** (1989) 1269-1310;

J. Fröhlich, *"Quantum Field Theories of Vortices and Anyons* ", Commun. Math. Phys. **121** (1989) 177-223;

I. J. R. Aitchison and N. E. Mavromatos, *"Spin-Monopoles and Anyon Statistics in Effective Gauge Theories of High T_C Superconductors* ", Mod. Phys. Lett. **B3** (1989) 1275-83;

S. Zhang, *"Gauge Invariance and Flux Quantization in the System of Fractional-Statistics Particles* ", Phys. Rev. **B40** (1989) 5219-5222;

A. M. Din, *"Interpolating Statistics and the Two-Skyrmion System* ", Nucl. Phys. **B330** (1990) 757-767;

D. Karabali, *"Soliton Operators of the Fractional Spin in 2+1 Dimensions* ", preprint, CCNY-HEP-90/5, City College, New York, 1990;

R. Jackiw, *"Topics in Planar Physics* ", preprint, MIT-CTP# 1824, Cambridge, 1990;

R. L. Davis, *"Magnus Forces and Statistics in 2+1 Dimensions* ", Mod. Phys. Lett. **A5** (1990) 853-862.

97. K. S. Gupta, G. Landi and B. Rai, *"Non-Abelian Chern-Simons Term and Statistics of the G/H Solitons* ", Mod. Phys. Lett. **A4** (1989) 2263-2281 and ibid. **A5** (1990) 225-226(E).

98. M. J. Bowick, D. Karabali and L. C. R. Wijewardhana, *"Fractional Spin Via Canonical Quantization of the O(3) Nonlinear Sigma Model* ", Nucl. Phys. **B271** (1986) 417-428.

99. B. I. Halperin, *"Statistics of Quasiparticles and the Hierarchy of Frac-*

tional Quantized Hall States ", Phys. Rev. Lett. **52** (1984) 1583-1586;

M. H. Friedman, J. B. Sokoloff, A. Widom and Y. N. Srivastava, "*Chiral Anomaly and the Rational Quantization of the Hall Conductance* ", Phys. Rev. Lett. **52** (1984) 1587-1589;

D. P. Arovas, F. Wilczek and J. R. Schrieffer, "*Fractional Statistics and the Quantum Hall Effect* ", Phys. Rev. Lett. **53** (1984) 722-723;

V. Kalmeyer and R. B. Laughlin, "*Equivalence of the Resonating-Valence-Bond and Fractional Quantum Hall States* ", Phys. Rev. Lett. **59** (1987) 2095-2098;

G. Morandi, "*Quantum Hall Effect* ", Monographs and Textbooks in Physical Science, Lecture Notes Number 10 (Bibliopolis, Naples, 1988).

R. B. Laughlin, "*The Relationship Between High-Temperature Superconductivity and the Fractional Hall Effect* ", Science **242** (1988) 525-533;

V. Kalmeyer and R. B. Laughlin, "*Theory of the Spin Liquid State of the Heisenberg Antiferromagnet* ", Phys. Rev. **B39** (1989) 11879-11899;

R. B. Laughlin, "*Superconducting Ground State of Noninteracting Particles Obeying Fractional Statistics* ", Phys. Rev. Lett. **60** (1988) 2677-2680; "*Spin Hamiltonian for which the Quantum Hall Wavefunction is Exact* ", Ann. Phys. (N.Y.) **191** (1989) 163-202;

C. B. Hanna, R. B. Laughlin and A. L. Fetter, "*Quantum Mechanics of the Fractional-Statistics Gas: Hartree-Fock Approximation* ", Phys. Rev. **B40** (1989) 8745-8758;

S. C. Zhang, T. H. Hansson and S. Kivelson "*Effective-Field-Theory Model for the Fractional Quantum Hall Effect* ", Phys. Rev. Lett. **62** (1989) 82-85;

Y.-S. Wu, "*Topological Aspects of the Quantum Hall Effect* ", preprint, IASSNS-HEP-90/33, Institute for Advanced Study, Princeton, 1990.

100. G. Baskaran and P. W. Anderson, "*Gauge Theory of High-Temperature Superconductors and Strongly Correlated Fermi Systems* ", Phys. Rev. **B37** (1988) 580-583;

I. Affleck, Z. Zou, T. Hsu and P. W. Anderson, "*SU(2) Gauge Symmetry of the Large-U Limit of the Hubbard Model* ", Phys. Rev. **B38** (1988) 745-747;

Z. Zou, "*SU(2) Gauge Symmetry and Anomaly of the S=1/2 Anitferromagnetic Heisenberg Model in 2+1 Dimensions* ", Phys. Lett. **A131** (1988) 197-202;

X. G. Wen and A. Zee, "*Quantum Statistics and Superconductivity in Two Spatial Dimensions* , preprint, NSF-ITP-89-155, Institute for Theoretical Physics, Santa Barbara, 1989;

C. Aneziris, A. P. Balachandran and A. M. Srivastava, "*The Half-Filled Hubbard Model and the 2+1 Dimensional Dirac Lagrangian* ", preprint, SU-4228-422, Syracuse University (1989) and Mod. Phys. Lett. **B** (in press);

Y.-H. Chen, F. Wilczek and E. Witten, "*On Anyon Superconductivity* ", Int. J. Mod. Phys. **B3** (1989) 1001-1067;

X. G. Wen, F. Wilczek and A. Zee, "*Chiral Spin States and Superconductivity* ", Phys. Rev. **B39** (1989) 11413-11423;

J. K. Jain and N. Read, "*Off-Diagonal Long-Range Order in Laughlin's States for Particles Obeying Fractional Statistics* ", Phys. Rev. **B40** (1990) 2723-2726;

B. Halperin, J. March-Russel and F. Wilczek, "*Consequences of Time-Reversal Symmetry Violation in Models of High T_c Superconductors* ", Phys. Rev. **B40** (1989) 8726-8744;

M. Greiter, F. Wilczek and E. Witten, "*Hydrodynamic Relations in Superconductivity* ", Mod. Phys. Lett. **B3** (1989) 903-918;

G. S. Canright and S. M. Girvin, "*Anyons, the Quantum Hall Effect, and Two-Dimensional Superconductivity* ", Int. J. Mod. Phys. **B3** (1989) 1943-1963;

Y. Hosotani and S. Chakravarty, "*Superconductivity With No Cooper Pairs* ", preprint, IASSNS-HEP-89/31, Institute for Advanced Study, Princeton, 1989;

D. H. Lee and M. P. A. Fisher, "*Anyon Superconductivity and the Fractional Quantum Hall Effect* ", Phys. Rev. Lett. **63** (1989) 903-906;

H. Aoki, "*Fractional Statistics and High T_C Superconductivity* ", Solid State Physics **24** (1989) 77-788;

S. Randjbar-Daemi, A. Salam and J. Strathdee, "*Chern-Simons Superconductivity at Finite Temperature* ", Nucl. Phys. **B340** (1990) 403-447 and "*Anyons and Chern-Simons Theory on Compact Spaces of Finite Genus* ", preprint, IC/90/11, Trieste, 1990;

S. Giddings and F. Wilczek, "*Spontaneous Fact Violation* ", Mod. Phys. Lett. **A5** (1990) 635-643;

T. Banks and J. Lykken, "*Landau Ginzburg Description of Anyonic Superconductors* ", Nucl. Phys. **B 336** (1990) 500-516;

J. Lykken, J. Sonnenschein and N. Weiss, "*Anyonic Superconductivity* ", preprint, Fermilab-Pub-89/231-T and UCLA/89/TEP/52, Fermilab, 1989; and "*Field Theoretical Analysis of Anyonic Superconductivity* ", preprint, Fermilab-Pub-90/50-T and UCLA/90/TEP/17, Fermilab, 1990;

E. Fradkin, "*Superfluidity of the Lattice Anyon Gas and Topological Invariance* ", preprint, P/89/12/173, Illinois, 1989;

P. Panigrahi, R. Ray and B. Sakita, "*Effective Lagrangian for a System of Non-Relativistic Fermions in 2+1 Dimensions Coupled to an Electromagnetic Field: Application to Anyonic Superconductors* ", preprint, CCNY-HEP-89-22, City College, New York 1989;

A. P. Balachandran, G. Landi and B. Rai, "*A Modified Chern-Simons Term: Chiral Spin Systems and Anyon Superconductivity* ", preprint, SU-4228-423, Syracuse University, 1989;

A. M. Din, "*Anyon Quarks and Superconductivity* ", Phys. Lett. **B231** (1989) 153-156;

F. Wilczek, "*Lectures on Fractional Statistics and Anyon Superconductivity* , in the Proceedings of the Second Ferrara School in Theoretical Physics, "*Anomalies, Phases and Defects* ", Ferrara 1989, Eds. M. Bregola, G. Marmo and G. Morandi (Bibliopolis, Naples, 1990);

M. Careba, T. E. Clark and C. E. M. Wagner, "*Superconductivity and Supersymmetric QED_3* ", preprint, PURD-TH-90-3, Purdue, 1990;

J. D. Lykken, *"Chern-Simons and Anyon Superconductivity"*, talk given at *"Strings 90"*, Texas A&M University, College Station, 1990, preprint, FERMILAB-CONF-90/72-T, Fermilab, 1990;
A. Zee, *"Semionics: A Theory of High Temperature Superconductivity"*, in *"High Temperature Superconducticity"*, Eds. K. Bedell, D. Pines and J. R. Schrieffer (Addison-Wesley, New York, 1990);
Ph. Gerbert, *"Anyons, Chern-Simons Lagrangians and Physics in 2+1 Dimensions"*, preprint, UCLA/90/TEP/36, Los Angeles, 1990.

101. E. Witten, *"Topological Quantum Field Theory"*, Commun. Math. Phys. **117** (1988) 353-386; *"Topological Sigma Models"*, Commun. Math. Phys. **118** (1988) 411-449; *"Topological Gravity"*, Phys. Lett. **206** (1988) 601-606.

102. E. Witten, *"Quantum Field Theory and the Jones Polynomial"*, talk presented at the IAMP Congress, Swansea, July, 1988, Eds. B. Simon, I. Davies and A. Truman, and in Commun. Math. Phys. **121** (1989) 351-399.

103. E. Witten, *"2+1 Dimensional Gravity as an Exactly Soluble System"*, Nucl. Phys. **B311** (1988/89) 46-78; *"Topology-Changing Amplitudes in 2+1 Dimensional Gravity"*, Nucl. Phys. **B331** (1989) 113-140; *"Quantization of Chern-Simons Gauge Theory With Complex Gauge Group"*, preprint, IASSNS-HEP-89/95, Institute for Advanced Study, Princeton, 1989;
S. Carlip, *"Exact Quantum Scattering in 2 + 1 Dimensional Gravity"*, Nucl. Phys. **B324** (1989) 106-122.

104. P. Salomonson, B.-S. Skagerstam and A. Stern, *"ISO(2,1) Chiral Models and Quantum Gravity in 2 + 1 Dimensions"*, preprint, UAHEP 897, University of Alabama, 1989 and Nucl. Phys. **B** (in press).

105. W. Siegel, *"Unextended Superfields in Extended Supersymmetry"*, Nucl. Phys. **B156** (1976) 135-143;

A. Achucarro and P. K. Townsend, "*A Chern-Simons Action for the Three-Dimensional Anti-Desitter Supergravity Theories* ", Phys. Lett. **180B** (1986) 89-92;

C. H. Tze and S. Nam, "*Global Dynamics of Electric and Magnetic Membrances on the Complex, Quaternionic and Octonionic Hopf Bundels* ", Phys. Lett. **B210** (1988) 76-84;

G. Moore and N. Seiberg, "*Taming the Conformal Zoo* ", Phys. Lett. **B220** (1989) 422-430;

S. Elitzur, G, Moore and N. Seiberg, "*Remarks on the Canonical Quantization of the Chern-Simons-Witten Theory* ", preprint, IASSNS-HEP-89/20, Institute for Advanced Study, Princeton, 1989;

G. V. Dunne, R. Jackiw and C. A. Trugenberger, "*Chern-Simons Theory in the Schrödinger Representation* ", Ann. Phys. (N.Y.) **194** (1989) 197-223;

H. Murayama, "*Explicit Quantization of the Chern-Simons Action* ", preprint, UT-542, Tokyo, 1989;

M. Bos and V. P. Nair, "*Coherent State Quantization of Chern-Simons Theory* ", Int. J. Mod. Phys. **A5** (1989) 959-988;

L. Smolin, "*Invariants of Links and Critical Points of the Chern-Simon Path Integrals* ", Mod. Phys. Lett. **A4** (1989) 1091-1112;

G. Chapline and B. Grossmann, "*Conformal Field Theory and a Topological Quantum Theory of Vortices and Knots* ", Phys. Lett. **B223** (1989) 336-342;

S. Axelrod, S. Della Pietra and E. Witten, "*Geometric Quantization of Chern-Simons Gauge Theory* ", preprint, IASSNS HEP-89/57, Institute for Advanced Study, Princeton, 1989;

H. Verlinde and E. Verlinde, "*Conformal Field Theory and Geometric Quantization* ", preprint, PUPT-89/1149 and IASSNS-HEP-89/58, Institute for Advanced Study, Princeton, 1989;

A. P. Polychronakos, "*Abelian Chern-Simons Theory in 2+1 Dimensions* ", Ann. Phys. (N.Y.) **203** (1990) 231-254;

I. J. R. Aitchison and N. E. Mavromatos, "*Effective 2+1 Dimensional*

Gauge Theory of Spin and Charge Degress of Freedom With Chern-Simons Coupling ", preprint, OUTP-89-12P, Oxford, 1989;

A. Stern, "*Pure Wess-Zumino Actions in 1+1 Dimensional Non-Linear σ-Models* ", Int. J. Mod. Phys. **5A** (1990) 415-426;

R. Floreanini, R. Percacci and E. Sezgin, "*Sigma Models With Purely Wess-Zumino-Witten Actions* ", Nucl. Phys. **B322** (1989) 255-276;

E. Guadagnini, M. Martellini and M. Mintchev, "*Chern-Simons Model and New Relations Between the HOMFLY Coefficients* ", Phys. Lett. **B228** (1989) 489-494;

G. Zembra, "*An Interpretation of the Abelian Chern-Simons Vacuum Holonomy* ", Int. J. Mod. Phys. **A5** (1990) 559-569;

Y.-S. Wu and K. Yamagishi, *Chern-Simons Theory and Kaufmann Polynomials* ", Int. J. Mod. Phys. **A5** (1990) 1165-1195;

E. Witten, "*Gauge Theories, Vertex Models, and Quantum Groups* ", Nucl. Phys. **B330** (1990) 285-346;

P. Cotta-Ramusino, E. Guadagnini, M. Martellini and M. Mintchev, "*Quantum Field Theory and Link Invariants* ", Nucl. Phys. **B330**, (1990) 557-574;

E. Guadagnini, M. Martellini and M. Mintchev, "*Wilson Lines in Chern-Simons Theory and Link Invariants* ", Nucl. Phys. **B330**, (1990) 575-607; "*Braids and Quantum Group Symmetry in Chern-Simons Theory* ", Nucl. Phys. **B336** (1990) 581-609;

L. Balieu, "*Chern-Simons 3-Dimensional and Yang-Mills 2-Dimensional Systems as 4-Dimensional Topological Quantum Field Theories* ", preprint, CERN-TH.5533/89 and PAR-LPTHE/8928, Geneva, 1989;

D. Boyanovsky, "*Chern-Simons With Matter Fields: Schrödinger Picture Quantization* ", preprint, PITT 89-12, Pittsburg, 1989;

D. Birmingham and S. Sen, "*Generalized Skein Relations from the Chern-Simons Field Theory* ", preprint, DIAS-STP-90-15, Dublin, 1990;

J. M. F. Labastida, P. M. Llatas and A. V. Ramallo, "*Knot Operators in Chern-Simons Gauge Theory* ", preprint, CERN-TH.5756/90, IEM-FT-20/90 and US-FT 7/90, Geneva, 1990;

T. P. Killingback, *"Quantization of SL(2,**R**) Chern-Simons Theory "*, preprint, CERN-TH.5698/90 and *"Chern-Simons Theory for Compact Non-Semisimple Lie Groups "*, preprint, CERN-TH-5699/90, Geneva, 1990.

106. B. Zumino, *"Chiral Anomalies and Differential Geometry "*, Lectures given at Les Houches, August 1983.
Also see B. Zumino, Y.-S. Wu and A. Zee, *"Chiral Anomalies, Higher Dimensions and Differential Geometry "*, Nucl. Phys. **B239** (1984) 477-507.

107. A. Stern, *"Frozen Solitons in a Two-Dimensional Ferromagnet "*, Phys. Rev. Lett. **59** (1987) 1506-1509. Also see A. Holz and C. Gong, *"Theory of q-Soliton and Phase Boundary Dynamics in a (2+1)-Dimensional O(3)-Antiferromagnet "*, Physica **A161** (1989) 474-507.

108. T. H. R. Skyrme, *"A Non-Linear Theory of Strong Interactions "*, Proc. Roy. Soc.(London **A247** (1958) 260-278; *"A Non-Linear Theory "*, Proc. Roy. Soc. (London) **A260** (1961) 121-138; *"Particle States of a Quantized Meson Field "*, Proc. Roy. Soc. (London) **A262** (1961) 237-245; *"A Unified Theory of Mesons and Baryons "*, Nucl. Phys. **31** (1962) 556-569; *"Kinks and the Dirac Equation "*, J. Math. Phys. **12** (1971) 1735-1743.

109. P. Jain, J. Schechter and R. Sorkin, *"Quantum Stabilization of the the Skyrme Soliton "*, Phys. Rev. **D39** (1989) 998-1001;
J. A. Mignaco and S. Wulck, *"Chiral Quantum Baryon "*, Phys. Rev. Lett. **62** (1989) 1449-1452.

110. G. 't Hooft, *"A Planar Diagram Theory for Strong Interactions"*, Nucl. Phys. **B72** (1974) 461-473; *"A Two-Dimensional Model for Mesons "*, Nucl. Phys. **B75** (1974) 461-470;
E. Witten, *"Baryons in the 1/N Expansion "*, Nucl. Phys. **B160** (1979) 57-115. For an introduction see e.g. S. Coleman in Ref. [72].

111. A. Kundu, Yu. P. Rubakov and V. I. Sanyuk, "*Topological Solitons in the Skyrme Model* ", Indian J. Pure and App. Phys. **17** (1979) 673-677.

112. D. Finkelstein and J. Rubinstein, "*Connection Between Spin and Statistics, and Kinks* ", J. Math. Phys. **9** (1968) 1762-1779.

113. J. G. Williams, "*Topological Analysis of a Nonlinear Field Theory* ", J. Math. Phys. **11** (1970) 2611-2616.

114. N. K. Pak and H. Ch. Tze, "*Chiral Solitons and Current Algebra* ", Ann. Phys. (N.Y.) **117** (1979) 164-194.

115. J. M. Gipson and H. Ch. Tze, "*Possible Heavy Solitons in the Strongly Coupled Higgs Sector* ", Nucl. Phys. **B183** (1981) 524-546.

116. G. S. Adkins, C. R. Nappi and E. Witten, "*Static Properties of Nucleons in the Skyrme Model* ", Nucl. Phys. **B228** (1983) 552-566.

117. See e.g. M. Gell-Mann and Y. Ne'eman, "*The Eightfold Way* " (W. A. Benjamin, Inc., New York, 1964).

118. E. Witten, "*Global Aspects of Current Algebra* ", Nucl. Phys. **B223** (1983) 422-432.

119. A. P. Balachandran, V. P. Nair, S. G. Rajeev and A. Stern, "*Exotic Levels from Topology in the Quantum-Chromodynamic Effective Lagrangian* ", Phy. Rev. Lett. **49** (1982) 1124-1127; "*Soliton States in the Quantum-Chromodynamic Effective Lagrangian* ", Phys. Rev. **D27** (1983) 1153-1164;
A. P. Balachandran, "*Solitons in $^3He - B$* ", Nucl. Phys. **B271** (1986) 227-252.

120. J. Goldstone and F. Wilczek, "*Fractional Quantum Numbers on Solitons* ", Phys. Rev. Lett.**47** (1981) 989-989.

121. J. M. Gipson, "*Quasi-Solitons in the Strongly Coupled Higgs Sector of the Standard Model* ", Nucl. Phys. **B231** (1984) 365-385.

122. C. R. Nappi, *"Skyrmion Phenomenology"* in *"Symposium on Anomalies, Geometry, Topology"*, Eds. W. A. Bardeen and A. R. White (World Scientific, Singapore,1985).

123. U.-G. Meissner, *"Low Energy Hadron Physics from Effective Chiral Lagrangians with Vector Mesons"* Phys. Rep. **161** (1988) 213-362.

124. L. D. Faddeev, *"Operator Anomaly for the Gauss Law"* Phys. Lett. **145B** (1984) 81-84;
R. Jackiw and R. Rajaraman, *"Vector-Meson Mass Generation by Chiral Anomalies"*, Phys. Rev. Lett. **54** (1985) 1219-1221;
E. D'Hoker and E. Farhi, *"Decoupling a Fermion in the Standard Electro-Weak Theory"*, Nucl. Phys. **B248** (1984) 77-89;
A. Niemi and G. W. Semenoff, *"Gauge Algebras in Anomalous Gauge-Field Theories"*, Phys. Rev. Lett. **56** (1986) 1019-1022;
P. Q. Hung and C. H. Tze, *"Can a Fourth Family Exist Without Its Quark Counterpart?"*, Z. Phys. **C39** (1988) 371-375.

125. Ö. Kaymakcalan, S. Rajeev and J. Schechter, *"Nonabelian Anomaly and Vector Meson Decays"*, Phys. Rev. **D30** (1984) 594-602;
H. Gomm, Ö. Kaymakcalan and J. Schechter, *"Anomalous Spin 1 Meson Decays from the Gauged Wess-Zumino Term"*, Phys. Rev. **D30** (1984) 2345-2355;
Ö. Kaymakcalan and J. Schechter, *"Chiral Lagrangians of Pseudoscalars and Vectors"*, Phys. Rev. **D31** (1985) 1109-1113.

126. R. V. Moody, *"Lie Algebras Associated With Generalized Cartan Matrices"*, Bull. Am. Math. Soc. **73** (1967) 217-221;
V. G. Kac, *"Infinite Dimensional Lie Algebras"* (Cambridge University Press, 1985).

127. P. Salomonson, B.-S.Skagerstam and A. Stern, *"Canonical Quantization of Chiral Bosons"*, Phys. Rev. Lett. **62** (1989) 1817-1820.

128. A. P. Balachandran, A. Barducci, F. Lizzi, V. G. J. Rodgers and A. Stern,

"*Doubly Strange Dibaryon in the Chiral Model* ", Phys. Rev. Lett. **52** (1984) 887-890;

A. P. Balachandran, F. Lizzi, V. G. J. Rodgers and A. Stern, "*Dibaryons as Chiral Solitons* ", Nucl. Phys. **B256** (1985) 525-556;

S. A. Yost and C. R. Nappi, "*Mass of the H Dibaryon in a Chiral model*", Phys. Rev. **D32** (1985) 816-818;

R. L. Jaffe and C. L. Korpa, "*Semiclassical Quantization of the Dibaryon Skyrmion* ", preprint, MIT-CTP#1233, Cambridge, 1985.

129. E. Rabinovici, A. Schwimmer and S. Yankielowicz, "*Quantization in the Presence of Wess-Zumino terms* ", Nucl. Phys. **B248** (1984) 523-535.

130. R. L. Jaffe, "*Perhaps a Stable Dihyperon* ", Phys. Rev. Lett. **38** (1977) 195-198; "*Multiquark Hadron: I. Phenomenology of $Q^2\overline{Q}^2$ Mesons* ", Phys. Rev. **D15** (1977) 267-280; "*Multiquark Hadrons. II. Methods* ", ibid. **D15** (1977) 281-289.

131. P. J. Mulders, A. T. Aerts and J. J. de Swart, "*Multiquark States. III. Q^6 Dibaryon Resonances* ", Phys. Rev. **D21** (1980) 2653-2671;

A. T. M. Aerts and C. B. Dover, "*S=-1 Dibaryon Production With Kaon Beams* ", Nucl. Phys. **B253** (1985) 116-148 and "*Narrow Dibaryons of Strangeness S=-1?* ", Phys. Lett. **146B** (1984) 95-100.

132. E. Guadagnini, "*Baryons as Solitons and Mass Formulae* ", Nucl. Phys. **B236** (1984) 35-47.

133. G. S. Adkins and C. R. Nappi, "*The Skyrme Model With Pion Masses* ", Nucl. Phys. **B233** (1984) 109-115.

134. M. Chemtob, "*Skyrme Model of Baryon Octet and Decouplet* " , Nucl. Phys. **256** (1985) 600-608;

M. Praszałowicz, "*A Comment on the Phenomenology of the SU(3) Skyrme Model* ", Phys. Lett. **158B** (1985) 264-269.

135. S. L. Glashow, "*Partial-Symmetries of Weak Interactions* , Nucl. Phys. **22** (1961) 579-588;

346

S. Weinberg, "*A Model of Leptons* , Phys. Rev. Lett. **19** (1976) 1264-1266;

A. Salam, "*Weak and Electromagnetic Interactions* " in "*Elementary Particle Theory*, Ed. N. Svartholm (Almqvist & Wiksell, Stockholm, 1969).

136. P. Langacker, "*Grand Unified Theories and Proton Decay* ", Phys. Rep. **72** (1981) 185-385.

137. J. Ambjørn and V. A. Rubakov, "*Classical Versus Semiclassical Electroweak Decay of the Techniskyrmion* ", Nucl. Phys. **B256** (1985) 434-448.

138. G. Eilam, D. Klabucar and A. Stern, "*Skyrmion Solution to the Weinberg-Salam Model* ", Phys. Rev. Lett. **56** (1986) 1331-1334.

139. G. Eilam and A. Stern, "*Soliton States in the Strongly Coupled Electroweak Theory* ", Nucl. Phys. **B294** (1987) 775-800.

140. T. Appelquist and C. Bernard, "*Strongly Interacting Higgs boson* ", Phys. Rev. **D22** (1980) 200-213.

141. Y. Brihaye and J. Kunz, "*A Sequence of New Classical Solutions in the Weinberg-Salam Model* ", Mod. Phys. Lett. **A4** (1989) 2723-2732;
"*Multi-Skyrmion Solutions in the Weinberg-Salam and Sakurai Model* ", Z. Phys. **C41** (1989) 663-666;
A. Dobado, "*On the Weak-Skyrmion Description of the Resonace in the Symmetry Breaking Sector of the Standard Model* ", talk presented at the "*International Europhysics Conference on High Energy Physics* ", Madrid, 1989, preprint, CERN 89-038, Geneva, 1989.

142. N. S. Manton, "*Topology in the Weinberg-Salam Theory* ", Phys. Rev. **D28** (1982) 2019-2026;
F. R. Klinkhamer and N. S. Manton, "*A Saddle-Point Solution to Weinberg-Salam Theory* ", Phys. Rev. **D30** (1984) 2212-2220.

143. G. 't Hooft, *"Symmetry Breaking Through Bell-Jackiw Anomalies"*, Phys. Rev. Lett. **37** (1976) 8-11; *"Computation of the Quantum Effects Due to a Four-Dimensional Pseudoparticle"*, Phys. Rev. **D14** (1976) 3432-3450; ibid. **D18** (1978) 2199-2200(E).

144. C. G. Callan, Jr., R. F. Dashen and D. J. Gross, *"The Structure of the Gauge Theory Vacuum"*, Phys. Lett. **63B** (1976) 334-340;
R. Jackiw and C. Rebbi, *"Vacuum Periodicity in a Yang-Mills Quantum Theory"*, Phys. Rev. Lett. **37** (1976) 172-175.

145. N. V. Krasnikov, V. A. Mateev and A. N. Tavkhelidze, *"The Problem of CP Invariance in Quantum Chromodynamics"*, Sov. J. Part. Nucl. **12** (1981) 38-47.

146. A. I. Vainstein, A. I. Zakarhov, V. A. Novikov and M. A. Shifman, *"ABC of Instantons"*, Sov. Phys. Usp. **25** (1982) 195-215.

147. For the analogues of θ states in the new variable approach to gravity, see A. Ashtekar, A. P. Balachandran and S. Jo, *"The CP Problem in Quantum Gravity"*, Int. J. Mod. Phys. **A4** (1989) 1493-1514.

148. For the analogues of gauge theory θ states for arbitrary manifolds and related considerations, see A. R. Shastri, J. G. Williams and P. Zwengrowski, *"Kinks in General Relativity"*, Int. J. Theor. Phys. **19** (1980) 1-23;
C. Isham and G. Kunstatter, *"Yang-Mills Canonical Vacuum Structure in a General Three Space"*, Phys. Lett. **102B** (1981) 417-420; *"Spatial Topology and Yang-Mills Vacua"*, J. Math. Phys. **23** (1982) 1668-1667;
C. Isham, *"Vacuum Tunneling In Static Space-Times"* in *"Old and New Questions in Physics, Cosmology, Philosophy and Theoretical Biology"*, Ed. by A. van Der Merwe (Plenum Press, New York, 1983); *"Topological an Global Aspects of Quantum Gravity"*, in the *"Proceedings of the 40th Summer School on Relativity, Groups and Topology"*, Les Houches, 1983, Eds. B. S. DeWitt and R. Stora (North-Holland, Amsterdam, 1984).

149. For the analogues of θ states in generally covariant gauge theories and in the vielbein approach to gravity, see A. P. Balachandran, S. Jo and A. M. Srivastava, "*What Becomes of the Vacuum Angle in Generally Covariant Gauge Theories and the Vielbein Approach to Gravity* ", Int. J. Mod. Phys. **A4** (1989) 2995-3014.

150. J. B. Hartle and D. M. Witt, "*Gravitational θ-States and the Wave Function of the Universe* ", Phys. Rev. **D37** (1988) 2833-2837.

151. J. L. Friedman and D. M. Witt, "*Internal Symmetry Groups of Quantum Geons* ", Phys. Lett. **120B** (1983) 324-328;
D. M. Witt, "*Symmetry Groups of State Vectors in Canonical Quantum Gravity* ", J. Math. Phys. **27** (1986) 573-592;
J. L. Friedman and D. M .Witt in "*Mathematics and General Relativity* ", Ed. by J. Isenberg (American Mathematical Society, Providence, 1988).

152. J. L. Friedman, N. Papastamatiou, L. Parker and Z. Huai, "*Non-Orientable Foam and an Effective Planck Mass for Point-Like Fermions* ", Nucl. Phys. **B309** (1988) 533-551.

153. A. P. Balachandran, "*Topological Aspects of Quantum Gravity* ", in the Proceedings of the "*1987 DST Workshop on Particle Physics-Superstring Theory* ", Eds. H. S. Mani and R. Ramachandran (World Scientific, Singapore, 1988).

154. C. Aneziris, A. P. Balachandran, M. Bourdeau, S. Jo, T. R. Ramadas and R. D. Sorkin, "*Statistics and General Relativity* ", Mod. Phys. Lett. **A4** (1989) 331-338; "*Aspects of Spin and Statistics in Generally Covariant Theories* ", Int. J. Mod. Phys. **A4** (1989) 5459-5510.

155. C. Aneziris, A. P. Balachandran, L. Kauffman and A. M. Srivastava, "*Novel Statistics for Strings and String "Chern-Simons Terms"* ", preprint, TPI-MINN-90/16-I and SU-4288-434, University of Minnesota, 1990 and Int. J. Mod. Phys. **A** (in press).

156. A. P. Balachandran, "*Statistics, Strings and Gravity* ", Lecture delivered at the "*IIIrd Regional Conference on Mathematical Physics* ", Eds. F. Hussain and A. Qadir (World Scientific, Singapore, 1990);
J. A. Harvey and J. Liu, "*Strings and Statistics* ", Phys. Lett. **B240** (1990) 369-374.

157. D. L. Goldsmith, "*The Theory of Motion Groups* ", Michigan Math. J. **28** (1981) 3-17; "*Motion of Links in the 3-Sphere* " Math. Scand. **50** (1982) 167-205.

158. Cf. H. Weyl, "*The Theory of Groups and Quantum Mechanics* " (Dover, 1931).

159. R. D. Sorkin (to be published).

160. A. M. Srivastava, "*Cosmological Consequences of Gravitational Interacting Planck-Mass Particles* ", Phys. Rev. **D36** (1987) 2368-2373.

161. A. P. Balachandran, A. Daughton, Z.-C. Gu, G. Marmo, R. D. Sorkin and A. M. Srivastava, "*A Topological Spin-Statistics Theorem or a Use of the Antiparticle* ", Mod. Phys. Lett. *A5* (1990) 1575-1585 and in the Proceedings of the Nobel Symposium "*The Birth and Early Evolution of Our Universe* ", Gräftåvallen, 1990, Eds. J. S. Nilsson, B. Gustafsson and B.-S. Skagerstam (World Scientific, to appear); "*A Spin-Statistics Theorem With No Relativity or Field Theory* ", preprint, SU-4288-433, Syracuse, 1990 and Int. J. Mod. Phys. **A** (in press);
A. P. Balachandran, W. D. McGlinn, D. O'Connor, L. O'Raifeartaigh, S. Sen and R. D. Sorkin (in preparation);
A. P. Balachandran, W. D. McGlinn, L. O'Raifeartaigh, S. Sen, R. D. Sorkin and A. M. Srivastava (in preparation).

162. C. Aneziris, A. P. Balachandran and D. Sen, "*Statistics in One Dimension* ", preprint, SU-4228-453, Syracuse, 1990;
A. P. Balachandran and G. Morandi (under preparation).

INDEX